Electric Power Conversion

Edited by Marian Găiceanu

Published in London, United Kingdom

IntechOpen

Supporting open minds since 2005

Electric Power Conversion
http://dx.doi.org/10.5772/intechopen.73992
Edited by Marian Găiceanu

Contributors
Mohamed Salem, Khalid Yahya, Brij Mohan Mundotiya, Wahdat Ullah, Krishna Kumar, R T Naayagi, Mahmoud Masoud, Mohamed Ebrahim, Jin Li, Boxue Du, Hucheng Liang, Carlos Alberto Reusser, Ivan Ramljak, Drago Bago, Farayi Musharavati, Kiran Singh, Marian Gaiceanu

First published in London, United Kingdom, 2019 by IntechOpen
IntechOpen is the global imprint of INTECHOPEN LIMITED, registered in England and Wales, registration number: 11086078, The Shard, 25th floor, 32 London Bridge Street
London, SE19SG – United Kingdom
Printed in Croatia

British Library Cataloguing-in-Publication Data
A catalogue record for this book is available from the British Library

Additional hard copies can be obtained from orders@intechopen.com

Electric Power Conversion Edited by Marian Găiceanu
p. cm.
Print ISBN 978-1-83880-355-1 Online ISBN
978-1-83880-356-8
eBook (PDF) ISBN 978-1-78984-046-9

For EU product safety concerns:
IN TECH d.o.o., Prolaz Marije Krucifikse Kozulić 3, 51000 Rijeka, Croatia, info@intechopen.com or visit our website at intechopen.com.

Meet the editor

Marian Găiceanu earned his PhD degree (Magna Cum Laude) in Electrical Engineering from the University of Galati, Romania, in 2002. He is currently with the Department of Automation and Electrical Engineering, Dunarea de Jos University of Galati, as an Academic Professor and PhD supervisor in Electrical Engineering, and he is also the Head of the Integrated Energy Conversion Systems and Advanced Control of Complex Processes Research Center, Romania. Dr. Găiceanu was with the Department of Regulation and Control, SIEI S.pA. (2002–2004) and Polytechnic of Turin, Italy (2002–2004 and 2006–2007) conducting several projects in the area of power converter systems for electrical drives, power quality, and PEM and SOFC fuel cell power converters for utilities and marine applications. Dr. Găiceanu has been an IEEE member since 2003 and a co-founder member of the IEEE Power Electronics Romania Chapter. He has been editor in chief of *The Annals of Dunarea de Jos* and *The International Symposium on Electrical and Electronics Engineering* journals. He is currently a reviewer of the *EPE* journal.

Contents

Preface

The purpose of this book is to integrate the technical knowledge obtained as a result of educational and research activities and to increase knowledge in the research area of electric power conversion. Electric power conversion systems are based on electromechanical conversion systems and static power converters.

In the first section, an incursion into the field of electric power conversion is presented.

Validation of the analytic mixed current-flux *d-q* mathematical model of a saturated, compensated, six-phase self-excited induction generator supplying a three-phase resistive or inductive load is shown in the second section. Both synchronous- and asynchronous-driven multiphase machines, with their main focus on multiphase power converter topologies associated with specific control, are presented.

Resonant power converters along with solutions to electromagnetic compatibility can be found in the third section. Power loss analysis based on experimental results of insulated-gate bipolar transistors in a high-voltage, high-power, dual-active bridge DC-DC converter for aerospace energy storage systems is presented. Both five-phase uncontrolled and controlled line commutated rectifiers are presented in a didactical manner.

In the fourth section, modern and efficient topologies and advanced control solutions to manage grid interactions of electric power conversion systems based on photovoltaic-delivered energy are envisaged. Moreover, two of the electric field applications are presented.

The applied methodology is based on knowledge developments and applicative research. Novice readers will learn how energy is converted adequately and adapted to different consumers. Advanced readers will discover different kinds of modern solutions and tendencies in the field of electric power conversion.

Professor Marian Găiceanu, PhD Habil. Eng.,
Dunărea de Jos University of Galați,
Galați, România

Section 1

Introduction

Chapter 1

Introductory Chapter: Electric Power Conversion

Marian Găiceanu

1. Chapter overview

The introductory chapter has in view an incursion in discovering electricity, how can be handled, and the future of it. The chapter starts with the ancient discovery of electricity. Starting from the Kite experiment to the energy use of lightning is mentioned in the *New Discoveries in the Electricity* section. Moreover, the current path from the electrostatic machine to ion wind propulsion system is mentioned in the same section. A short history of energy conversion technology is described in the forthcoming section. Different electric power conversion technologies are mentioned. The possible pathways of the future electric power conversion are mentioned. Some ideas about electric power conversion development are mentioned at the end of this chapter.

2. The ancient discovery of the electricity

Electricity is a natural phenomenon. One of the first known discovered natural electricity generations is the *electric fish* (discovered in twenty eighth century BC by the Egyptian). These types of fish are named *electrogenic* (one of the most dangerous fish is the *electric eel—electrophorus electricus*—due to the lethal electric shocks generation), being capable to generate the electric fields through electric organ discharge (EOD). Nowadays, the *bioelectrogenesis* (generation of the electricity by the living organisms) phenomenon is studied by the *electrophysiology* science branch. The receiver fish of the electric field is named *electroreceptive*, having the ability to receive the electric field through the *electroreception* feature [1].

The Thales of Miletus (around 600 BC) remarked the attraction of lightweight material by rubbed amber (ἤλεκτρον in Greeks or ēlektron), discovering static electricity [2]. However, as the archeological discoveries are on-going the researchers make hypothesis of electric light existence in ancient Egypt [3–6].

3. New discoveries in the electricity

In 1600 the scientist William Gilbert published the on the *Magnet treatise*, becoming the *Father of Electricity* [7]. In his published work *Pseudodoxia Epidemica* (1646), second book—*Tenets Concerning Mineral and Vegetable Bodies*, Thomas

Browne used the word *Electricity*. The inventor of the *Electricity* (1663) was declared Otto von Guericke by producing static electricity by friction of a sulfur ball [8].

The first discover of conductors and insulators were made by the Stephen Gray (1731). Charles François de Cisternay du Fay discovered the existence of two electricity types: vitreous and resinous (1733), which were renamed positive and negative electricity by Benjamin Franklin in 1750.

Through the bolt of lightning Benjamin Franklin discovers the connection between lightning and electricity, as observed in the *Kite Experiment* (1752) [9].

The bolt of lightning can be intra-cloud (**Figure 1a**), between clouds (**Figure 1b**) or between cloud and ground (**Figure 1c**) [10].

A massive steel sculpture was performed by Isamu Noguchi in Philadelphia [11], as a memorial to Benjamin Franklin (**Figure 1d**) [12].

An average value of the energy released during thunderstorm is about 10,000,000 kilowatt-hours (3.6 × 1013 joule) [13].

Taking into account such huge quantity of energy provided in a very short time, the scientists propose the use of the captured lightning energy as alternative energy in different kinds of patents [14, 15] in order to use the stored energy lately.

In the future, as the nanotechnology penetrates the scientific knowledge environment the lightning energy could be a real alternative energy solution.

In 1741 William Watson used a vacuum glow-lamp supplied by an electrostatic machine.

Recently, at the end of 2018 the MIT engineers have been discovered the first no moving parts aircraft by using the electrostatic energy (**Figure 2**) [16]. They used the electro aerodynamic thrust principle to design ion wind propulsion system.

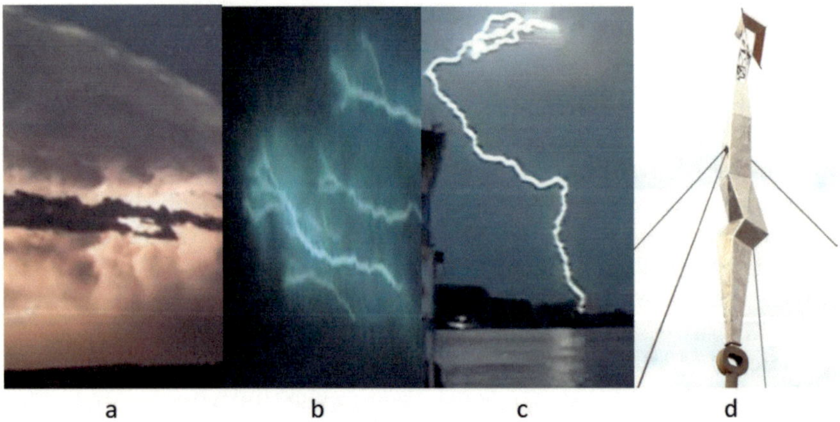

Figure 1.
The bolt of lightning (a) intracloud (b) between vertical clouds (c) between cloud and ground (d) bolt of lightning.

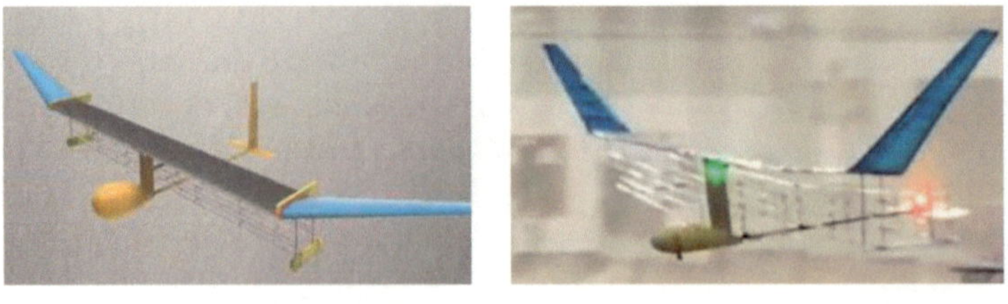

Figure 2.
The first no moving parts aircraft by using the electrostatic energy [17].

The bioelectricity was discovered by Luigi Galvani in 1780. The voltaic pile was made in 1800 by Alessandro Volta (the international unit for electric potential difference—voltage—is volt). This invention was a huge step regarding the source of the electricity, being more reliable than the electrostatic generator.

The actual technologies in electrical energy storage could be reviewed in [18].

The NAWA Technologies introduced in mass production the new technology as ultra fast carbon battery which is carbon-based ultracapacitor with vertically aligned carbon nanotube (VACNT) [19].

The future of the energy storage could be graphene-based supercapacitors. This new type delivers four times more energy density than the current supercapacitors, maintaining high power and long operational life (Meilin Liu) [20]. In 1802, Humphry Davy invented the incandescent lamp.

4. A short history of energy conversion technology

The energy conversion technology starts with biomass (primarily wood: dried plants) conversion to heat. The mechanical energy was obtained to replace the animal or human power. In order to obtain the mechanical energy as primary energy sources wind (air flowing by using windmills) or flowing water (by using waterwheels) were used. The first notes of windmills were from Hero of Alexandria, in the first century of common era (CE). The ocean tides were the primary source of energy for tidal mills since 1086. The idea of conversion steam power to move a piston into cylinder (earlier pump) was delivered by Denis Papin (1679). The conversion of the thermal energy (steam) into hydraulic energy was patented by Thomas Savery (1698). Thomas Newcome, together with Savery invented the first piston-operated steam pump (1712). James Watt was the Father of the Watt steam engine (1765), a modern way of the steam engine.

The first steam powered locomotive was provided by Richard Trevithick (1803) [21]. Robert Stirling patented in 1816 the engine without high-pressure boiler (for safety reason). The mechanical energy conversion into electricity was discovered by Michael Faraday by using electrical generator (1830s).

5. Energy conversion technologies

The total energy of one system, according to the first thermodynamics law, is composed by heat and capacity to do the work. The standard unit of energy is *joule* (after the name of scientist James Prescott Joule). The efficiency of the energy conversion is defined by the second law of thermodynamics [22]. The energy cannot be created or destroyed, only transformed it from one form to another (law of conservation of energy). In other words, the energy within a reference frame the energy is always the same.

The energy per unit time is power. The unit *watt* was introduced by Carl William Siemens (1882) for *real power* in AC system. The *watt* name is provided after the James Watt.

According to [23] (the *metric EU directive*), the symbol of the *reactive power* is "*var*" and it was proposed in 1929 by the Prof. Constantin Budeanu and introduced in 1930 by the International Electrotechnical Commission in Stockholm, as international unit for reactive power [24].

There are two main states of energy potential (store) and kinetic (in motion, working). There are several forms of energy: chemical, nuclear, gravitational, elastic (stored), electrical, electromagnetic (radiant, light), mechanical, thermal (heat), ionization, sound.

The energy conversion takes place through power converters. Power converters can be rotating (electromechanical) or static.

5.1 Electromechanical power converters

In 1821 Michael Faraday invented the electric motor (a primitive version).

The relationship between the current and the voltage, known as Ohm's law, was quantified by Georg Ohm in 1827. Four years later, in 1831 Michael Faraday discovered the electromagnetic induction and the electromechanical conversion of the energy. Through the experiments, Faraday demonstrated that the electricity is obtained by friction, electromagnetic induction, chemical or thermoelectric. Werner von Siemens demonstrated the obtaining of the electric light from dynamo generators (1867). Siemens laid the groundwork of the modern electric generators. In treatise on electricity and magnetism (1873), James Clerk Maxwell published the unified theory of electricity and magnetism through Maxwell's equations.

Thomas Alva Edison found in 1870 the electric lighting solution by using incandescent light bulb. In 1882 Edison opened the Electric Light Company, the first public electric power plant to supply 110 volts direct current (DC) for the lights. In Westinghouse the first hydroelectric power plant opened in 1882.

Nikola Tesla invented the Tesla coil in 1883. In order to transmit the electric power through this type of transformer the low voltage is changed to high voltage. An induction motor supplied from the alternating current was developed by Tesla in 1887.

George Westinghouse developed the first multiple-voltage alternative current (AC) power distribution system by using the transformers, in 1886. The electric energy was produced from the hydroelectric generator. Westinghouse imported the transformers and Siemens AC alternators (generators) from Europe and formed Westinghouse Electric & Manufacturing Company in 1886, a power distribution plant.

Galileo Ferraris was the "Father of three-phase current" (1885). In 1888, Galileo Ferraris published the work on the AC polyphase motor.

In 1889, Mikhail Dolivo-Dobrovolsky invented the cage version of the three-phase induction motor and in 1890 the wound rotor version of the actual three-phase induction motor.

Nowadays, the power transmission over long distance is reliable in AC power system. It requires high voltage obtained by using transformers conducting to low currents; therefore, low transmission losses occur. However, by introducing the static power converters, the DC transmission becomes a challenge.

The generation of the electricity mostly takes place into a power station by using the electromechanical generators. The primary sources of the generators are heat engines fueled by fossil sources or nuclear fission but also by renewable energy: wind, water motion energy, solar or geothermal power.

5.2 Static power converters

The development of power electronics starts with French scientist J. Jasmin. Jasmin discovered in 1882 that the mercury electric arc assures conduction in one way [25]. By using this property, the rectifier had been born transforming alternating current (AC) in direct current (DC). In 1892, L. Arons invented the vacuum valve with mercury arc. In 1906, J. A. Fleming invented the vacuum diode. The silicon valve was invented by G. W. Pickard in 1906. The vacuum triode was invented by L. de Forest in 1907. The first production of ignitrons was made by Westinghouse Company in 1933. Starting with the transistor invention in 1948, the next generation of power converters arises. The silicon controlled rectifier (SCR) or thyristor, was invented in 1956 (John Moll). During 1956–1975, the power converters based on SCRs had been developed.

The new generation (1975–1990) of the power converters starts with new invented power switches: MOSFETs (metaloxide-semiconductor field-effect transistor, 1980), bipolar transistors and power bipolar junction transistors (BJT), the power gate turn-off (GTO) thyristors. The next generation of the power converters starts with the microprocessors, and application specified integral circuits (ASIC). The automatic control of power converters improves significantly the performances. The introduction of the intelligent power modules (IPM) boosts the performances of the power converters.

The main types of power converters are: rectifier, dc-dc chopper, inverters, direct (cycloconverters and matrix converters) and indirect (back-to-back) frequency converters. From commutation point of view, the power converters are natural commutated and force commutated. The phase-controlled rectifiers are natural commutated and used at high power. Due to the power quality problems, the performances are lower than of the force commutated power converters.

Besides of the development of the power devices and of the drivers, the modulation strategy is a key factor of efficiency improvements allowing the increased power capacity. The most common modulation strategy is pulse width modulation (PWM).

In order to obtain a variable speed, the alternating machines (synchronous and asynchronous) are combined with power converters. The most advanced control methods of the alternating machine are scalar control, vector control (field oriented), direct torque control, and sensorless control. The general exploration of the man-machine visual interactivity in the sense of the metaheuristic search algorithms can be a modern way for the energy domain exploitation. The modern algorithms based on artificial intelligence to find the optimal solution are also applied in this stringent energy area.

The features of the modern drives include remote control, networking, human-machine interface functions [26].

The new power converters should be protected against threats due to the digital communication and control. Therefore, the cybersecurity for the smart power converters should be taken into account [27]. Artificial intelligence and machine learning algorithms will be part of modern drive control systems.

5.3 Overview of the of the energy system

By knowing the input (energy production) and the output (energy consumption) of the the energy system the adequate mathematical model can be deducted. Taking into account the energy production (**Figure 3**—net electricity generation was 3.10 million GWh in 2016 [28]) from primary sources and the final energy consumption (**Figure 4**—EU 2016 energy consumption 2.78 million GWh) the main *energy conversion technologies* can be described briefly.

The hot water from the geothermal natural sources is used in steam power plant to obtain the electricity. The kinetic and potential energy from the rivers are transformed in electric energy through the hydro power station. The windmill captures the power of air flow to obtain the mechanical power. The mechanical power is converted into electricity by using the electric generators. The nuclear power plant used the nuclear energy to obtain the electricity. The photovoltaic cells are used to convert the solar energy into electricity. The most pollutant power conversion technologies are based on the fossil fuels. The transport sector is the most dominant consumption one where the fossil fuels are dominant, followed by the industry sector and households (**Figure 4**) [29]. Therefore, alternative energy solutions as

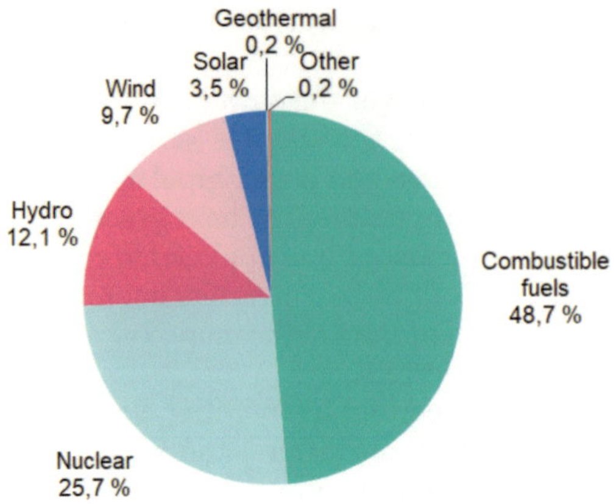

Figure 3.
Energy production [28] (% of total, based on GWh), EU-28, 2016.

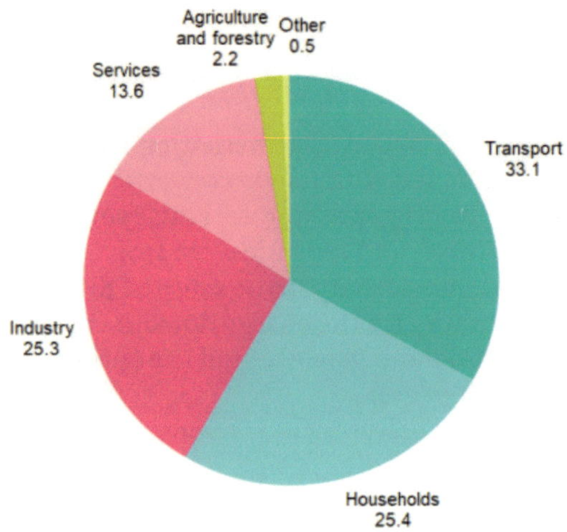

Figure 4.
Energy consumption [29] (% of total, based on GWh), EU-28, 2016.

environmentally friendly way should be taken into account. The prediction of the world energy consumption between 2010 and 2040 is an increase by 56% [30]. Currently, almost the entire world energy production is covered by hydropower plants, thermal power stations, and nuclear power plants. The environmentally friendly renewable energy will expand in the future.

The first optical amplification of the light based on stimulated emission of electromagnetic radiation, known as light amplification by stimulated emission of radiation (LASER) was invented in 1960 by Theodore H Maiman. The application of laser could be in energy transmission between the space stations and earth or could be used for the propulsion of the spatial vehicle [31].

Recently (2018), the Lawrence Livermore National Laboratory (LLNL) has obtained record of 2.15 megajoules (MJ) of energy with the National Ignition Facility (NIF) laser system [32]. The main purpose of NIF is to produce "green" electric power by fusion without producing radioactive waste.

Nobel prize in physics 2014 was awarded for the invention of the blue light-emitting diodes (LEDs), helped to create energy-efficient light sources in a completely new way. The impact of discovering white LED lamps will decrease significantly the world's electricity used for lighting [33].

The extreme light infrastructure—nuclear physics (ELI-NP) pan European project has the main outcome the highest intensity laser system consisting of two 10 PW laser arms. The laser system will develop laser beam of 10^{23} W/cm^2 power intensity and up to 10^{15} V/m electrical fields [34]. The current record is HERCULES Petawatt laser at the University of Michigan, USA. The focused laser beam has the intensity of 2×10^{22} W/cm^2 [35, 36]. For a comparison, the strength intensity of the lightning electric field is 3×10^6 V/m (30 kV/cm) under dry air condition, at atmospheric pressure [37, 38]. The methods of measuring of these strength fields are described in [39]. The project ELI-NP was promoted by Professor Gérard Mourou in 2006 [40], one of the winners of the Nobel prize in physics 2018 [41]. By using laser technology, laser power transfer will be a solution to power the space vehicles or interplanetary communications.

6. Cyber physical systems

In order to increase the safety, security and reliability of the electricity the entire systems of energy domain should integrate the modern cyber-security features. Cyber physical systems (CPS) are based on the special architecture [42].

Two key enabled factors conduct to the design of modern energy plant infrastructures:

1. due to the limited flexibility and vulnerability of the ancient bulk centralized power systems;

2. protecting the bidirectional (receiving the adequate signals from the devices, transmitting the adequate control signals) communication data.

According to [43], there are two main causes that enable major changes to the electricity infrastructure:

- Environmental sustainability

- Effective management of pervasive data and extracted information

Two deterministic CPS models are described in [44], and used for practical realization of the distributed CSP through two projects: PRET (precision timed (PRET) computation in cyber-physical systems) [45], and Ptides (programming temporally-integrated distributed embedded systems) [46].

The pathways of the cyber physical systems developing should conduct to open CPS (cyber-physical systems) platforms [47]. The vision or short-term research strategy of any country should include the integration into Continental Cyber Physical Platforms (CCPP), respectively Intercontinental Cyber Physical Platforms (ICCPP). One development direction already started through the Open Source Cyber Platform [47]. At the same time, one of the European initiatives is made by the European Strategy Forum on Research Infrastructures [48].

Some of the main CPS design features could be mentioned as follows: to be modular, easily interconnecting with other CPS (plug and play), communication, self-healing, restoration, resiliency, autonomy, robustness, sensor-connectivity, and security assurance.

In order to ensure world protection against internal or external attacks, the synergies between the local CPS should be enabled, integrating in one intercontinental platform (ICCPP). One of the European projects has as main objective to disperse the CPSs to market through the seven platforms [49].

According to [50], the information about the current continental research platform in Europe is available. The European research infrastructures in Energy area could be finding on [51].

7. Conclusion

As the electric energy has the great importance for human development, in order to obtain it, any research strategy should include the electric power conversion development. Nowadays, the symbiosis between energy and information and communications technology is already taken into consideration. However, one key success of fast technology developing is the sharing of knowledge through open source. All the budget funded research projects should provide back the accumulated research results to the entire world. A win-to-win mechanism to access the high-tech laboratories by the worldwide researchers has been already created at least by the European Parliament and US officials. Another key-enabled of fast technology developing is the adequate training of the human resources through the advanced and innovative teaching, and researching methods. The curriculum development and adequate tools for education should be adapted very fast from the Digital Era [52] to the forthcoming Cyborg Era [53].

Acknowledgements

This work was supported by a grant of the Romanian National Authority for Scientific Research, CNDI—UEFISCDI, project number PN-II-PT-PCCA-2011-3.2-1680.

Author details

Marian Găiceanu[1,2]

1 Dunarea de Jos University of Galati, Galati, Romania

2 Integrated Energy Conversion Systems and Advanced Control of Complex Processes Research Center, Galati, Romania

*Address all correspondence to: marian.gaiceanu@ieee.org

References

[1] Benedict K, Yuzhi H, Long John A. Electroreception in early vertebrates: Survey, evidence and new information. Palaeontology. 2018;**61**(3):325-358. DOI: 10.1111/pala.12346

[2] Keyser PT. The purpose of the parthian galvanic cells: A first century A.D. electric battery used for analgesia. Journal of Near Eastern Studies. 1993;**52**(2):81-83

[3] The Library of Congress. Everyday Misteries [Internet]. 2017. Available from: http://www.loc.gov/rr/scitech/mysteries/static.html [Accessed: January 10, 2019]

[4] Ancient Code, Ivan Petricevic. Did Ancient Egyptians Have Electricity? 2012. Available from: https://www.ancient-code.com/did-ancient-egyptians-have-electricity [Accessed: January 10, 2019]

[5] Larry Brian Radka. A Short History of Ancient Electricity. 2006. Available from: https://www.bibliotecapleyades.net/ciencia/ciencia_hitech05.htm [Accessed: January 10, 2019]

[6] Frank Dörnenburg, "Electric Lights in Egypt?" Available from: http://WorldMysteries.com, http://old.world-mysteries.com/sar_lights_fd1.htm [Accessed: January 10, 2019]

[7] Available from: https://www.smashinglists.com/10-first-electricity-milestones/ [Accessed: January 10, 2019]

[8] Guarnieri M. Electricity in the age of enlightenment. IEEE Industrial Electronics Magazine. 2014;**8**(3):60-63. DOI: 10.1109/MIE.2014.2335431

[9] National Archives, The Kite Experiment; 19 October 1752 [Internet]. 2017. Available from: https://founders.archives.gov/documents/Franklin/01-04-02-0135 [Accessed: January 10, 2019]

[10] Lightning and Thunder. Encyclopedia Britannica [Internet]. Available from: https://www.britannica.com/science/lightning-meteorology/media/340767/194043 [Accessed: January 10, 2019]

[11] The Electric Ben Franklin [Internet]. 1995. Available from: http://www.ushistory.org/franklin/philadelphia/lightning.htm [Accessed: January 10, 2019]

[12] Bolt of Lightning. A Memorial to Benjamin Franklin #1695, Direct Line Development [Internet]. Available from: http://freephillyphoto.com/photos/1695/ [Accessed: January 10, 2019]

[13] Atmospheric Electricity [Internet]. 2019. Available from: https://en.wikipedia.org/wiki/Atmospheric_electricity [Accessed: January 10, 2019]

[14] Lightning Energy Storage System. US20140042987A1. 2014; Northern Lights Semiconductor Corp. Available from: https://patents.google.com/patent/US20140042987A1/en [Accessed: January 10, 2019]

[15] Collecting and Storage Device of Lightning. CN106787229A. 2017; Anyang Teachers' College. Available from: https://patents.google.com/patent/CN106787229A/en [Accessed: January 10, 2019]

[16] Xu H, He Y, Strobel KL, Gilmore CK, Kelley SP, Hennick CC, et al. Flight of an aeroplane with solid-state propulsion. Nature. 2018;**563**:532-535. Available from: https://www.nature.com/articles/s41586-018-0707-9 [Accessed: January 10, 2019]

[17] Available from: http://news.mit.edu/2018/first-ionic-wind-plane-no-moving-parts-1121 [Accessed: January 10, 2019]

[18] Luo X, Wang J, Dooner M, Luo JCX, et al. Overview of current development in electrical energy storage technologies and the application potential in power system operation. Applied Energy. 2015;**137**:511-536. Table 10. Technical characteristics of electrical energy storage technologies

[19] Available from: http://www.nawatechnologies.com/en/nawa-technologies-revolutionizes-energy-storage-with-new-ultra-fast-carbon-battery/ [Accessed: January 10, 2019]

[20] Available from: http://www.rh.gatech.edu/features/12-emerging-technologies-may-help-power-future [Accessed: January 10, 2019]

[21] Available from: https://www.britannica.com/biography/Richard-Trevithick [Accessed: January 10, 2019]

[22] Michaelides EE. The Second Law of Thermodynamics as Applied to Energy Conversion Processes. Wiley Online Library; 1984. DOI: 10.1002/er.4440080306

[23] EUR-Lex. Document 31980L0181. 1979. [Internet]. Available from: https://eur-lex.europa.eu/legal-content/EN/ALL/?uri=celex:31980L0181 [Accessed: January 10, 2019]

[24] Available from: https://en.wikipedia.org/wiki/Volt-ampere_reactive

[25] Introduction to Electronic Engineering. ISBN 978-87-7681-539-4. 2010; Valery Vodovozov & Ventus Publishing ApS

[26] Available from: http://www.ene.ttu.ee/elektriajamid/oppeinfo/materjal/AAV0020/1Introduction1.pdf [Accessed: January 10, 2019]

[27] Qi J, Hahn A, Lu X, Wang J, Liu C-C. Cybersecurity for distributed energy resources and smart inverters. IET Cyber-Physical Systems: Theory & Applications. 2016;**1**(1):28. DOI: 10.1049/iet-cps.2016.0018. Available from: https://ieeexplore.ieee.org/stamp/stamp.jsp?arnumber=7805366

[28] Net Electricity Generation, EU-28 (% of total, based on GWh), Source: Eurostat (nrg_105a). 2016. Available from: https://ec.europa.eu/eurostat/statistics-explained/index.php/Electricity_production_consumption_and_market_overview [Accessed: January 10, 2019]

[29] Available from: http://appsso.eurostat.ec.europa.eu/nui/submitViewTableAction.do [Accessed: January 10, 2019]

[30] Today in Energy, U.S. Energy Information Administration [Internet]. 2017. Available from: https://www.eia.gov/todayinenergy/detail.php?id=32912

[31] Available from: http://adsabs.harvard.edu/abs/1975AsAer..13...56B

[32] Available from: https://phys.org/news/2018-07-national-ignition-facility-laser-energy.html

[33] The Nobel Prize in Physics 2014. NobelPrize.org. Nobel Media AB 2019. 2019. Available from: https://www.nobelprize.org/prizes/physics/2014/summary/ [Accessed: January 10, 2019]

[34] Gales S, Tanaka KA, Balabanski DL, Negoita F, Stutman D, Tesileanu O, et al. The extreme light infrastructure-nuclear physics (ELI-NP) facility: New horizons in physics with 10 PW ultra-intense lasers and 20 MeV brilliant gamma beams. Reports on Progress in Physics. 2018;**81**(9):094301. DOI: 10.1088/1361-6633/aacfe8. Available

from: https://www.ncbi.nlm.nih.gov/pubmed/29952755 [Accessed: January 10, 2019]

[35] Bahk S-W, Rousseau P, Planchon TA, Chvykov V, Kalintchenko G, Maksimchuk A, et al. Generation and characterization of the highest laser intensities (10^{22} W/cm^2). Optics Letters. 2004;**29**(24):2837. DOI: 10.1364/ol.29.002837

[36] Available from: http://www.guinnessworldrecords.com/world-records/highest-intensity-focused-laser/ [Accessed: January 10, 2019]

[37] Sabu S, Srichitra S, Joby NE, Premlet B. Electric field characteristics during a thunderstorm: A review of characteristics of electric field prior to lightning strike. In: IEEE International Conference on Signal Processing, Informatics, Communication and Energy Systems (SPICES). 2017. DOI: 10.1109/spices.2017.8091357

[38] Rakov VA, Uman MA. Lightning: Physics and Effects. New York: Cambridge University Press; 2003. ISBN: 9780521583275

[39] Iarossi S, Poscolieri M, Rafanelli C, Franceschinis D, Rondini A, Maggi M, et al. The measure of atmospheric electric field. Lecture Notes in Electrical Engineering. 2011;**91**(3):175-179

[40] Available from: http://www.eli-np.ro/index.php

[41] Available from: https://www.nobelprize.org/uploads/2018/10/press-physics2018.pdf

[42] Yang L, Yu P, Wang B, Yao S, Liu Z. Review on cyber-physical systems. IEEE/CAA Journal of Automatica Sinica. 2017;**4**(1):27-40

[43] Kezunovic M, Annaswamy A, Dobson I, Grijalva S, Kirschen D, Mitra J, et al. Energy Cyber-Physical Systems: Research Challenges and Opportunities. 2016; National Science Foundation

[44] Lee EA, Cheng AMK. The past, present and future of cyber-physical systems: A focus on models. Sensors. 2015;**15**(3):4837-4869. DOI: 10.3390/s150304837

[45] Available from: https://ptolemy.berkeley.edu/publications/papers/07/PRET/ [Accessed: January 10, 2019]

[46] Available from: https://ptolemy.berkeley.edu/projects/summaries/11/PTIDES.htm [Accessed: January 10, 2019]

[47] Jorge Juan Rodríguez Vázquez. Open CPS Platforms. 2015. DOI: 10.13140/RG.2.1.3161.4561. Available from: https://www.researchgate.net/publication/280095935_Open_CPS_Platforms [Accessed: January 10, 2019]

[48] Available from: http://roadmap2018.esfri.eu/[Accessed: January 10, 2019]

[49] Available from: https://www.eurocps.org/ [Accessed: January 10, 2019]

[50] Available from: http://roadmap2018.esfri.eu/media/1066/esfri-roadmap-2018.pdf [Accessed: January 10, 2019]

[51] Available from: http://ec.europa.eu/research/infrastructures/index.cfm [Accessed: January 10, 2019]

[52] Shepherd J. What Is the Digital Era? Source Title: Social and Economic Transformation in the Digital Era2004. DOI: 10.4018/978-1-59140-158-2.ch001. Available from: https://www.igi-global.com/chapter/digital-era/29024 [Accessed: January 10, 2019]

[53] Robert F. Service, BIOELECTRONICS. The Cyborg Era Begins, Science. 07 Jun 2013;**340**(6137):1162-1165. DOI:10.1126/science.340.6137.1162

Section 2

Electromechanical Power Conversion

Chapter 2

Analysis of Compensated Six-Phase Self-Excited Induction Generator Using Double Mixed State-Space Variable Dynamic Model

Kiran Singh

Abstract

In this article, a mixed current-flux d-q modeling of a saturated compensated six-phase self-excited induction generator (SP-SEIG) is adopted during the analysis. Modeling equations include two independent variables namely stator current and magnetizing flux rather than single independent variables either current or flux. Mixed modeling with stator current and magnetizing flux is simple by having only four saturation elements and beneficial in study of both stator and rotor parameters. Performance equations for the given machine utilize the steady-state saturated magnetizing inductance (Lm) and dynamic inductance (L). Validation of the analytical approach was in good agreement along with three-phase resistive or resistive-inductive loading and also determined the relevant improvement in voltage regulation of machine using series capacitor compensation schemes.

Keywords: mixed double state space variables, self-excitation, six-phase, compensation, induction generator, non-conventional energy

1. Introduction

Traditionally, synchronous generators have been used for power generation, but induction generators are increasingly being used these days because of their relative advantageous features over conventional synchronous generators. The need for external reactive power limits the application of an induction generator as isolated unit. The use of SEIG, due to its reduced unit cost, simplicity in operation and ease of maintenance are most suited in such system. These entire features facilitate the operation of induction generator in stand-alone mode to supply far-flung areas where extension of grid is economically not viable. The stand-alone SEIG can be used with conventional as well as non-conventional energy sources to feed remote single family, village community, etc. in order to expedite the electrification of rural and remote locations. A detailed dynamic performance of induction generator 'IG' operating in different modes i.e. isolated and grid-connected is necessary for the optimum utilization of its various favorable features. The investigations spread over last two decades also indicate the technical and economic vitality of using number

 IntechOpen

of phases higher than three in AC machines for applications in marine ships, thermal power plant to drive induced draft fans, electric vehicles and circulation pumps in nuclear power plants etc. In this area, research is still in its early stage, yet some extremely great authority's findings have been reported in the previous literatures indicating the general expediency of multi-phase systems. The literature regarding multi-phase IG is nearly not available since it has only three findings before 2004. The first article on multi-phase induction generator is appeared in 2005, along with rest of theoretical and practical works on SP-SEIG using single state- space variables either stator and rotor d-q axis currents or stator and rotor d-q axis flux linkages so far as reported [1]. On the basis of previous article reviews, before companion paper of 2015 [1], there were no literatures on modeling and analysis of SP-SEIG, using d-q axis components of stator current and magnetizing flux as mixed state–space variables. In the view of novelty, such mathematical modeling and analysis were carried out in detail for three-phase SEIG only, in a very few available literatures and some of which are mentioned by [2]. The purpose of this article is also to accomplish a similar task for modeling and analysis of SP-SEIG with series compensation scheme using double state-space variables as were proposed by companion paper without compensation. The simulation is performed on series compensated SP-SEIG by using 4th order Runge-Kutta subroutine in Matlab software.

2. Modeling description

Concerns about mathematical modeling of SP-SEIG, short-shunt series compensation capacitors and static resistive 'R' and balanced three-phase reactive 'R-L'

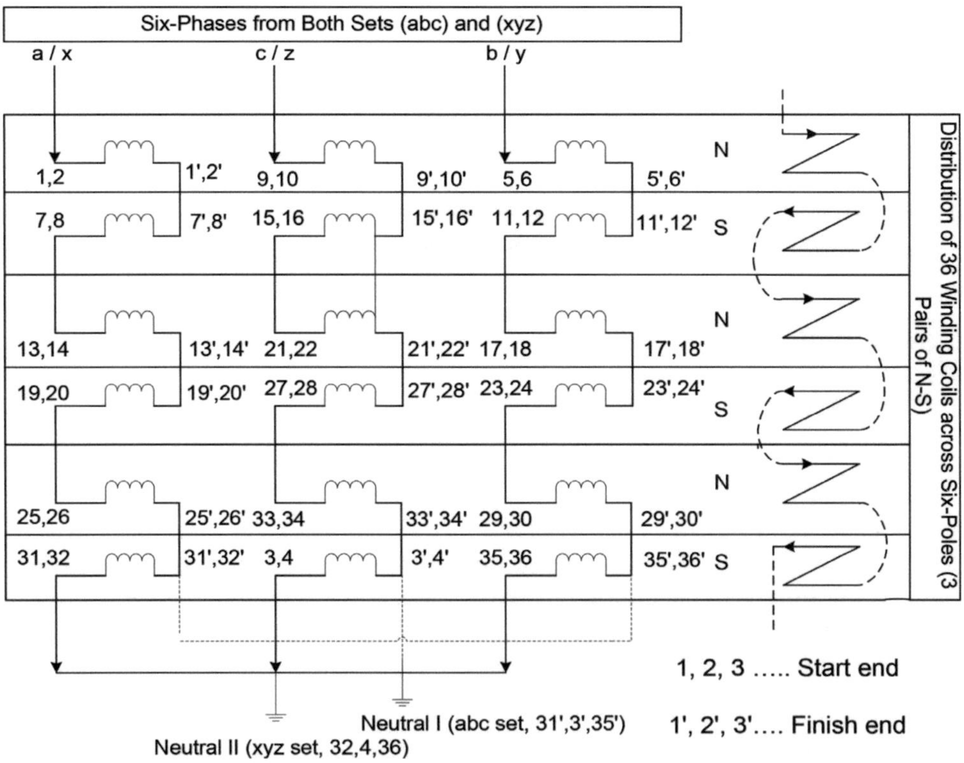

Figure 1.
Distribution of 6-phases in 36 slots of 6 pole induction machine.

loads as previously discussed in [1, 3] are not described in this section, only briefly summarized along with newly addition of long-shunt series compensation capacitors.

2.1 SP-SEIG model

A basic two-pole, six-phase induction machine is schematically described by its stator and rotor axis [1]. In which, six stator phases of both sets, *a, b, c* and *x, y, z* (set I and II, respectively) are arranged to form two sets of uniformly distributed star configuration, displaced by an arbitrary angle of 30 electrical degree gravitate asymmetrical winding structure. The distribution of 6-phases in 36 slots of 6 pole induction machine is also shown in **Figure 1**. Previously, voltage equations using single state-space variable namely flux linkage were used in the expanded form for six-phase induction machine.

After simplification of voltage equations using double mixed state-space variables namely stator currents and magnetizing fluxes in the machine model following form occurs from Eqs. (1)–(39) of [1]:

$$[V_{dq}] = [H][dX_{dq}]/dt + [J][X_{dq}] \tag{1}$$

where $[V_{dq}] = [V_{d1}\ V_{q1}\ V_{d2}\ V_{q2}\ 0\ 0]^t$, $[X_{dq}] = [i_{d1}\ i_{q1}\ i_{d2}\ i_{q2}\ \psi_{dm}\ \psi_{qm}]^t$ and matrices [H] and [J] are given by Eqs. (2) and (3), respectively.

$$7\begin{bmatrix} -L_{\sigma 1} & 0 & 0 & 0 & 1 & 0 \\ 0 & -L_{\sigma 1} & 0 & 0 & 0 & 1 \\ 0 & 0 & -L_{\sigma 2} & 0 & 1 & 0 \\ 0 & 0 & 0 & -L_{\sigma 2} & 0 & 1 \\ L'_{\sigma r} & 0 & L'_{\sigma r} & 0 & \left(1+\frac{L'_{\sigma r}}{L'_{dd}}\right) & \frac{L'_{\sigma r}}{L_{dq}} \\ 0 & L'_{\sigma r} & 0 & L'_{\sigma r} & \frac{L'_{\sigma r}}{L_{dq}} & \left(1+\frac{L'_{\sigma r}}{L_{qq}}\right) \end{bmatrix} \tag{2}$$

$$\begin{bmatrix} -r_1 & wL_{\sigma 1} & 0 & 0 & 0 & -w \\ -wL_{\sigma 1} & -r_1 & 0 & 0 & w & 0 \\ 0 & 0 & -r_2 & wL_{\sigma 2} & 0 & -w \\ 0 & 0 & -wL_{\sigma 2} & -r_2 & w & 0 \\ r'_r & -(w-w_r)L'_{\sigma r} & r'_r & -(w-w_r)L'_{\sigma r} & \frac{r'_r}{L_m} & -(w-w_r)\left(1+\frac{L'_{\sigma r}}{L_m}\right) \\ (w-w_r)L'_{\sigma r} & r'_r & (w-w_r)L'_{\sigma r} & r'_r & -(w-w_r)\left(1+\frac{L'_{\sigma r}}{L_m}\right) & \frac{r'_r}{L_m} \end{bmatrix} \tag{3}$$

The nonlinear equations of voltage and current across the shunt excitation capacitor and series compensation capacitors (short-shunt and long-shunt) can be transformed into d-q axis by using reference frame theory, i.e. Park's (d_{q0}) transformation [4], are given by Section 2.2 of [1] and (Section 2.3 of [1] and by following Section 2.2), respectively. Modeling of static loads is also given in Section 3 of [1].

2.2 Modeling of long-shunt capacitors

Current through series capacitors Cls1 and Cls2 (in case of long shunt), connected in series with winding set I and II, respectively, is same as the machine

ψ_{dm}, ψ_{qm}, $1/L_{dd}$, $1/L_{qq}$, $1/L_{dq}$	d- and q-axis magnetizing flux linkages, saturation dependent coefficients of system matrix A
p, σ, P, J, θr, ω, ω_b, ω_r	differentiation w. r. t. time, index for constant, number of pole pairs, moment of inertia, electrical angular displacement of the rotor, reference frame speed, base speed and rotor speed
L_m, L, L_{dq}, cosμ, sinμ	steady-state saturated magnetizing inductance, dynamic inductance, cross-saturation coupling between the d- and q-axis of stator, angular displacements of the magnetizing current space vector with respect to the d-axis of the common reference frame
C_{ls1} C_{ls2}	long-shunt capacitors across the stator winding set I and II

Table 1.
Machine model symbols.

current. The machine current along with series capacitance determine the voltage across series long-shunt capacitor and when transformed in to d-q axis by using Park's transformation is given in Eqs. (4) and (5) [5–7].

$$\begin{aligned} \rho V_{q1ls} &= i_{q1}/C_{ls1} \\ \rho V_{d1ls} &= i_{d1}/C_{ls1} \\ \rho V_{q2ls} &= i_{q2}/C_{ls2} \\ \rho V_{d2ls} &= i_{d2}/C_{ls2} \end{aligned} \tag{4}$$

and the load terminal voltage is expressed as

$$\begin{aligned} V_{Lq1} &= V_{q1} + V_{q1ls} \\ V_{Ld1} &= V_{d1} + V_{d1ls} \\ V_{Lq2} &= V_{q2} + V_{q2ls} \\ V_{Ld2} &= V_{d2} + V_{d2ls} \end{aligned} \tag{5}$$

The remaining symbols of machine model have their usual meanings from Ref. [1] and **Table 1**.

3. Methodology

In this section, a numerical method is introduced to the solution of Eqs. (1)–(3); where double mixed current flux state space model is discussed by [1]. The ordinary linear differential equations can be solved by the analytical technique rather than approximation method. Eq. (1) is non-linear differential and cannot be solved exactly with high expectations, only approximations are estimated numerically by computer technique using 4th order Runge-Kutta method or classical Runge-Kutta method or often referred as "RK4" as so commonly used [8]. The analytical response of compensated SP-SEIG in only single operating mode is carried out under significant configuration using RK4 subroutine implemented in Matlab M-file. The dynamic performances were determined under no load, R load and R-L loading condition in only the single mode of excitation capacitor bank, and in both modes of compensating series capacitor bank. The following analytical dynamic responses of series compensated SP-SEIG is considered for the validity of proposed approaches in this chapter.

- During R and R-L loading with short shunt compensation.
- During R and R-L loading with long shunt compensation.

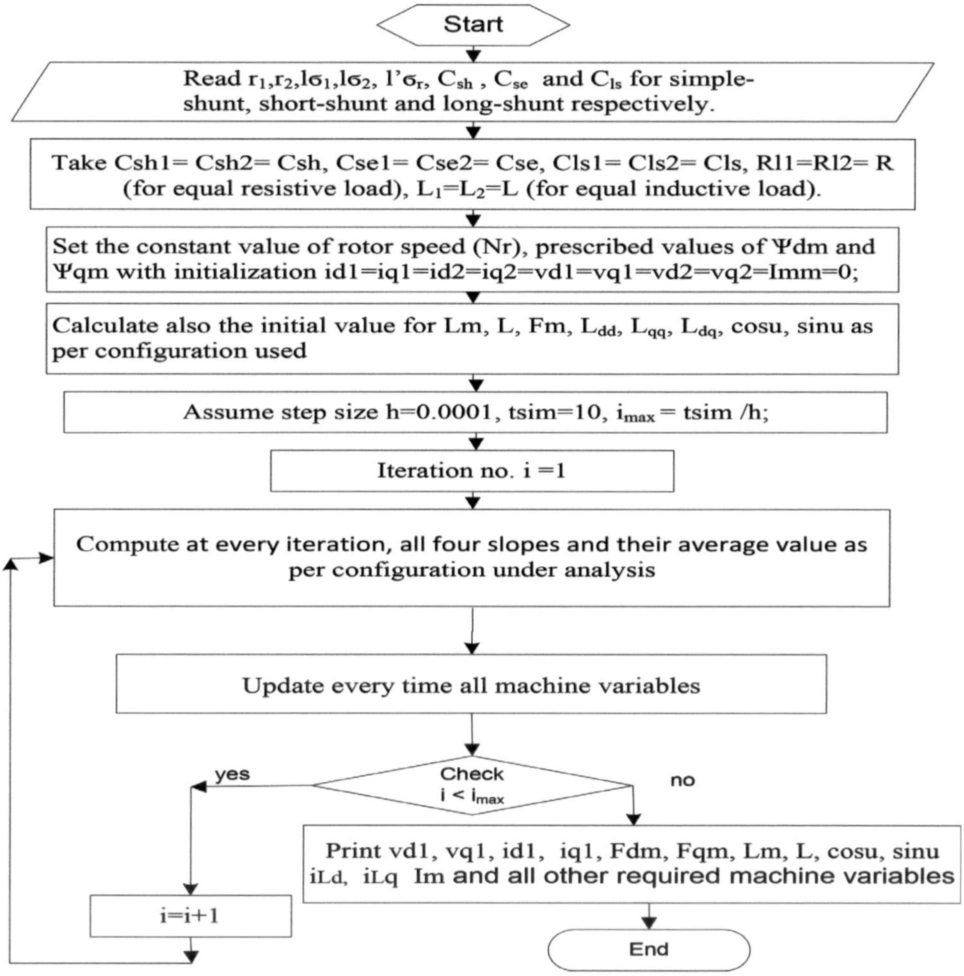

Figure 2.
Algorithm for Runge-Kutta method implemented for SP-SEIG under constant rated speed at 1000 RPM.

Both analytical responses are well detailed in Section 4. The analytical study of compensated SP-SEIG is given in Section 4 by using an explicit MATLAB program incorporates the RK4 method. An Algorithm of RK4 method for the analysis of compensated SP-SEIG is also shown in **Figure 2**. The parameters of studied machine and saturation dependent coefficients of system matrix [H] are also reported by [1] for further dynamic analysis of saturated compensated SP-SEIG using double mixed state space variable model under constant rotor speed along with appropriate initial estimated variables values which are also responsible in the development of rated machine terminal voltage and it depends on other machine variables.

4. Analytical response

Analytical performances of compensated SP-SEIG are illustrated in **Figures 3–6** along with sudden switching of R load of 200 Ω and a balanced three-phase R-L load (200 Ω in series with 500 mH) at t = 2 s when short-shunt and long-shunt compensation along both three-phase winding sets. These waveforms are

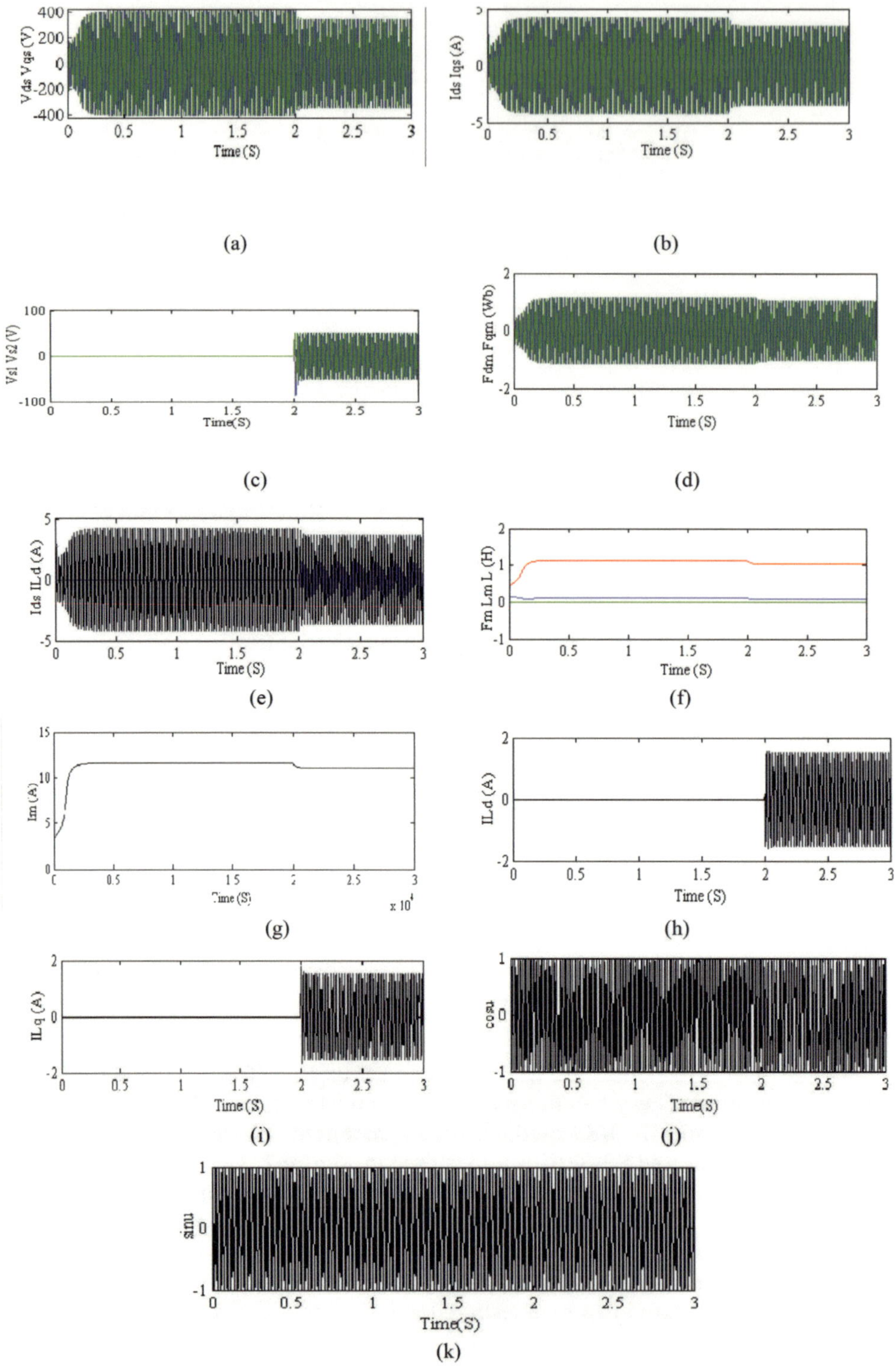

Figure 3.
Analytical waveforms during sudden switching of R load of 200 Ω at t = 2 s.

correspond to d-q axis stator voltage, d-q axis stator current, d-q axis load current, d-q axis magnetizing flux, d-q axis series capacitor voltage and magnetizing current. In the case of R-L load, generated voltage and current amplitude has been dropped few more volts and amperes, respectively, compared to R load. Combined

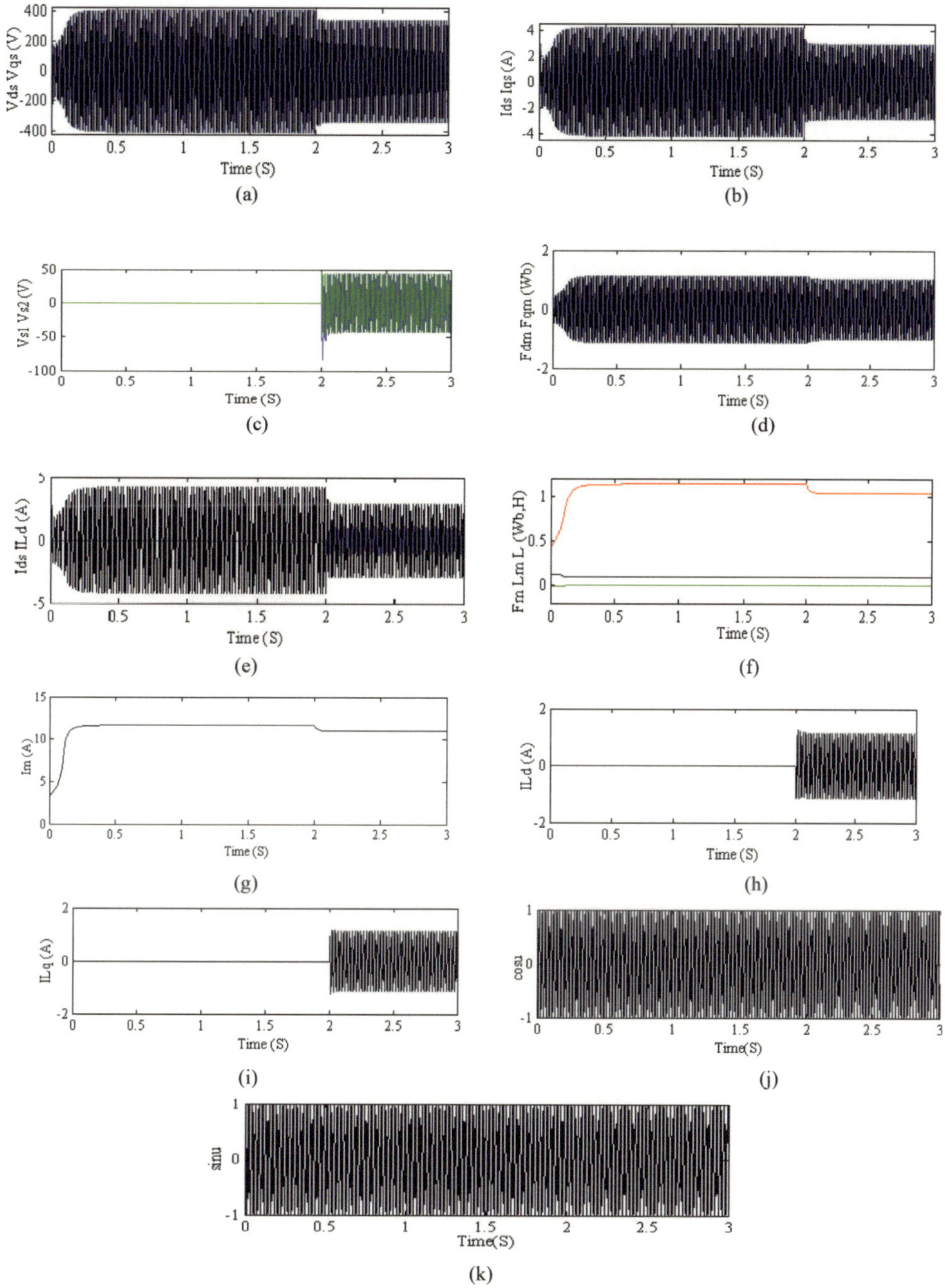

Figure 4.
Analytical waveforms during sudden switching of RL load at t = 2 s.

amplitude of magnetizing flux, steady-state saturated magnetizing inductance and dynamic (tangent slope) inductance along with the combined amplitude waveform of d-axis stator current and load current during no-load and sudden switching of R load of 200 Ω at t = 2 s are also shown in **Figures 3–6**. As per discussion of [1], generated RMS value of steady state terminal voltage and current is about 225 V line to line and 4.46 A at rated speed of 1000 RPM with the value of excitation capacitance of 38.5 μF per phase in simple-shunt configuration for the selected machine in this article. In proposing stator current and magnetizing flux mixed variable model,

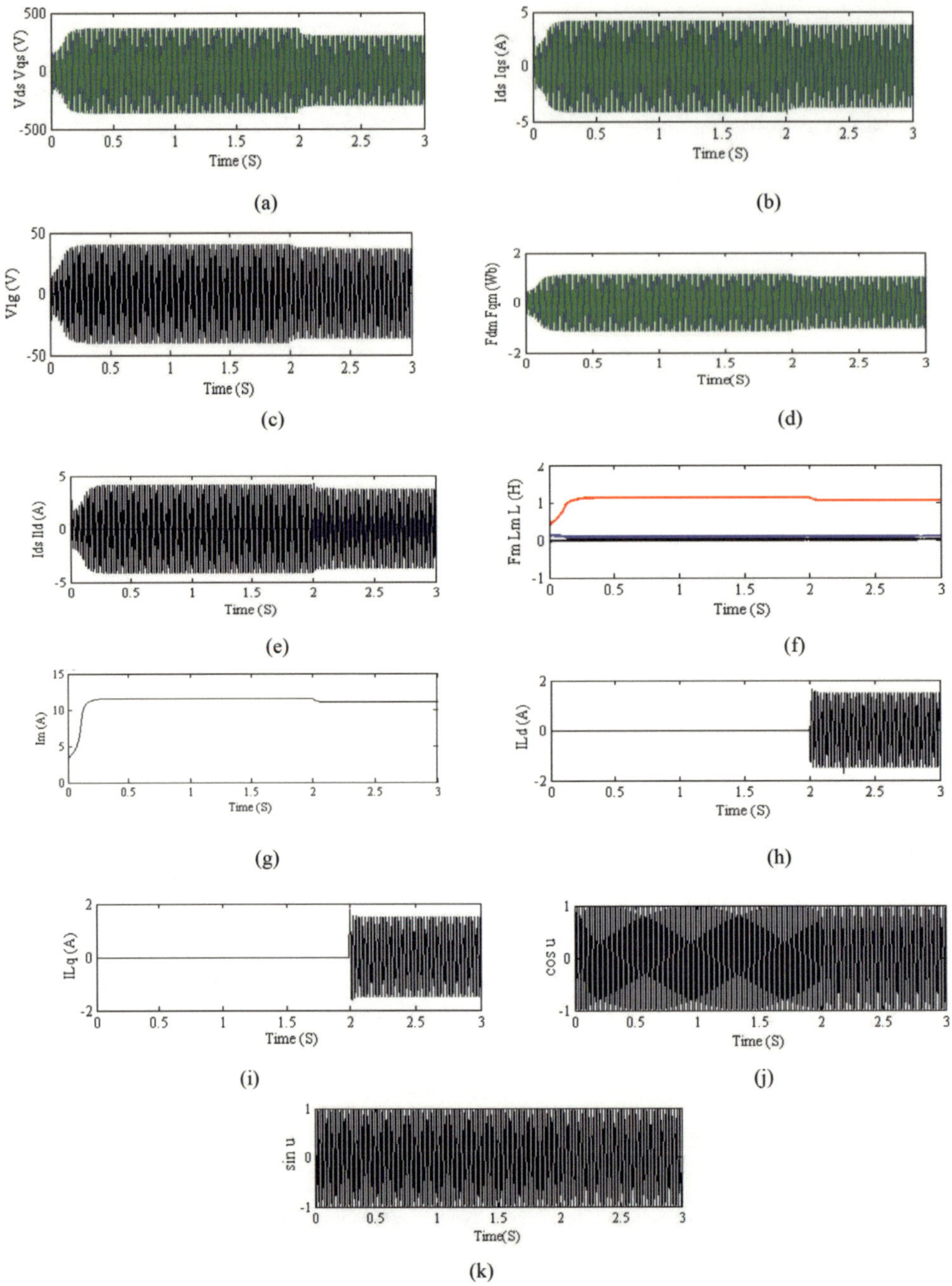

Figure 5.
Analytical waveforms during sudden switching of R load of 200 Ω at t = 2 s.

it has been observed that SP-SEIG terminal voltage and current build-up from their initial values to final steady state values entirely depends upon its initial few Weber values of d-and q-axis magnetizing flux at constant rated speed of 1000 RPM.

4.1 Short-shunt series compensated SP-SEIG

Performance of short-shunt SP-SEIG in only its single mode of operation using the values of excitation and series capacitor banks of 38.5 and 108 μF per phase, respectively, has been predicted from the built explicit MATLAB program using RK4 subroutine. Computed waveforms are shown in **Figures 3** and **4**. Application

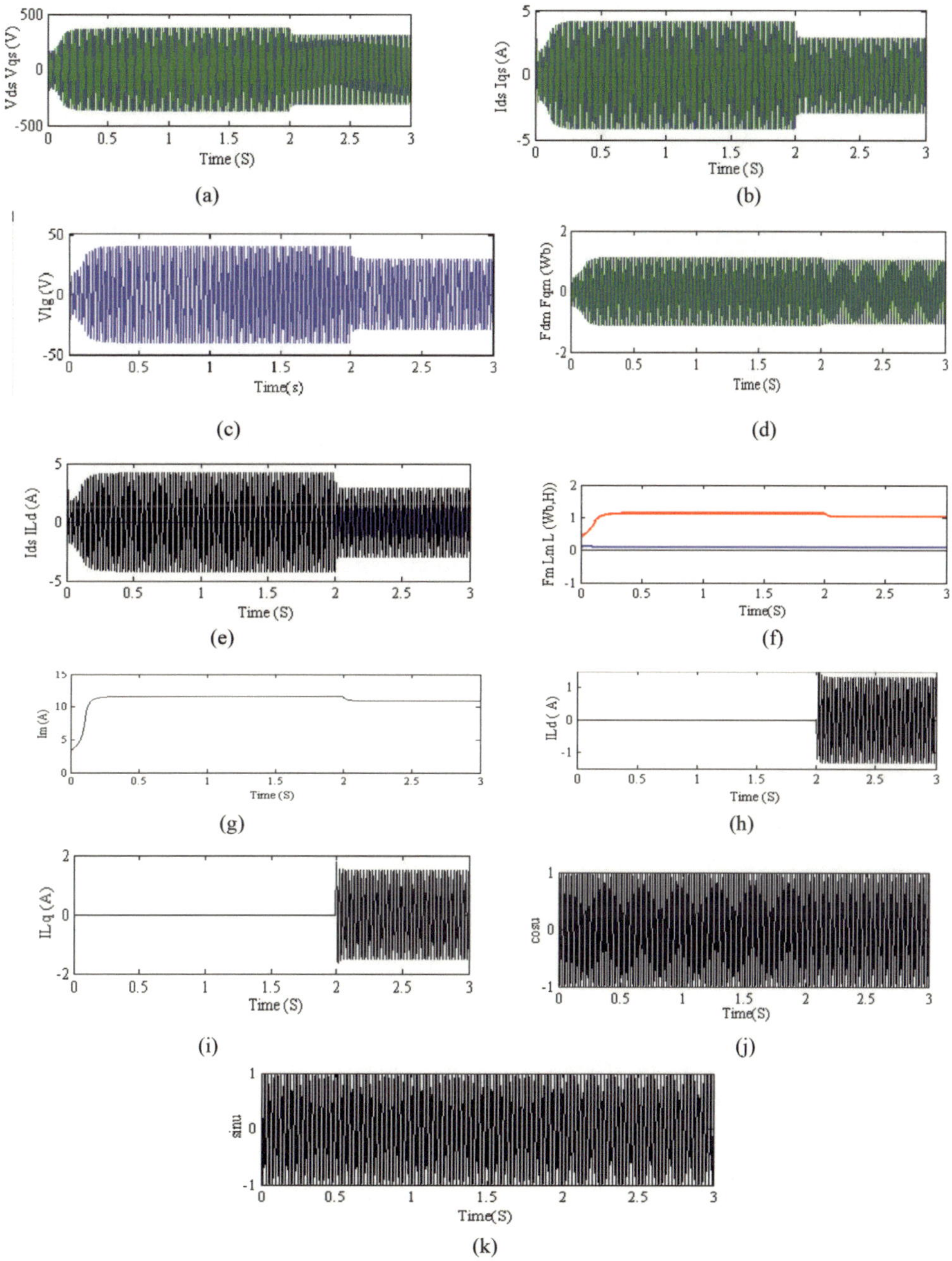

Figure 6.
Analytical waveforms during sudden switching of R-L load at t = 2 s.

of short-shunt scheme results in overvoltage across the generator terminals as shown in **Figure 3a**, the per phase voltage level is more than the voltage level of **Figure 5a** and it is illustrated in **Figure 3c**.

4.1.1 When both three-phase winding sets are connected in short-shunt configuration with independent R loading

The analytical d-q waveform of voltage, current, magnetizing flux and magnetizing current during no-load and sudden switching of R load of 200 Ω at t = 2 s are shown in corresponding **Figure 3a, b, d** and **g**. In addition, combined amplitude waveforms of magnetizing flux, steady-state saturated magnetizing inductance and dynamic (tangent slope) inductance is shown in **Figure 3f**. The angular displacements of the magnetizing current space vector with respect to the d-axis of the common reference frame are also shown in **Figures 3j** and **k**. The combined d-q axis voltage drop across the series short-shunt capacitor is given in **Figure 3c** and combined amplitude waveform of d-axis stator current and load current during no-load and sudden switching of R load of 200 Ω at t = 2 s is given in **Figure 3e**. The d- and q-axis load currents are also depicted in **Figure 3h** and **i**, respectively, during sudden switching of R load of 200 Ω at t = 2 s.

4.1.2 When both three-phase winding sets are connected in short-shunt configuration with independent R-L loading

In the same order of figure numbers, all computed waveforms are shown in **Figure 4** with sudden switching of R-L load (200 Ω resistance in series with 500 mH inductor) at t = 2 s. Analytical generated RMS steady state voltage and corresponding current at rated speed of 1000 RPM are shown in **Figure 4a** and **b**. Drops in d-q-axis generating and lagging load currents are shown in **Figure 4b, h** and **i**.

4.2 Long-shunt series compensated SP-SEIG

It is also seen that like short-shunt compensation, long-shunt compensation is also self- regulating in nature. In the same manner, analysis of long-shunt SP-SEIG along with a single mode of excitation and the series capacitor banks of 38.5 and 350 F respectively, have also been computed and predicted by using the RK4 subroutine and illustrated in **Figures 5** and **6**. Here, the value of series capacitor is more than the twice of short-shunt series capacitor.

4.2.1 When both three-phase winding sets are connected in long-shunt configuration with independent R loading

The analytical waveform of voltage, current, magnetizing flux and magnetizing current during no-load and sudden switching of R load of 200 Ω at t = 2 s with long-shunt compensation along both three-phase winding sets are respectively shown in **Figure 5a, b, d** and **g**. Combined amplitude waveforms of magnetizing flux, steady-state saturated magnetizing inductance and dynamic (tangent, slope) inductance is shown in **Figure 5f**. The angular displacements of the magnetizing current space vector with respect to the d-axis of the common reference frame are also shown in **Figure 5j** and **k**. The application of the long-shunt scheme results in less overvoltage or reduced terminal voltage across the generator terminals. As it is shown in **Figure 5a**, the per phase voltage level is less than the voltage level of **Figure 3a**. It gives evidence that long-shunt SP-SEIG is able to deliver output power at reduced terminal voltages, as shown in **Figure 5c**. The combined d-q axis voltage drop across the series short-shunt capacitor is given in **Figure 5c** and combined amplitude waveform of d-axis stator current and load current during no-load and sudden switching of resistive load of 200 Ω at t = 2 s is given in **Figure 5e**. The d- and q-axis load currents are also depicted in **Figure 5h** and **i** respectively, during sudden switching of resistive load of 200 Ω at t = 2 s. The steady state no-load voltage is generated at rated speed of 1000 RPM.

4.2.2 When both three-phase winding sets are connected in long-shunt configuration with independent R-L loading

In the same order of figure numbers, all computed waveforms are also shown in **Figure 6** with sudden switching of R-L load (200 Ω resistance in series with 500 mH inductor) at t = 2 s. The RMS analytical value of steady state voltage is generated at rated speed of 1000 RPM. Same like as Section 4.1, the d-q axis drops in generating and lagging load currents are also given in **Figure 6b, h** and **i**.

5. Discussion

In the machine model, two mixed state space variables have been chosen in place of single state space variable (general case). Double mixed stator current and air-gap flux state space model belongs to one of the more complex model types compared to remaining mixed variable model of stator flux linkages-stator current, rotor flux linkages-rotor current, rotor flux linkages-stator current, stator flux linkages-rotor current, air-gap flux-rotor current, magnetizing current-stator flux linkages and rotor flux linkages-magnetizing current. On the other perspective, considerably simpler than d-q axis winding current model and has simple matrix model. Short-shunt compensation results in overvoltage across the generator terminals during no-load. Whereas, long-shunt compensation gives reduced terminal voltage as compared to short-shunt compensation. In long-shunt configuration, deep saturation provides higher level of magnetizing (or stator) current against large load current at sudden switching of R (or RL) load. While, sometimes, in long-shunt compensation scheme, even reduced generated no load terminal voltage as compared to short-shunt scheme, can be capable of almost same generated total output power. In this fashion, when one moves from simple-shunt to short-shunt (or long-shunt) compensated SP-SEIG, voltage regulation has to be improved by maintaining almost similar magnitude of voltage response after load, as was in simple-shunt SP-SEIG at no-load in **Figure 5a** of Ref. [1]. Proposed saturated machine model will be applied in air-gap flux field orientation vector control strategy.

6. Conclusion

Mixed stator current and air-gap flux as a double state-space variables model preserves information about both stator and rotor parameters. A careful value selection of the combination, i.e. shunt and series capacitors may avoid the excessive voltage across the terminals of SP-SEIG during sudden switching of machine load. It is noticed that involvement of extra supplied reactive power as per self-regulating nature of short-shunt as compared to long-shunt series capacitors in each lines, retains the similar output profile of simple-shunt scheme, when machine load is suddenly switched on after few seconds. In both cases (R and R-L loading), the little bit of marked voltage drops were occurred during R-L loading compared to the R loading when variation from no load to full load.

Author details

Kiran Singh
Department of Electrical Engineering, Indian Institute of Technology, Roorkee, Uttarakhand, India

*Address all correspondence to: kiransinghiitr@gmail.com

References

[1] Singh K, Singh GK. Modelling and analysis of six-phase self-excited induction generator using mixed stator current and magnetizing flux as state-space variables. Electric Power Components and Systems. 2015;**43**: 2288-2296

[2] Liao YW, Levi E. Modelling and simulation of a stand-alone induction generator with rotor flux oriented control. Electric Power Systems Research. 1998;**46**:141-152

[3] Singh K, Singh GK. Stability assessment of isolated six-phase induction generator feeding static loads. Turkish Journal of Electrical Engineering & Computer Sciences. 2016;**24**:4218-4230

[4] Krause PC. Analysis of Electric Machinery. New York: McGraw Hill Book Company; 1986

[5] Khan MF, Khan MR. Generalized model for investigating the attributes of a six-phase self-excited induction generator over a three-phase variant. International Transactions on Electrical Energy Systems. 2018;**28**:e2600

[6] Khan MF, Khan MR. Performance analysis of a three phase self-excited induction generator operating with short shunt and long shunt connections. IEEE Biennial International Conference on Power and Energy Systems: Towards Sustainable Energy (PESTSE). 2016:1-6

[7] Chermiti D, Abid N, Khedher A. Voltage regulation approach to a self-excited induction generator: Theoretical study and experimental validation. International Transactions on Electrical Energy Systems. 2017;**27**(5):e2311

[8] Kohler Werner E, Johnson Lee W. Elementary Differential Equations with Boundary Value Problems. 2nd ed. Addison-Wesley; 2006

Chapter 3

Power Converter Topologies for Multiphase Drive Applications

Carlos A. Reusser

Abstract

The yet growing demand for higher demanding industrial applications and the global concern about harmful emissions in the atmosphere have increased the interest for new developments in electric machines and power converters. To meet these new requirements, multiphase machines have become a very attractive solution, offering potential advantages over three-phase classical solutions. Multiphase machine's power demand can be split over more than three phases, thus reducing the electric field stress on each winding (protecting the insulation system) and the requirements on maximum power ratings, for semiconductor devices. Moreover, only two degrees of freedom (i.e. two independently controllable currents) are required for independent flux and torque control. Due to the previous facts, the use of multiphase drives has become very attractive for applications and developments in areas such as electric ship propulsion, more-electric aircraft, electric and hybrid electric road vehicles, electric locomotive traction and in renewable electric energy generation. As a consequence of this multiphase drive tendency, the development of power converter topologies, capable of dealing with high power ratings and handling multiphase winding distributions, has encourage the development of new converter topologies, control strategies and mathematical tools, to face this new challenge.

Keywords: multiphase AC drive, neutral-point clamped, cascaded H-bridge, nine-switch converter, 11-switch converter, field-oriented control (FOC), direct torque control (DTC)

1. Introduction

Multiphase variable-speed drives, based on multiphase AC machines, are nowadays the most natural solution for high-demanding industrial, traction and power generation and distribution applications [1].

The types of multiphase machines for variable-speed applications are in principle the same as their three-phase counterparts. These are synchronous machines, which depending on the excitation can be subclassified into wound rotor, permanent magnet or reluctance machines and induction machines.

In a multiphase winding, the stator winding distribution becomes more concentrated, rather than distributed, as in the case of three-phase windings. This fact and the particularity of using quasi-sinusoidal, rather than sinusoidal voltages because of the inverting process in the power conversion stage, have several advantages that can be summarized as lower field harmonics content and better fault tolerance, because of extra degrees of freedom and less susceptibility to torque pulsations, due to an excitation field with less harmonic content [2].

As a consequence, the use of a multiphase winding configuration improves the MMF spatial distribution, by reducing its harmonic content and losses, due to flux leakage, and increasing the machine's efficiency. These facts have increased the interest for research and development for transportation applications, such as cargo ships, aircraft and road vehicles, thus also contributing to the reduction of greenhouse effect emissions.

2. Mathematical modeling of multiphase machines

All electrical machines can be found to be variations of a common set of fundamental principles, which apply alike the number of phases of which the machine is constructed.

In this context, multiphase machines can be treated as belonging to an n-dimensional space, corresponding to the respective state variables. The machine model, on its original phase-variable form, can be transformed into $n/2$ two-dimensional subspaces, for a machine with an even number of phases. If the number of phases is odd, then the original n-dimensional subspace can be decomposed into a $(n-1)/2$ two-dimensional subspaces and a single-dimensional quantity, which corresponds to a common mode subspace.

Each new two-dimensional subspace is orthogonal to each other, so there is no mutual coupling and they are represented in a stationary reference frame, respecting all other state variables. These new two-dimensional subspaces are could be denoted as $u_k v_k$, where k stands for the respective new orthogonal subspace.

Let us consider an arbitrary state variable λ, defined as $\boldsymbol{\lambda} = [\lambda_1 \lambda_2 \dots \lambda_n]^T$, and let $T\langle\,\rangle$ be a linear operator, which transforms the n-dimensional space into $n/2$ two-dimensional subspaces, for n even, as in Eq. (1):

$$T\langle\,\rangle = \sqrt{\frac{2}{n}} \begin{bmatrix} 1 & \cos(\alpha) & \cos(2\alpha) & \dots & \cos(2\alpha) & \cos(\alpha) \\ 0 & \sin(\alpha) & \sin(2\alpha) & \dots & -\sin(2\alpha) & -\sin(\alpha) \\ 1 & \cos(2\alpha) & \cos(4\alpha) & \dots & \cos(4\alpha) & \cos(2\alpha) \\ 0 & \sin(2\alpha) & \sin(4\alpha) & \dots & -\sin(4\alpha) & -\sin(2\alpha) \\ \vdots & \vdots & \vdots & \vdots & \vdots & \vdots \\ 1 & \cos\left(\frac{n}{2}\right)\alpha & \cos 2\left(\frac{n}{2}\right)\alpha & \dots & \cos 2\left(\frac{n}{2}\right)\alpha & \cos\left(\frac{n}{2}\alpha\right) \\ 0 & \sin\left(\frac{n}{2}\right)\alpha & \sin 2\left(\frac{n}{2}\right)\alpha & \dots & -\sin 2\left(\frac{n}{2}\right)\alpha & -\sin\left(\frac{n}{2}\right)\alpha \end{bmatrix} \quad (1)$$

In the case of n odd, then the corresponding transformation is given as in Eq. (2):

$$T\langle\,\rangle = \sqrt{\frac{2}{n}} \begin{bmatrix} 1 & \cos(\alpha) & \cos(2\alpha) & \cdots & \cos(2\alpha) & \cos(\alpha) \\ 0 & \sin(\alpha) & \sin(2\alpha) & \cdots & -\sin(2\alpha) & -\sin(\alpha) \\ 1 & \cos(2\alpha) & \cos(4\alpha) & \cdots & \cos(4\alpha) & \cos(2\alpha) \\ 0 & \sin(2\alpha) & \sin(4\alpha) & \cdots & -\sin(4\alpha) & -\sin(2\alpha) \\ \vdots & \vdots & \vdots & \vdots & \vdots & \vdots \\ 1 & \cos\left(\frac{(n-1)}{2}\right)\alpha & \cos 2\left(\frac{(n-1)}{2}\right)\alpha & \cdots & \cos 2\left(\frac{(n-1)}{2}\right)\alpha & \cos 2\left(\frac{(n-1)}{2}\right)\alpha \\ 0 & \sin\left(\frac{(n-1)}{2}\right)\alpha & \sin 2\left(\frac{(n-1)}{2}\right)\alpha & \cdots & -\sin 2\left(\frac{(n-1)}{2}\right)\alpha & -\sin 2\left(\frac{(n-1)}{2}\right)\alpha \\ \sqrt{\frac{1}{2}} & \sqrt{\frac{1}{2}} & \sqrt{\frac{1}{2}} & \cdots & \sqrt{\frac{1}{2}} & \sqrt{\frac{1}{2}} \end{bmatrix} \quad (2)$$

The last row of $T\langle\,\rangle$ in Eq. (2) corresponds to the projection of the state variable onto the common mode subspace.

Using the above-presented mathematical decomposition, any machine can be represented by an equivalent two-axis idealized machine, called the Kron's primitive machine. The Kron's primitive machine corresponds to the representation of the n-dimensional space, in the u_1v_1 or $\alpha\beta$ subspace, using the linear transformation $T\langle\,\rangle_{n\to\alpha\beta}$; this definition can be found as the generalized theory of electrical machines.

The mathematical model describing a generalized n-dimensional electrical machine is given in Eqs. (3) and (4), for dynamics, respectively:

$$\boldsymbol{v_s} = [R_s]\,\boldsymbol{i_s} + \frac{d}{dt}\boldsymbol{\psi_s} \tag{3}$$

$$\boldsymbol{v_r} = [R_r]\,\boldsymbol{i_r} + \frac{d}{dt}\boldsymbol{\psi_r} \tag{4}$$

$$\boldsymbol{v_s} = [v_{s1}v_{s2}\dots v_{sn}]^T \qquad \boldsymbol{i_s} = [i_{s1}i_{s2}\dots i_{sn}]^T \qquad \boldsymbol{\psi_s} = [\psi_{s1}\psi_{s2}\cdots\psi_{sn}]^T \tag{5}$$

$$\boldsymbol{v_r} = [v_{r1}v_{r2}\dots v_{rn}]^T \qquad \boldsymbol{i_r} = [i_{r1}i_{r2}\dots i_{rn}]^T \qquad \boldsymbol{\psi_r} = [\psi_{r1}\psi_{r2}\cdots\psi_{rn}]^T \tag{6}$$

Considering symmetrical windings for both stator and rotor, then $[R_s]$ and $[R_r]$ are diagonal $n\times n$ matrices, thus $[R_s] = diag\,(R_s)^{(n)}$ and $[R_r] = diag\,(R_r)^{(n)}$. The stator and rotor flux linkages can be found as in Eqs. (7) and (8):

$$\boldsymbol{\psi_s} = [L_s]\,\boldsymbol{i_s} + [L_{sr}]\,\boldsymbol{i_r} \tag{7}$$

$$\boldsymbol{\psi_r} = [L_r]\,\boldsymbol{i_r} + [L_{rs}]\,\boldsymbol{i_s} \tag{8}$$

where due to the machine's symmetry $[L_{rs}] = [L_{sr}]^T$; the corresponding stator and rotor inductance matrices are described as follows:

$$[L_s] = \begin{bmatrix} L_{s11} & L_{s12} & \dots & L_{s1(n-1)} & L_{s1n} \\ \vdots & \vdots & \vdots & \vdots & \vdots \\ L_{sn1} & L_{sn2} & \dots & L_{sn(n-1)} & L_{snn} \end{bmatrix} \tag{9}$$

$$[L_r] = \begin{bmatrix} L_{r11} & L_{r12} & \dots & L_{r1(n-1)} & L_{r1n} \\ \vdots & \vdots & \vdots & \vdots & \vdots \\ L_{rn1} & L_{rn2} & \dots & L_{rn(n-1)} & L_{rnn} \end{bmatrix} \tag{10}$$

It has to be noticed that for both stator and rotor windings, $L_{11} = L_{22} = \dots = L_{nn}$ which correspond to the winding self-inductance $\overline{L}$; for the corresponding stator and rotor self-inductances within each one, due to the machine's symmetry, it fulfills that $L_{jk} = L_{kj} = L_m \quad \forall j, k\; j \neq k$. Then for the stator and rotor winding self-inductances, it can be stated that $L_{jj} = L_\sigma + L_m$ where L_σ corresponds to the stator or rotor winding leakage inductance. On the other hand, the stator-to-rotor mutual inductance matrix $[L_{sr}]$ can be found as in Eq. (11):

$$[L_{sr}] = M \begin{bmatrix} \cos(\theta_r) & \cos(\theta_r-(n-1)\alpha) & \cos(\theta_r-(n-2)\alpha) & \dots & \cos(\theta_r-\alpha) \\ \cos(\theta_r-\alpha) & \cos(\theta_r) & \cos(\theta_r-(n-1)\alpha) & \dots & \cos(\theta_r-2\alpha) \\ \cos(\theta_r-2\alpha) & \cos(\theta_r-\alpha) & \cos(\theta_r) & \dots & \cos(\theta_r-3\alpha) \\ \vdots & \vdots & \vdots & \dots & \vdots \\ \cos(\theta_r-(n-1)\alpha) & \cos(\theta_r-(n-2)\alpha) & \cos(\theta_r-(n-3)\alpha) & \dots & \cos(\theta_r) \end{bmatrix} \tag{11}$$

The resulting generalized machine representation is obtained by applying the linear transformation operator $T\langle\,\rangle_{n\to\alpha\beta}$ to Eqs. (3) and (4), resulting in the following dynamic representation, in the $\alpha\beta$ subspace, in Eqs. (12) and (13):

$$\boldsymbol{v}_s^{(\alpha\beta)} = R_s \boldsymbol{i}_s^{(\alpha\beta)} + \frac{d}{dt}\boldsymbol{\psi}_s^{(\alpha\beta)} \tag{12}$$

$$\boldsymbol{v}_r^{(\alpha\beta)} = R_s \boldsymbol{i}_r^{(\alpha\beta)} + \frac{d}{dt}\boldsymbol{\psi}_r^{(\alpha\beta)} - j\omega_r \boldsymbol{\psi}_r^{(\alpha\beta)} \tag{13}$$

The corresponding stator and rotor flux linkages given in Eqs. (7) and (8) are transformed into the $\alpha\beta$ subspace as in Eqs. (14) and (15):

$$\boldsymbol{\psi}_s^{(\alpha\beta)} = [L_s]\,\boldsymbol{\psi}_s^{(\alpha\beta)} + [L_{sr}]\,\boldsymbol{i}_r^{(\alpha\beta)} \tag{14}$$

$$\boldsymbol{\psi}_r^{(\alpha\beta)} = [L_r]\,\boldsymbol{i}_r^{(\alpha\beta)} + [L_{rs}]\,\boldsymbol{i}_s^{(\alpha\beta)} \tag{15}$$

The presented methodology considers a n-phase winding with uniform distribution that is $\alpha = \frac{2\pi}{n}$; in the case of machine with multiple groups of three-phase windings, α is the shifting angle between each group of three-phase windings. It has to be noted that for each three-phase winding transformation, an additional row has to be included in Eq. (2).

The developed electromechanical torque can be expressed in terms of the electromechanical energy conversion as in Eqs. (16) and (17):

$$T_e = \frac{\partial}{\partial\theta_r} W_{fld}\left(\boldsymbol{i}_s^{(\alpha\beta)}, \boldsymbol{\psi}_s^{(\alpha\beta)}, \theta_r\right) \tag{16}$$

$$T_e = \frac{1}{2}\left[\boldsymbol{i}^{(\alpha\beta)}\right]\frac{\partial}{\partial\theta_r}[L]\left[\boldsymbol{i}^{(\alpha\beta)}\right]^T \tag{17}$$

$$\left[\boldsymbol{i}^{(\alpha\beta)}\right] = \left[\boldsymbol{i}_s^{(\alpha\beta)} \quad \boldsymbol{i}_r^{(\alpha\beta)}\right]^T \tag{18}$$

$$[L] = \begin{bmatrix} [L_s] & [L_{sr}] \\ [L_{sr}]^T & [L_r] \end{bmatrix} \tag{19}$$

Following the electromechanical torque can be expressed as in Eq. (20):

$$T_e = \frac{1}{2}\boldsymbol{i}_s^{(\alpha\beta)}\frac{\partial}{\partial\theta_r}[L_{sr}]\,\boldsymbol{i}_r^{(\alpha\beta)^T} \tag{20}$$

The generalized machine representation given by Eqs. (12), (13) and (20) can be expressed in an arbitrary orthogonal synchronous reference frame dq, which rotates at synchronous speed ω_k; thus the corresponding rotation into the dq state variables is given by Eq. (21):

$$\theta_k = \int \omega_k\, dt \tag{21}$$

Rotation into the dq subspace is given by the unitary rotation matrix U defined in Eq. (22):

$$U = \begin{bmatrix} \cos\theta_k & \sin\theta_k \\ -\sin\theta_k & \cos\theta_k \end{bmatrix} \tag{22}$$

The generalized machine dynamic representation in the dq reference frame, as consequence of the rotation of Eqs. (12), (13) and (20), is given in Eqs. (23)–(25), representing the Kron's primitive machine model in a generalized synchronous reference frame:

$$\boldsymbol{v}_s^{(dq)} = R_s \boldsymbol{i}_s^{(dq)} + \frac{d}{dt}\boldsymbol{\psi}_s^{(dq)} + j\omega_k \boldsymbol{\psi}_s^{(dq)} \tag{23}$$

$$\boldsymbol{v}_r^{(dq)} = R_s \boldsymbol{i}_r^{(dq)} + \frac{d}{dt}\boldsymbol{\psi}_r^{(dq)} + j(\omega_k - \omega_r)\boldsymbol{\psi}_r^{(dq)} \tag{24}$$

$$T_e = p\frac{L_m}{L_r}\left(\psi_r^d i_s^q - \psi_r^q i_s^d\right) \tag{25}$$

Equations (23) and (24) can be written in their matrix form as in Eq. (26), becoming a generalized impedance model, as described in Eq. (27):

$$\begin{bmatrix} v_s^d \\ v_s^q \\ v_r^d \\ v_r^q \end{bmatrix} = \begin{bmatrix} R_s + L_s\frac{d}{dt} & -\omega_k L_s & L_m\frac{d}{dt} & -\omega_k L_m \\ \omega_k L_s & R_s + L_s\frac{d}{dt} & \omega_k L_m & L_m\frac{d}{dt} \\ L_m\frac{d}{dt} & -(\omega_k - \omega_r)L_m & R_r + L_r\frac{d}{dt} & -(\omega_k - \omega_r)L_r \\ (\omega_k - \omega_r)L_m & L_m\frac{d}{dt} & (\omega_k - \omega_r)L_r & R_r + L_r\frac{d}{dt} \end{bmatrix} \begin{bmatrix} i_s^d \\ i_s^q \\ i_r^d \\ i_r^q \end{bmatrix} \tag{26}$$

$$\left[\boldsymbol{v}^{(dq)}\right] = [z]\left[\boldsymbol{i}^{(dq)}\right] \tag{27}$$

2.1 Multiphase synchronous machines

The use of multiphase synchronous machines has been focused on its application for medium- and high-power generation systems, being their primary use, on wind energy conversion systems (WECS). The previous statement is based on the fact that most of wind energy conversion systems operate in the low-voltage range, principally due to restrictions of winding insulation. This fact has stimulated the development of multiphase generator topologies, thus gaining increasing interest in the research for new converter topologies [3, 4]. Some of their main advantages are the following:

- The total power can be divided into lower power converters.
- Due to that each converter is insulated from each other, there is no circulating current between converters, which leads to no power derating for the converters.
- The increase of the number of phases in the generator voltages is phase-shifted so that low-order harmonics are reduced and consequently smaller filters can be used.
- Reduced torque pulsations.
- Fault-tolerant redundancy under winding fault conditions.
- Additional degrees of freedom which can be used to improve the machine performance.

Arrangement of multiple three-phase windings has become a very popular construction technique, for multiphase machines. In this field, the most common configuration is the six-phase machine, based on two independent three-phase windings, which are spatially shifted in 30 degrees, as shown in **Figure 1**.

The use of multiple three-phase windings has the advantage of guaranteeing full decoupling under faulty conditions, thus preventing the circulation of common mode currents and pulsating torque.

The linear transformation operator $T\langle\,\rangle$ is defined as in Eq. (28), with $\varphi = \frac{\pi}{6}$:

$$T\langle\,\rangle = \frac{1}{\sqrt{3}}\begin{bmatrix} 1 & \cos 4\varphi & \cos 8\varphi & \cos\varphi & \cos 5\varphi & \cos 9\varphi \\ 0 & \sin 4\varphi & \sin 8\varphi & \sin\varphi & \sin 5\varphi & \sin 9\varphi \\ 1 & \cos 8\varphi & \cos 4\varphi & \cos 5\varphi & \cos\varphi & \cos 9\varphi \\ 0 & \sin 8\varphi & \sin 4\varphi & \sin 5\varphi & \sin\varphi & \sin 9\varphi \\ 1 & 1 & 1 & 0 & 0 & 0 \\ 0 & 0 & 0 & 1 & 1 & 1 \end{bmatrix} \tag{28}$$

The equivalent model in the *dq* synchronous subspace can be derived from Eq. (26); in this case it has to be noted that the synchronous reference frame *dq* coincides with the rotor natural reference frame, thus $\omega_k = \omega_r$, becoming Eq. (29) and (30) for the stator:

$$v_s^d = R_s i_s^d + \frac{d}{dt}\psi_s^d - \omega_k \psi_s^q \tag{29}$$

$$v_s^q = R_s i_s^q + \frac{d}{dt}\psi_s^q + \omega_k \psi_s^d \tag{30}$$

Equations (31) and (32) correspond to the damping winding effect:

$$0 = R_r i_r^d + \frac{d}{dt}\psi_r^d \tag{31}$$

$$0 = R_r i_r^q + \frac{d}{dt}\psi_r^q \tag{32}$$

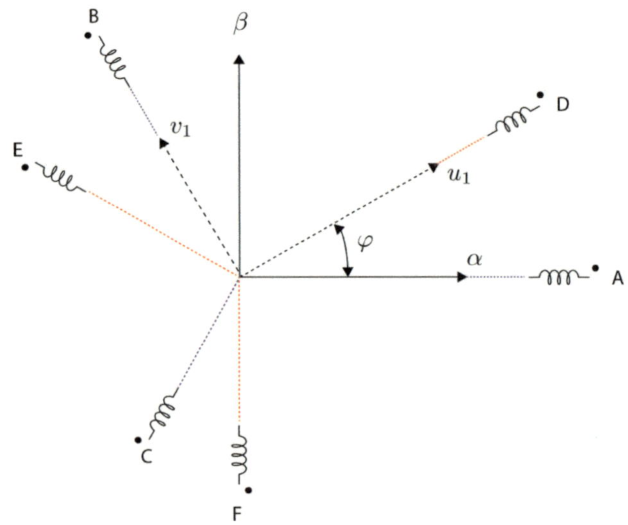

Figure 1.
Dual three-phase stator winding configuration.

The corresponding flux linkages, considering the general case of an anisotropic machine, are given in the following equations:

$$\psi_s^d = L_{sd}\, i_s^d + L_{md}\, i_r^d + \psi_m \tag{33}$$

$$\psi_s^q = L_{sq}\, i_s^q + L_{mq}\, i_r^q \tag{34}$$

$$\psi_r^d = L_{rd}\, i_r^d + L_{md}\, i_s^d + \psi_m \tag{35}$$

$$\psi_r^q = L_{rq}\, i_r^q + L_{mq}\, i_s^q \tag{36}$$

and the electromechanical torque is given in Eq. (37):

$$T_e = p\left[\psi_m\, i_s^q + \left(L_{md}\, i_r^d\, i_s^q - L_{mq}\, i_r^q\, i_s^d\right)\right] + p\left[L_{md} - L_{mq}\right] i_s^q\, i_s^d \tag{37}$$

2.2 Multiphase induction machines

Modern industrial high-demanding processes are commonly based on induction machines. They are very attractive for these kind of applications because of their simplicity and capability to work under extreme torque demanding conditions [5].

Within the previous context, multiphase induction machines have become very popular for applications where high redundancy and power density are required. In particular, the use of multiple three-phase windings (six and nine phases) in naval propulsion systems has aroused much interest and encouraged the research of new multiphase converter topologies and control schemes [6, 7].

Based on the Kron's primitive machine model developed in Eq. (26), the dynamical model of the multiphase induction machine is given in Eqs. (38)–(41):

$$v_s^d = R_s\, i_s^d + \frac{d}{dt}\psi_s^d - \omega_k\, \psi_s^q \tag{38}$$

$$v_s^q = R_s\, i_s^q + \frac{d}{dt}\psi_s^q + \omega_k\, \psi_s^d \tag{39}$$

$$0 = R_r\, i_r^d + \frac{d}{dt}\psi_r^d \tag{40}$$

$$0 = R_r\, i_r^q + \frac{d}{dt}\psi_r^q \tag{41}$$

Due to the isotropy of the induction machine and the absence of an MMF source in the rotor (permanent magnets or field winding), the corresponding flux linkages are given in Eqs. (42)–(45):

$$\psi_s^d = L_s\, i_s^d + L_m\, i_r^d \tag{42}$$

$$\psi_s^q = L_s\, i_s^q + L_m\, i_r^q \tag{43}$$

$$\psi_r^d = L_r\, i_r^d + L_m\, i_s^d \tag{44}$$

$$\psi_r^q = L_r\, i_r^q + L_m\, i_s^q \tag{45}$$

The electromechanical torque expression can be derived from Eq. (37) by considering $\psi_m = 0, L_{md} = L_{mq} = L_m$, becoming Eq. (46):

$$T_e = p L_m\left[i_r^d\, i_s^q - i_r^q\, i_s^d\right] \tag{46}$$

3. Multiphase power converters

As stated previously, multiphase machines have many advantages over traditional three-phase-based machine drives, by reducing the impact of low-frequency torque pulsations and the dc-link current harmonic content. They also, due to the nature of their winding configuration, improve the system reliability, by introducing redundant operation conditions. As a consequence the use of multiphase converter topologies with multiphase drive arrangements has been proved as a viable approach for its application in high-demanding industrial applications.

In this field, the development of power converters capable to deal with the multiphase machine structure has capture much attention in the recent years; thus several topologies have been introduced in the last decades. This topologies can consist of arrangements of conventional two-level three-phase voltage source converters (2LVSC), multilevel converters (MLVSC) or in more specialized and dedicated topologies such as multiphase matrix converters.

3.1 Classical topologies

Classical topologies for multiphase converters are commonly based in arrangements of parallel-connected fundamental cells (multiple legs), or in multiple channel configuration, of voltage source converter topologies.

Commonly used topologies for multiphase applications are H-bridge converter (HBC), neutral-point clamped converter (NPC) and the cascaded H-bridge (CHB) topology, which are shown in **Figure 2**.

Classical multiphase VSC topologies consist of an arrangement of n individual half-cells connected in parallel to a single dc-link, as shown in **Figure 3**.

Modulation of each individual half-cell circuit is implemented in such a way, to obtain n voltage output signals shifted by $2\pi/n$. Generation of the required voltage signals is supported by classical carrier-based strategies, such as sinusoidal pulse width modulation (SPWM) and space vector pulse width modulation (SVPWM).

SPWM methods can be implemented for two-level or multilevel half-cells. In the first case, only one high-frequency carrier is required for the respective switching signals per phase S_x, as given in Eqs. (46) and (48):

$$S_x = \begin{cases} 1 & |u_{xN}^*| \geq |u_c| \\ 0 & |u_{xN}^*| < |u_c| \end{cases} \tag{47}$$

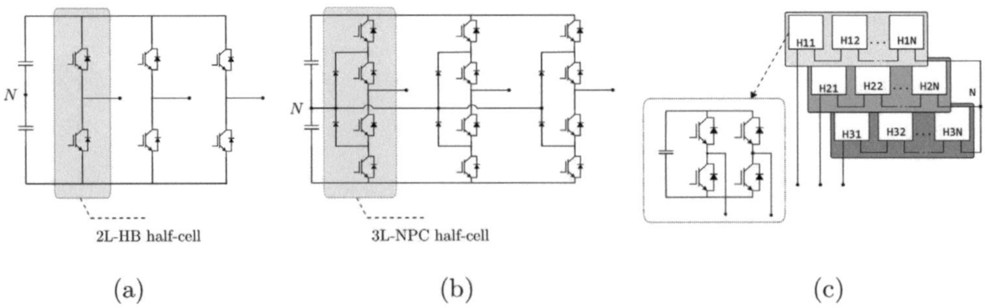

Figure 2.
Classical VSC topologies: (a) two-level bridge, (b) neutral-point clamped and (c) cascaded H-bridge.

$$u_{xN}^{*} = v_{xN}^{*} - \frac{1}{2}\left\{min\left(v_{aN}^{*}, v_{bN}^{*}, \dots, v_{nN}^{*}\right) + max\left(v_{aN}^{*}, v_{bN}^{*}, \dots, v_{nN}^{*}\right)\right\} \tag{48}$$

where x stands for the corresponding phase $x = [1, \dots, n]$, u_{xN}^{*} is the reference to be synthesized for the corresponding phase, and u_c stands for the carrier wave.

For multilevel parallel arrangements (NPC or CHB), the use of multiple carries, as an extension of the two-level PWM methods, has been proven as a suitable solution. Level-shifted PWM (LSPWM) has become a very popular modulation technique, because it fits for any multilevel converter topology and ensures low harmonic distortion. The corresponding switching states are given in Eq. (49):

$$S_{xm} = \begin{cases} 1 & |u_{xN}^{*}| \geq |u_{cm}| \\ 0 & |u_{xN}^{*}| < |u_{cm}| \end{cases} \tag{49}$$

where x stands for the corresponding phase $x = [1, \dots, n]$, m to the corresponding level and, u_{cm} is the carrier for the matched m level.

The main advantages of this topology are its simplicity by using half-bridges for each leg (phase), and the requirement of a single dc-link the feed the n-phase inverting stage. The use of a parallel arrangement of fundamental half-cells generates $(n-1)$ dimensional spaces, which can be decomposed into $(n-1)/2$ subspaces, ensuring redundant switching states.

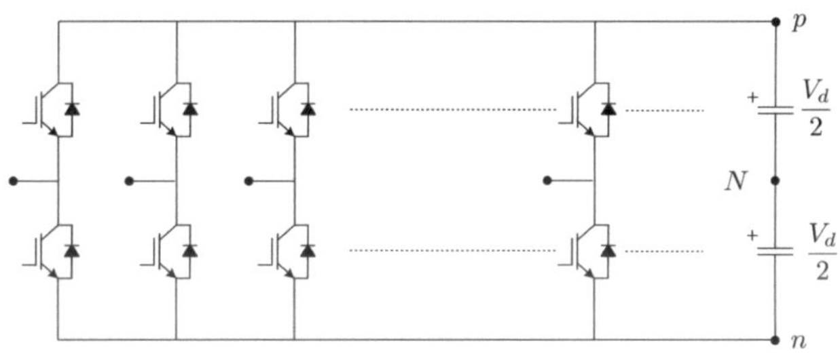

Figure 3.
Multiphase multicell topology.

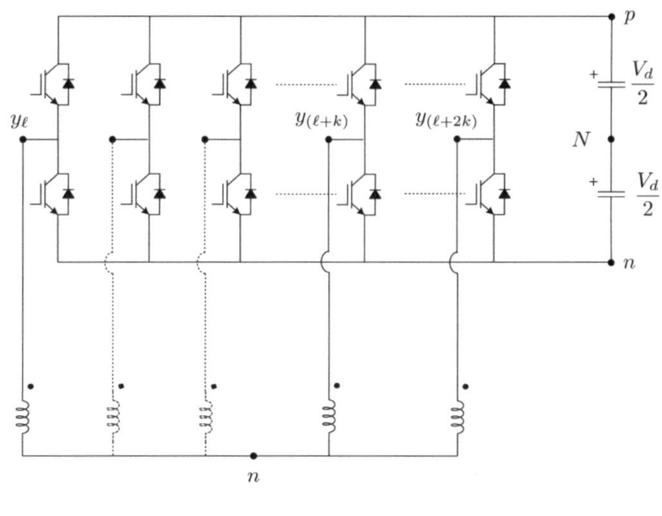

Figure 4.
Common connected load arrangement.

Multicell topology also enables the possibility to have common or split connected loads. In the first case, each phase in the load side is connected to a common neutral point N as presented in **Figure 4**; the main drawback with this configuration is the circulation of zero sequence currents, since the neutral points $N - n$ are not insulated, thus establishing a common mode voltage $v_{Nn} \neq 0$ in the dq synchronous reference frame.

Split connected load arrangement is possible, if $n = 3k | k = [1, 2, \ldots]$, so the n phase system can be divided into k three-phase insulated independent subsystems, as shown in **Figure 5**, with symmetric or asymmetric electrical shifting δ. It has to be noted that independently on the symmetric or asymmetric configuration, the

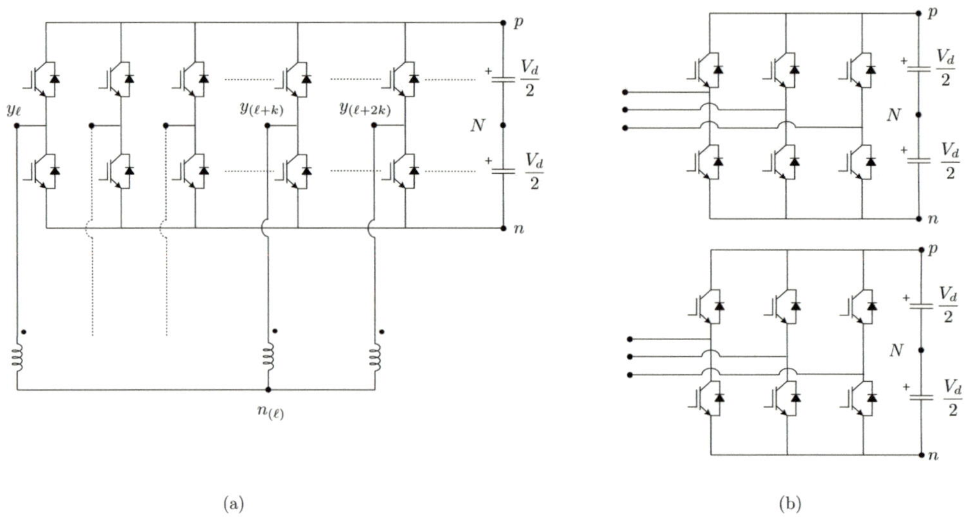

Figure 5.
Split connected load arrangements. (a) Single-channel topology and (b) multichannel topology.

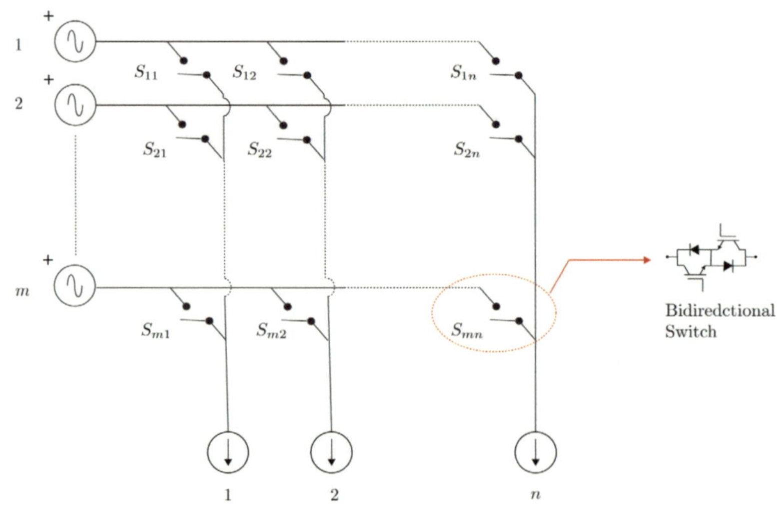

Figure 6.
mxn matrix converter equivalent model.

electrical shifting between each phase on a single three-phase group is still $2\pi/3$, constituting each individual symmetrical three-phase system.

Figure 6a shows a split-phase single-channel arrangement where the corresponding phase for each half-cell is derived as $y_\ell, y_{\ell+k}, y_{\ell+2k}$ with $\ell = [1, \dots, k]$. A multichannel arrangement is shown in **Figure 6b**, which introduces some advantages, like full load dynamic decoupling and lower dc-link power rating. On the other hand, additional dc-link capacitors are required.

For both previously described topologies, SPWM is achieved by implementing the modulation as given in Eqs. (47) and (48) considering the asymmetry (if required) φ for each reference as $u_{xN}^*(\varphi)$.

In the case of space vector modulation for multiphase converters, it is necessary to extend the space vector representation into its corresponding subspaces. So as presented previously in this chapter, the output voltage space vector can be decomposed into $n/2$ orthogonal subspaces for n even and into $(n-1)/2$ orthogonal subspaces and single-dimensional quantity (common mode component) for n odd. Under this formulation, the reference voltage space vector to be synthesized can be expressed as in Eqs. (50) and (51) for n even and odd, respectively:

$$\boldsymbol{u}_\mathbf{x}^* = \left[u^{(1)}\, u^{(2)} \dots, u^{(n)}\right]^T \tag{50}$$

$$\boldsymbol{u}_\mathbf{x}^* = \left[u^{(1)}\, u^{(2)} \dots, u^{(n-1)}\right]^T \tag{51}$$

For each orthogonal subspace, there exits two active vectors, $\overline{v}_{uk}$ and $\overline{v}_{vk}$, corresponding to k orthogonal subspace. Each active vector is applied for a duty-cycle δ to synthesize the corresponding reference space vector, as in Eqs. (52) and (53):

$$\boldsymbol{u}_\mathbf{x}^* = \delta_{u1}\overline{v}^{(u1)} + \delta_{v1}\overline{v}^{(v2)} + \delta_{u2}\overline{v}^{(u2)} + \delta_{v2}\overline{v}^{(v2)} + \dots + \delta_{u(n/2)}\overline{v}^{(u(n/2))} + \delta_{v(n/2)}\overline{v}^{(v(n/2))} \tag{52}$$

$$\boldsymbol{u}_\mathbf{x}^* = \sum_{j=1}^{n/2} \delta_{uj}\overline{v}^{(uj)} + \sum_{j=1}^{n/2} \delta_{vj}\overline{v}^{(vj)} \tag{53}$$

In the case of n odd, an additional degree of freedom is introduced by the common mode voltage v_{cm} which represents a single quantity and not a space vector as in Eq. (54):

$$\boldsymbol{u}_\mathbf{x}^* = \sum_{j=1}^{(n-1)/2} \delta_{uj}\overline{v}^{(uj)} + \sum_{j=1}^{(n-1)/2} \delta_{vj}\overline{v}^{(vj)} + \delta_{cm} v_{cm} \tag{54}$$

Eqs (53) and (54) can be interpreted as the partitioning of the $\alpha\beta$ stationary reference frame, into $n!$ or $(n-1)!$ adjacent sectors, and an additional common mode component, orthogonal to the $\alpha\beta$ subspace. Sector identification can then be achieved by implementing Eq. (55) where S corresponds to the given sector within the $\alpha\beta$ subspace:

$$S = \left(\frac{\theta}{M!}\right) + 1 \quad M = \begin{cases} n & \text{even number of subspaces.} \\ (n-1) & \text{odd number of subspaces.} \end{cases} \tag{55}$$

3.2 Matrix converter

The matrix converter is a direct AC-AC converter, which uses an arrangement of bidirectional switches, to connect each input phase, with a single corresponding output phase, thus generating an arrangement of *mxn* power switches, where *m* is the number of input phases and *n* the number of output phases. In **Figure 6,** the equivalent model of a *mxn* matrix converter is shown.

Due to the absence of a dc-link stage, the output voltages should be synthesized by selecting segments of the input voltages, by generating the adequate switching states. However, some switching state restrictions have to be taken into account, because of the particular topology of the matrix converter. Let's consider a generalized switching state S_{jk}, such that

$$S_{jk} = \begin{cases} 1 & \text{switch}\, S_{jk}\, \text{is in on state.} \\ 0 & \text{switch}\, S_{jk}\, \text{is in off state.} \end{cases} \quad \forall j = 1, 2, \ldots, m \quad k = 1, 2, \ldots, n \tag{56}$$

For every instant t, the switching state S_{jk} must comply with both conditions stated in Eqs (57) and (58), thus meaning that for all t only one input phase is connected to one output phase, avoiding short-circuit condition, and also all output phases are connected to at least one input phase, ensuring no open-circuit condition. This last condition is intended to protect the bidirectional switches that cannot handle reverse current flow, due to the inductive energy discharge:

$$\sum_{j=1}^{n} S_{jk} = 1 \tag{57}$$

$$S_{jk} = 1 \quad \forall k = 1, 2, \ldots, n \tag{58}$$

Sinusoidal PWM scheme is implemented via the *Venturini* method [8], which is based on the solution of the relational input-output equations of the matrix converter, given in Eqs. (59)–(61):

$$[\boldsymbol{V_o}] = [M]\,[\boldsymbol{V_i}] \quad M = \begin{bmatrix} \delta_{11} & \delta_{12} & \ldots & \delta_{1m} \\ \delta_{21} & \delta_{22} & \ldots & \delta_{2m} \\ \vdots & \vdots & \ldots & \vdots \\ \delta_{n1} & \delta_{n2} & \ldots & \delta_{nm} \end{bmatrix} \tag{59}$$

$$[\boldsymbol{I_i}] = [M]^T\,[\boldsymbol{I_o}] \tag{60}$$

$$P_i = P_o \tag{61}$$

where $[\boldsymbol{V_o}]$ and $[\boldsymbol{V_i}]$ stand for the output and input voltage space vectors, respectively, $[\boldsymbol{I_o}]$ and $[\boldsymbol{I_i}]$ for the output and input current space vectors, $[M]$ is the low-frequency matrix or modulation index transfer matrix, and P_o and P_i correspond to the output and input active power.

On the other hand, space vector PWM formulation has no difference, as it is implemented in VSC. However, its complexity lies in the fact in that the absence of a nearly constant dc-link voltage, from which the reference voltage space vectors are synthesized. So both space vectors are to be composed using the input and output, voltage and current space vectors simultaneously. Moreover, the total number of possible switching states is $2^{(nm)}$, from which the forbidden conditions have to be considered. Space vector implementation is largely explained in literature [9].

The matrix converter has several advantages over multiphase voltage source converters, for multiphase drive applications. It is capable to synthesize nearly sinusoidal output voltage and currents, with low-order harmonics, thus improving the MMF distribution in machine air gap, eliminating torque ripple and preventing mechanical stresses on the output shaft. It provides bidirectional energy flow and provides full power factor control. Moreover, due to the absence of dc-link, it presents more power density, because of the lack of large capacitors, becoming an alternative for integrated drive converter configurations [10, 11]. However the complexity in the implementation of SPWM and SVPWM schemes, and the complex commutation strategy for the bidirectional switches, makes the matrix converter less attractive than the voltage source-based multiphase solutions.

There exist several variations of the direct matrix converter topology presented previously, such as the indirect and sparse matrix converter topologies, which are extensively described in literature [12]. These topological variations relay on the same basis but differ in the number and type of power switches used.

3.3 Nine-switch converter

The topology is derived from two three-phase voltage source converters that share a positive and a negative busbar, respectively, as shown in **Figure 7**.

The nine-switch converter has the ability to operate in back-to-back mode, as rectifier (A stage input) and inverter (B stage output), as two channel rectifier (A and B stages input) and as two stage inverter (A and B stages output), enabling the converter to handle a six-phase systems with just one channel, in spite of commonly 12-switch back-to-back topologies.

However, this topological advantage introduces some drawbacks to the nine-switch topology, because of the fact that S_x, S_y and S_z switches are shared by both converter stages A and B. Hence 3 switches are eliminated of the 12 needed for

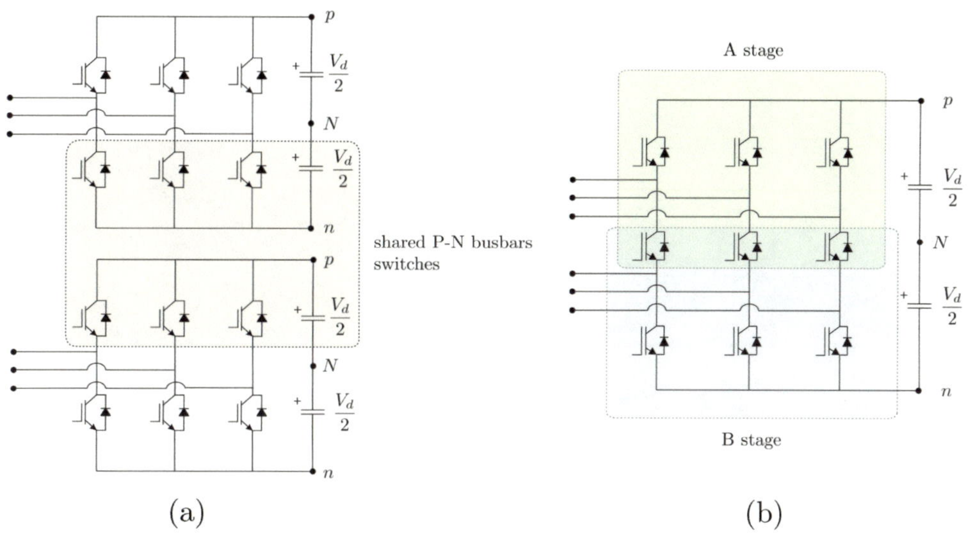

Figure 7.
(a) Classical arrangement with 12 switches. (b) Nine-switch converter topology.

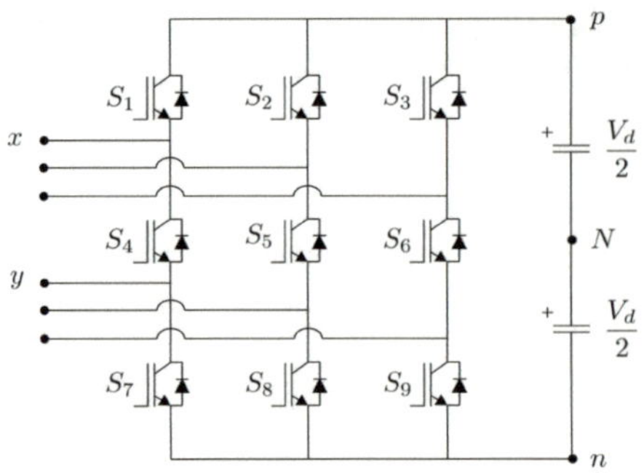

S_1	S_4	S_7	v_{xN}	v_{yN}
1	1	0	$V_d/2$	$V_d/2$
0	1	1	$-V_d/2$	$-V_d/2$
1	0	1	$V_d/2$	$-V_d/2$

Figure 8.
Nine-switch converter allowed switching states.

multiphase operation as in multichannel topology, some forbidden switching states are introduced, thus remaining 27 allowed states, and also the maximum output voltage gain is limited [13]. In **Figure 8,** the allowed switching states per leg are shown.

Carrier-based PWM modulation schemes such as sinusoidal PWM (SPWM), space vector modulation (SVM) and min-max (third harmonic injection) PWM [14] can be applied to the nine-switch converter, using two independent SPWM modulation schemes, one for each converter stage, as in the 12-switch back-to-back converter. However, because of the restrictions introduced to the modulation, pattern by the middle switches S_4, S_5 and S_6, its switching pattern can be obtained as in Eq. (62):

$$S_{k+3} = \overline{S_k \cdot S_{k+6}} \quad k = 1, \ldots, 3 \tag{62}$$

The implementation of SPWM considers the use of two voltage reference space vectors x^* and y^*, for each three-phase winding group. The NSC can operate either in constant (CFM) or variable frequency (VFM) modes, depending on each space vector angular frequency. However, due to the nature of the particular application in multiphase drives, the NSC is to be operated in the constant frequency mode, as presented in **Figure 9**.

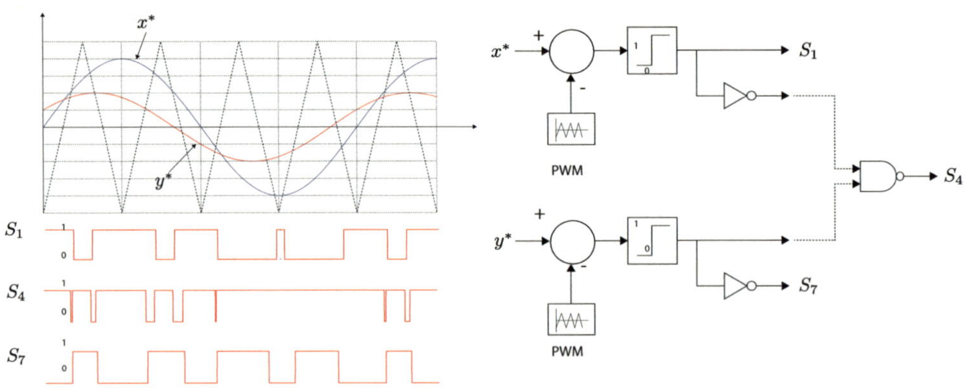

Figure 9.
CFM modulation.

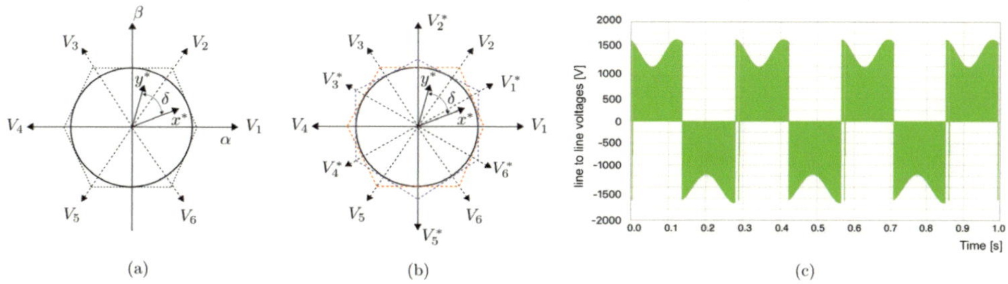

Figure 10.
Space vector decomposition. (a) Single carrier, (b) shifted multi-carrier and (c) line-line voltage.

For the constant frequency mode operation (CFM), both reference space vectors x^* and y^* have the same angular frequency ω_k, preserving its spatial shifting δ, so they are to be found at least in contiguous sectors of the voltage hexagon, as shown in **Figure 10a**, being the maximum voltage gain g given in Eq. (63):

$$g = \frac{1}{\sqrt{3}\left(\frac{\delta}{2} + \frac{\pi}{3}\right)} \tag{63}$$

By introducing a phase shifting of $\pi/6$ between both PWM carriers, a second group of active vectors is introduced as shown in **Figure 10b**, which results in a voltage gain as in Eq. (64):

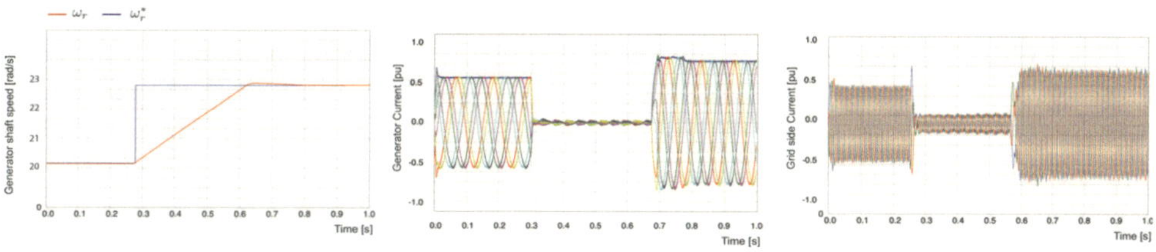

Figure 11.
Simulation results for a back-to-back NSC for wind energy conversion.

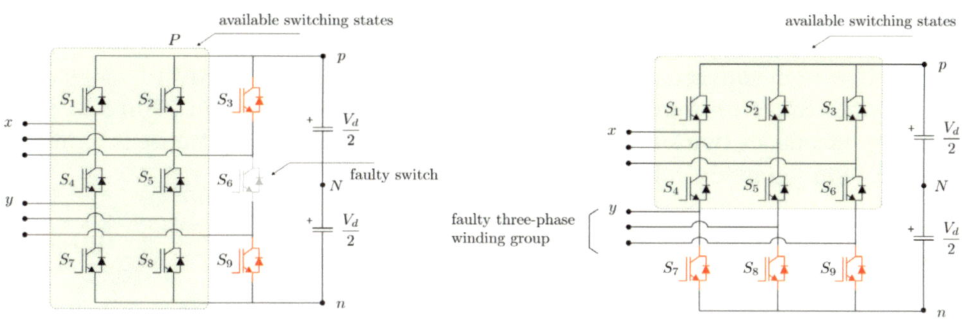

S_2	S_5	S_8	v_{xN}	v_{yN}	v_{pN}	v_{qN}
1	1	0	$V_d/2$	$-V_d/2$	$V_d/2$	$V_d/2$
0	1	1	$V_d/2$	$-V_d/2$	$-V_d/2$	$-V_d/2$
1	0	1	$V_d/2$	$-V_d/2$	$V_d/2$	$-V_d/2$

S_1	S_4	S_7	v_{xN}	v_{yN}
0	1	1	$-V_d/2$	$-V_d/2$
1	0	1	$V_d/2$	$-V_d/2$

Figure 12.
Nine-switch converter reconfiguration under faulty operation.

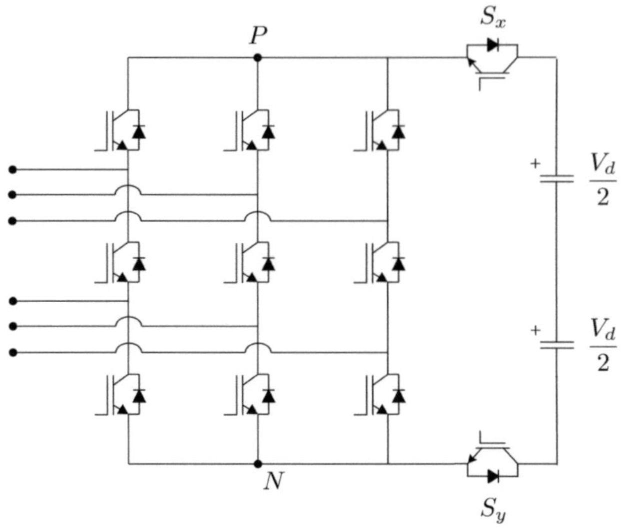

Figure 13.
11-switch converter.

$$g = \frac{1}{\sqrt{3}\left(\delta + \frac{\pi}{6}\right)} \tag{64}$$

Recently, the NSC has been gaining much attention in various applications like isolated wind-hydro hybrid power system, power quality enhancement and hybrid electric vehicles because of its ability to interconnect multiphase power systems in common or split configurations and also independent three-phase-based systems, in constant or variable frequency modes. **Figure 11** shows simulation results for a wind energy conversion system (WECS) based on a nine-switch back-to-back converter topology [15].

Another important feature of the NSC is the capability to rearrange its switching states under single- or multiple-switch faulty conditions [16], as shown in **Figure 12**.

3.4 11-switch converter

The 11-switch converter, presented in [17], consists of a modified topology of the 9-switch converter topology, previously discussed. As shown in **Figure 13**, this topology introduces two additional switches S_x, S_y, whose main propose is to mitigate the common mode voltage during the zero switching states.

4. Control of multiphase electric drives

As discussed previously in this chapter, the electromechanical torque developed by the multiphase machine depends only on the state variables in $\alpha\beta$ subspace, thus meaning that the other subspaces do not contribute to the energy conversion process but only losses. This fact makes possible the implementation of oriented control schemes, such as field-oriented control (FOC), direct torque control (DTC) and model-based predictive control (MBPC), in a synchronous reference frame dq, by rotating the state variables in $\alpha\beta$ subspace.

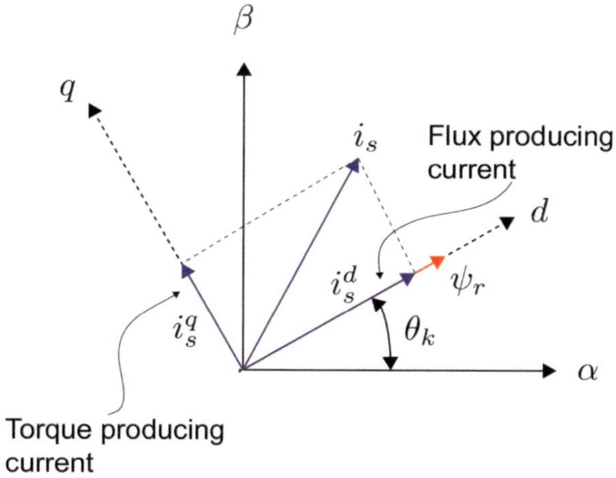

Figure 14.
dq synchronous reference frame rotation.

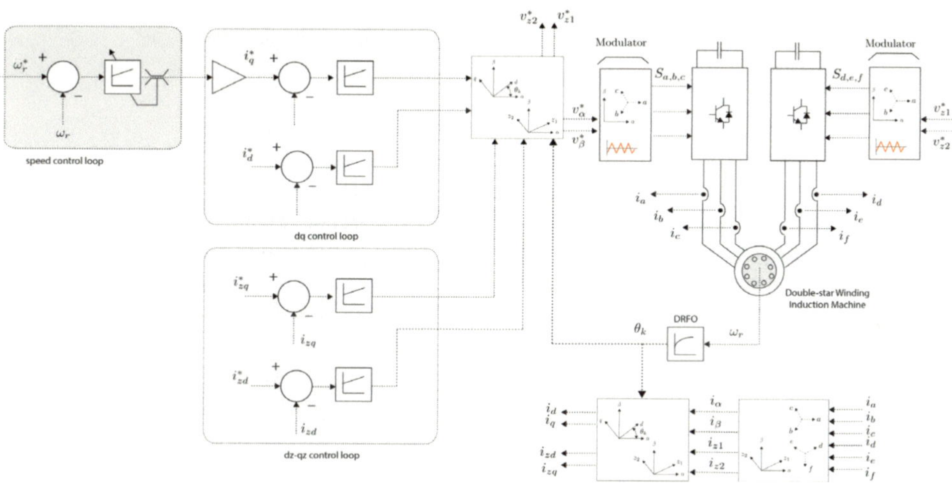

Figure 15.
Field-oriented control scheme of a dual three-phase winding induction machine.

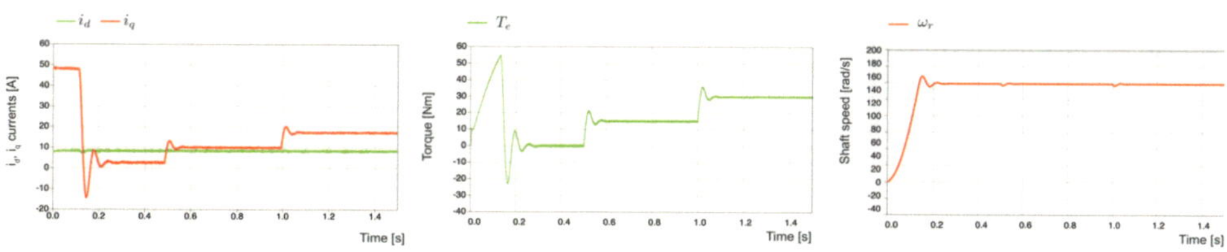

Figure 16.
Dual three-phase winding induction machine FOC scheme simulation results.

An additional control loop in the *dq* reference frame can be implemented, to compensate the losses in the remaining subspaces. Control goals for the multiphase drive can be summarized as maximum torque per Ampere operation, control of nominal flux and control of rotor speed/torque.

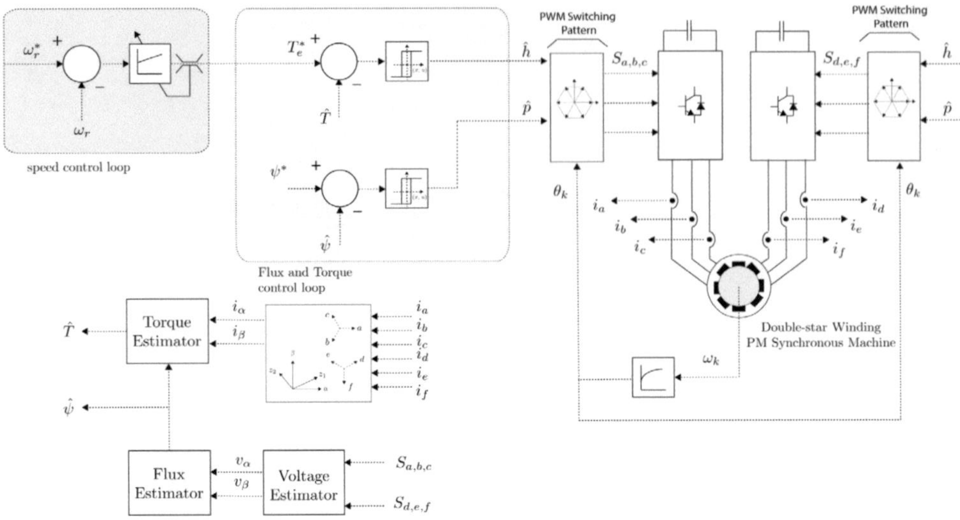

Figure 17.
Direct torque control scheme of a dual three-phase winding PM synchronous machine.

4.1 Field-oriented control

The characteristics of field-oriented control (FOC) have made this control strategy the most widely used for high-demanding industrial applications. Field-oriented control is based in the decoupling of the current space vector, into a flux-producing component and a torque-producing component. This is achieved by rotating the current space vector from $\alpha\beta$ subspace, into the synchronous rotating reference frame dq, oriented with respect to the rotor flux linkage space vector, as shown in **Figure 14**.

In this way the magnetizing flux can be controlled, so that the machine operates with nominal flux under any condition, and also torque can be controlled only by i_s^q, because $\psi_r^q = 0$, due to the orientation of ψ_r which only exists in the direct axis orientation. Implementation of the FOC scheme for a multiphase AC drive is shown in **Figure 15**.

Figure 16 shows simulation results for a dual three-phase induction machine drive using FOC scheme, under load impact.

The main drawbacks of FOC are the requirement in the estimation of the rotor flux linkage space vector and the rotation of the state space variables into this synchronous reference frame, becoming a very complex process. Also, as the dynamic response, this control strategy is limited by maximum bandwidth achievable for the PI controllers, which represents one of the fundamental limitations of linear controllers.

4.2 Direct torque control

Direct torque control (DTC) is based on the estimation of torque and flux directly from the state variables of the AC machine. The torque and flux can be controlled by applying the suitable voltage vector, synthesized by the available switching states of the converter.

The required voltage vector is chosen via a switching table, as function of the actuation of the torque and flux loop hysteresis controllers (in terms of increasing or decreasing flux or torque for a certain operational point). Implementation of the DTC scheme for a multiphase AC drive is shown in **Figure 17**.

The main characteristics of DTC are its simple implementation and a fast dynamic response achieved by using hysteresis controllers. Furthermore, the required switching states are directly assigned from the switching table algorithm, so no modulator is needed.

5. Conclusions

In this chapter, the main advantages of multiphase machine drives for its application in high-demanding industrial processes, traction and renewable energy grid interfacing were presented, with their main focus on multiphase power converter topologies. Various technical issues related to classical multiphase converters, based on multicell arrangements, were discussed, as well as some new converter topologies, such as the nine-switch converter (NSC) and 11-switch converter (ESC), were introduced.

The main advantages of classical multiphase converter topologies, based on voltage source converters (VSC), are mostly referred to their topological simplicity and their capability to implement conventional sinusoidal-PWM-based techniques. On the other hand, the increasing number of semi-conductors and dc-link capacitors (in the case of multichannel arrangements) and the need for common mode current compensation are their major drawbacks. In this field of application, matrix converters arise as suitable alternative, which enables the possibility to handle multiple output phases. However, maximum voltage gain limitations and the complexity of the modulation and commutation strategies are their main disadvantages, when compared to classical topologies.

Nine-switch and eleven-switch converters appear as a middle-point alternative between multicell and matrix converter topologies. These topologies allow the use of sinusoidal-PWM-based modulation techniques, without the need of complex modulation and commutation strategies (as in the case of the matrix converter), using a single dc-link stage.

Implementation of control strategies for multiphase drive, such as field-oriented control (FOC) and direct torque control (DTC), has been easily achieved by using multi-space decomposition of the each state space vectors (representing a state space variable).

Author details

Carlos A. Reusser
Department of Electronics, Universidad Tecnica Federico Santa Maria, Valparaiso, Chile

*Address all correspondence to: carlos.reusser@usm.cl

References

[1] Levi E. Multiphase electric machines for variable-speed applications. IEEE Transactions on Industrial Electronics. 2008;**55**(5):1893-1909

[2] Levi E. Advances in converter control and innovative exploitation of additional degrees of freedom for multiphase machines. IEEE Transactions on Industrial Electronics. 2016;**63**(1):433-448

[3] Zhang Z, Matveev A, Ovrebo S, Nilssen R, Nysveen A. State of the art in generator technology for offshore wind energy conversion systems. In: Electric Machines Drives Conference (IEMDC), 2011 IEEE International; May 2011. pp. 1131-1136

[4] Li H, Chen Z. Design optimization and evaluation of different wind generator systems. In: International Conference on Electrical Machines and Systems, 2008. ICEMS 2008; Oct 2008. pp. 2396-2401

[5] Levi E, Bojoi R, Profumo F, Toliyat HA, Williamson S. Multiphase induction motor drives—A technology status review. IET Electric Power Applications. 2007;**1**(4):489-516

[6] Banerjee A, Tomovich MS, Leeb SB, Kirtley JL. Control architecture for a doubly-fed induction machine propulsion drive. In: 2013 Twenty-Eighth Annual IEEE Applied Power Electronics Conference and Exposition (APEC); March 2013. pp. 1522-1529

[7] Leon JI, Kouro S, Franquelo LG, Rodriguez J, Wu B. The essential role and the continuous evolution of modulation techniques for voltage-source inverters in the past, present, and future power electronics. IEEE Transactions on Industrial Electronics. 2016;**63**(5):2688-2701

[8] Venturini M, Alesina A. The generalised transformer: A new bidirectional, sinusoidal wave- form frequency converter with continuously adjustable input power factor. In: 1980 IEEE Power Electronics Specialists Conference; June 1980. pp. 242-252

[9] Lee MY, Wheeler P, Klumpner C. Space-vector modulated multilevel matrix converter. IEEE Transactions on Industrial Electronics. 2010;**57**(10): 3385-3394

[10] Wheeler PW, Clare JC, Apap M, Lampard D, Pickering SJ, Bradley KJ, et al. An integrated 30kw matrix converter based induction motor drive. In: 2005 IEEE 36th Power Electronics Specialists Conference; June 2005. pp. 2390-2395

[11] Abebe R, Vakil G, Calzo GL, Cox T, Lambert S, Johnson M, et al. Integrated motor drives: State of the art and future trends. IET Electric Power Applications. 2016;**10**(8):757-771

[12] Pena R, Cardenas R, Reyes E, Clare J, Wheeler P. A topology for multiple generation system with doubly fed induction machines and indirect matrix converter. IEEE Transactions on Industrial Electronics. 2009;**56**(10): 4181-4193

[13] Liu C, Wu B, Zargari N, Xu D, Wang J. A novel three-phase three-leg ac/ac converter using nine igbts. IEEE Transactions on Power Electronics. 2009;**24**(5):1151-1160

[14] Hava A, Kerkman R, Lipo T. Simple analytical and graphical tools for carrier based pwm methods. In: Power Electronics Specialists Conference, 1997. PESC '97 Record, 28th Annual IEEE, Vol. 2; Jun 1997. pp. 1462-1471

[15] Reusser CA, Kouro S, Cardenas R. Dual three-phase pmsg based wind energy conversion system using 9-switch dual converter. In: 2015 IEEE Energy Conversion Congress and Exposition (ECCE); Sep. 2015. pp. 1021-1022

[16] Reusser CA. Full-electric ship propulsion, based on a dual nine-switch inverter topology for dual three-phase induction motor drive. In: 2016 IEEE Transportation Electrification Conference and Expo (ITEC); June 2016. pp. 1-6

[17] Kumar EA, Satyanarayanat G. An eleven-switch inverter topology supplying two loads for common-mode voltage mitigation. In: 2017 IEEE PES Asia-Pacific Power and Energy Engineering Conference (APPEEC); Nov 2017. pp. 1-6

Section 3

Static Power Conversion

Chapter 4

Five-Phase Line Commutated Rectifiers

Mahmoud I. Masoud

Abstract

Multiphase systems including multiphase generators or motors, especially five-phase, offer improved performance compared to three-phase counterpart. Five-phase generators could generate power in applications such as, but not limited to, wind power generation, electric vehicles, aerospace, and oil and gas. The five-phase generator output requires converter system such as AC-DC converters. This chapter introduces the basic construction and performance analysis for uncontrolled/controlled five-phase line commutated rectifier guided by numerical examples. This rectifier is suitable for wind energy applications to be the intermediate stage between five-phase generator and DC load or inverter stage. The filtration for AC side and effects of source inductances are detailed in other references. Here, this chapter gives the reader a quick idea about the analysis and performance of multiphase line commutated rectifiers and specifically five phase.

Keywords: five phase, multiphase generators, wind turbine, controlled and uncontrolled rectifiers, AC-DC converters, line commutated rectifiers

1. Introduction

Nowadays, energy generation using renewable energy sources is considered as contemporary issue. Many countries are experiencing economic pressures because of the reduction in oil prices. Moreover, energy that is produced by burning fossil fuels is mainly responsible for the global air pollution problem, and consequently, it is not environmentally friendly. As a consequence, many countries are impelled to concentrate on renewable energy sources including generation, management, and planning. Renewable energy share, of the total energy consumption by the end of 2016, is 24.5% where it was 3.3% in 2010 [1]. This number is tremendously increasing each year. Accordingly, most research institutes foster research in the energy sector to go to clean and green energy generation. From renewable energy sources, wind energy is considered to be a pillar using appropriate generator. By the end of 2016, 29 countries have more than 1 GW in operation from wind energy, and the total capacity generated for the world is 487 GW. Only in 2016 a 23.4 GW is added due to wind energy generation [1]. Incontrovertibly, wind power becomes a big industry and a big topic for research. It includes mechanical parts and electrical parts. Accordingly, a generator that converts wind mechanical power to electrical power is a must. But, the power generated using wind turbines is not constant and could fluctuate. Also, the load profile fluctuates with respect to time. To accomplish this, specially designed power systems for meeting these demands are built which can feed these drastic load changes. In all cases, the generator output voltage should be rectified. Owing to the numerous advantages when compared to its three-phase counterpart such as high fault-tolerant capability, five-phase direct-drive permanent magnet generators are a key area of focus in power generation with renewable and wind energy systems [2–6]. Hence, the generator output requires a converter, such as an AC-DC rectifier, to match load requirements. Rectifiers are broadly classified into two types, namely, line-commutated rectifiers (LCRs) and power factor control rectifiers. LCRs are either uncontrolled (using diodes) or controlled (using thyristors). This type of rectifiers operates on power frequency or generated frequency and has the advantages of simple construction and low cost. In a controlled LCR, the DC-link voltage can be controlled by adjusting the firing angle, thus facilitating high power capability, easy control, and a simple gating system [7, 8]. However, LCRs still have a low input power factor, high input current total harmonic distortion (THD), and unidirectional current flow [9, 10]. To address such problems, power factor control rectifiers, which operate at a high switching frequency, such as pulse-width modulated (PWM) rectifiers are used where the filter size for PWM rectifiers is smaller owing to the higher switching frequencies [9, 11, 12]. However, one of the main drawbacks of using PWM rectifiers is the limitation of the switch current and voltage, which restricts the use of such rectifiers in high-power-scale applications [3, 5, 13, 14].

This chapter introduces uncontrolled/controlled line-commutated five-phase rectifier as we target high-power wind energy applications. The study includes the analysis of both AC side and DC side. Moreover, the gating signal generation is introduced. The study includes ideal source to represent the PM generator. Effect of source inductance and filtration of AC side are introduced in Refs. [3, 5].

2. Five-phase source

The phase voltage of the five-phase source, which is the output of the five-phase generator, can be written as

$$v_{jn} = V_{ph\_\max} \sin\left(\omega t - \frac{2\pi k}{5}\right) \tag{1}$$

where j represents phases a, b, c, d, and e, respectively, and k equals 0, 1, 2, 3, and 4, respectively. The five-phase source has two line voltages (adjacent and nonadjacent) [3, 5]. The line voltages can be calculated using phasor diagram shown in **Figure 1**.

The line voltage for adjacent phases has amplitude 1.1756 V_{ph_max} and leads to the phase voltage by 54°; see **Figure 1a**. The line voltage for nonadjacent phases has amplitude 1.902 V_{ph_max} and leads to the phase voltage by 18°; see **Figure 1b**.

Example 1: *For a 200 V (peak value-phase voltage) five-phase source, calculate and plot the phase and line voltage waveforms.*

Solution

Phase voltages are

$$v_a = 200\sin\ (\omega t),\quad v_b = 200\sin\left(\omega t - \frac{2\pi}{5}\right),\quad v_c = 200\sin\left(\omega t - \frac{4\pi}{5}\right),$$
$$v_d = 200\sin\left(\omega t + \frac{4\pi}{5}\right),\quad v_e = 200\sin\left(\omega t + \frac{2\pi}{5}\right)\quad V$$

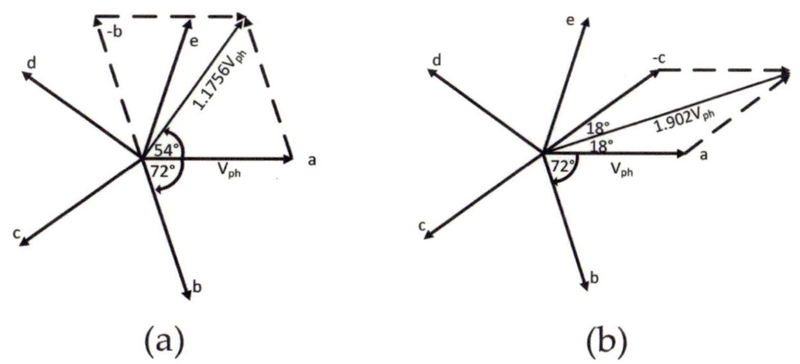

Figure 1.
Phasor diagram for five-phase source: (a) line voltage for adjacent phases and (b) line voltage for nonadjacent phases.

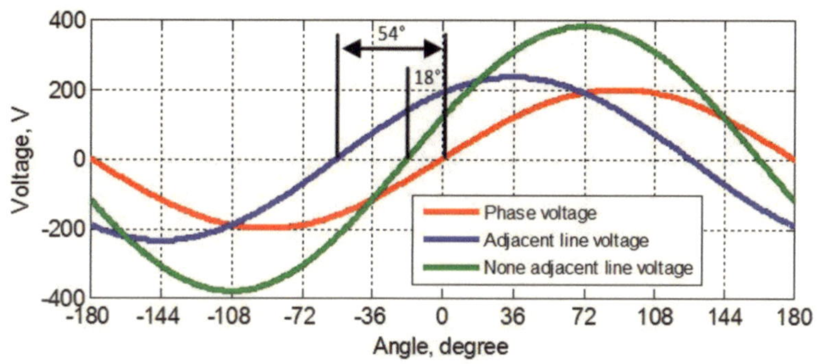

Figure 2.
Phase voltage, adjacent line voltage, and non-adjacent line voltage.

Figure 2 shows the line voltages for adjacent and non-adjacent phase voltages. The line voltage for adjacent phases is

$$v_{ab} = 200^{*}1.1756 \sin\left(\omega t + \frac{3\pi}{10}\right) = 235.12 \sin\left(\omega t + \frac{3\pi}{10}\right) \quad V$$

The line voltage for nonadjacent phases is

$$v_{ac} = 200^{*}1.902 \sin\left(\omega t + \frac{\pi}{10}\right) = 380.4 \sin\left(\omega t + \frac{\pi}{10}\right) \quad V$$

3. Uncontrolled five-phase rectifier

The five-phase uncontrolled rectifier is shown in **Figure 3**. It consists of 10 diodes where the load is fed from five-phase half wave connection, five switches *D1–D5*, and return path being via another half wave connection, five switches *D6–D10*, to one of the five supply lines.

The load voltage or output DC voltage can be calculated by the addition of the two five-phase half-wave voltages, relative to the supply neutral point n, appearing at the positive and negative sides of the load, respectively. The voltage is 10 pulses in nature having its maximum instantaneous value of that of line voltages for nonadjacent phases and leads the phase voltage by $\pi/10$. The load voltage follows in turn 10 sinusoidal voltages during one cycle, those being v_{ac}, v_{ad}, v_{bd}, v_{be}, v_{ce}, v_{ca}, v_{da}, v_{db}, v_{eb},

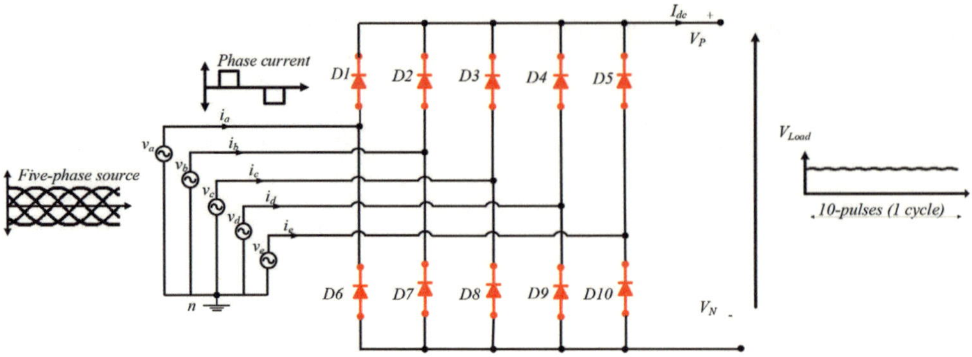

Figure 3.
Five-phase uncontrolled line commutated rectifier.

and v_{ec}. The average voltage can be calculated with the aid of **Figure 4**. The repeated period accommodates 36° and the limits are −18° to +18°.

The average or mean load voltage can be calculated as

$$V_{mean} = \frac{1}{\pi/5}\left[\int_{-\frac{\pi}{10}}^{\frac{\pi}{10}} 1.902V_{ph\_\max}\cos\omega t\, d\omega t\right] \tag{2}$$

$$V_{mean} = 1.87\, V_{ph\_\max} \tag{3}$$

The waveform of supply voltage (phase voltage), positive voltage with respect to neutral (V_P), negative voltage with respect to neutral (V_N), load voltage, diode 1 voltage, diode 1 current, and supply current are shown in **Figure 5** parts 'a' to 'e'. The peak reverse voltage for one diode is the line voltage of nonadjacent phases and as indicated in **Figure 5c**.

The ripple voltage for DC side is

$$\%\ Voltage\ \ ripple = \frac{Maximum\ \ value - Minimum\ \ value}{V_{mean}} * 100 \tag{4}$$

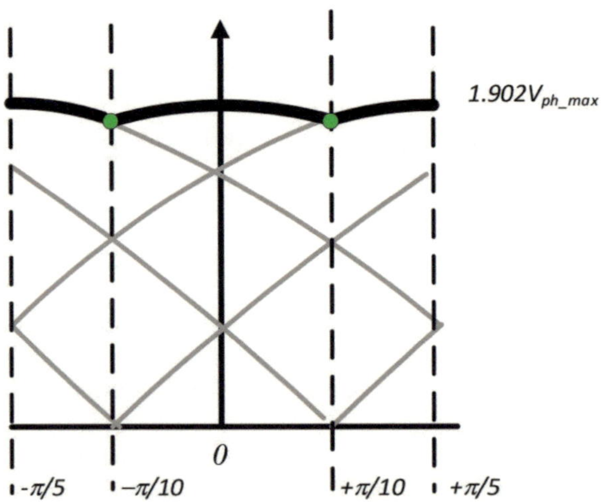

Figure 4.
The load voltage limits.

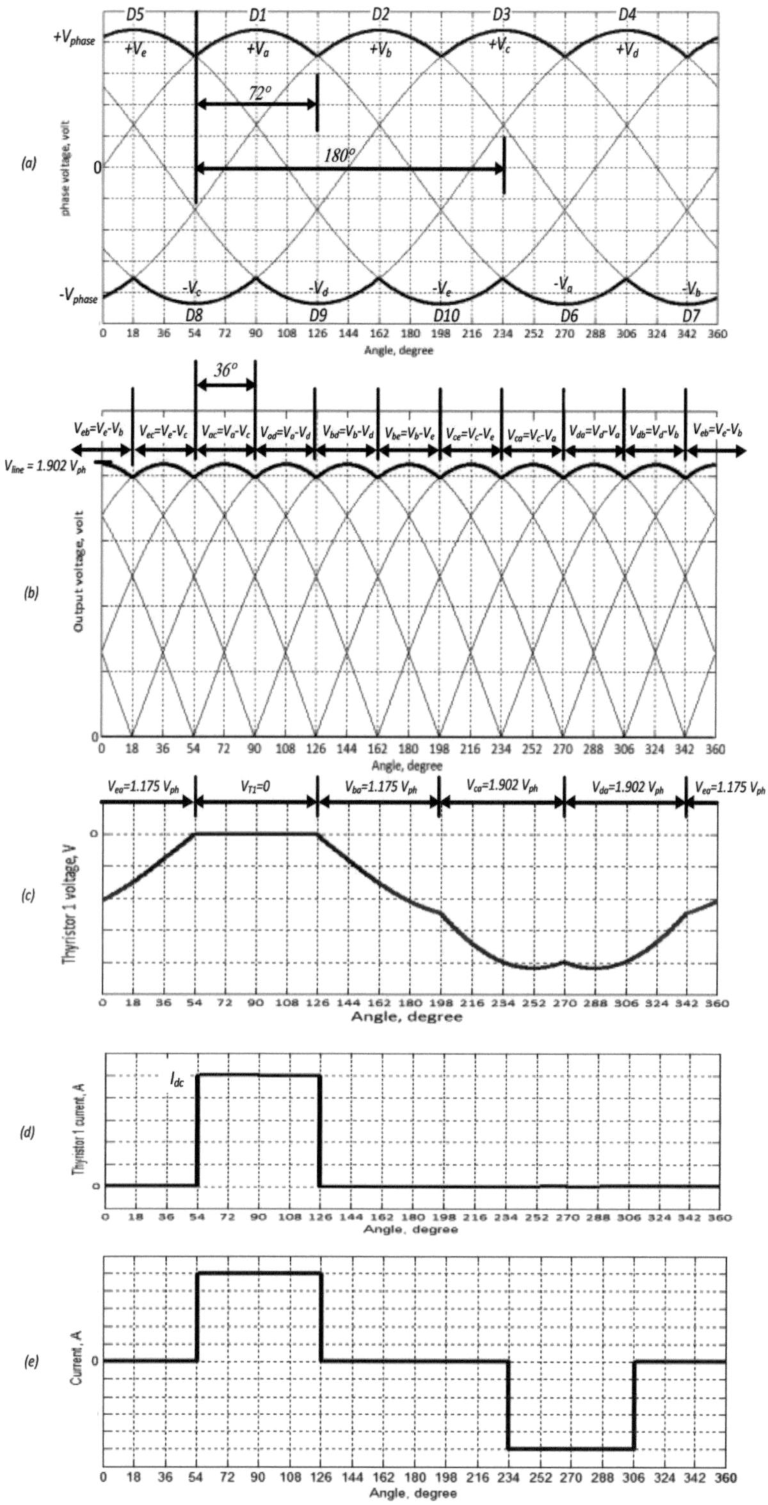

Figure 5.
Five-phase uncontrolled rectifier waveforms: (a) supply voltage, positive voltage with respect to neutral VPN, and negative voltage with respect to neutral VNN, (b) output voltage waveform, (c) diode 1 current, (d) diode 1 current, and (e) supply current for phase 'a'.

The DC-side harmonics are generated at $10kf$, where f is the fundamental frequency of the AC source or generated power frequency and k equals (1, 2, 3, etc.). The lowest order harmonic represents 2.1% of the average DC value as shown in **Figure 6**.

The supply current is a quasi-square wave in its nature as shown in **Figure 4e**. The load current is constant as the load is considered highly inductive. Switch current operates for only $2\pi/5$ from one cycle; accordingly, the switch average current can be calculated as in Eq. (5) which represents 20% of the DC-link current.

$$I_{av\_switch} = \frac{1}{2\pi}\left[\int_{\frac{3\pi}{10}}^{\frac{7\pi}{10}} I_{dc}\ d\omega t\right] = 0.2\ I_{dc} \tag{5}$$

This means five-phase rectifier could have switches with lower ratings than three-phase counterpart which feed the same load as the average value of switch current in three-phase system is 33.3% [8, 15]. Part 'e' in **Figure 5** shows that supply current is a quasi-square wave where input current comprises $2\pi/5$ alternating polarity blocks, with each phase displaced to others by $2\pi/5$. The input phase current can be expressed as [5, 8, 11, 15].

$$i_s(t) = \sum_{n=1,\,3,\,5,\,\ldots}^{\infty} \frac{4I_{dc}}{n\pi} \sin\frac{n\pi}{5} \sin\ (n\omega t) \tag{6}$$

Substituting $n = 5$ (or any odd multiple of 5) into Eq. (6) results in $i_{s5}(t) = 0$. The fundamental input current ($n = 1$) is

$$i_{s1}(t) = 0.7484 I_{dc}\ \sin\ (\omega t) \tag{7}$$

and rms value of fundamental current is

$$I_{s1} = 0.529 I_{dc} \tag{8}$$

The rms value of supply current is calculated as

$$I_s = \sqrt{\frac{1}{\pi}\int_{3\pi/10}^{7\pi/10} I_{dc}^2\, d\omega t} = \sqrt{\frac{2}{5}} I_{dc} = 0.6324 I_{dc} \tag{9}$$

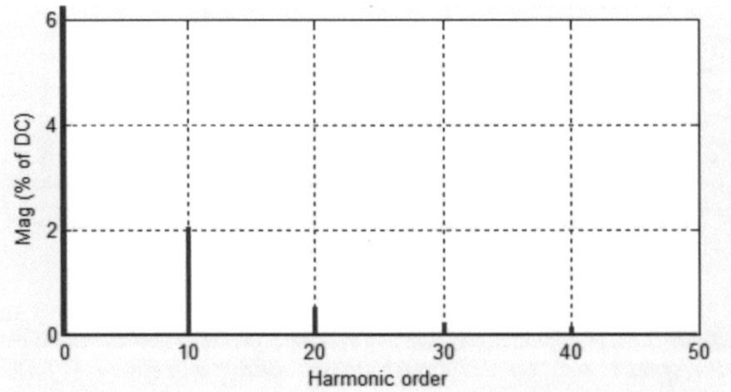

Figure 6.
Harmonic spectrum for load voltage percentage of average DC value.

The harmonic contents of the supply current normalized to the fundamental component (I_{sn}/I_{s1}) are shown in **Figure 7**. The harmonic factor (HF_n) is defined as

$$HF_n\% \quad = \quad \frac{I_{sn}}{I_{s1}} * 100 \qquad (10)$$

Example 2: *A 240 V, 50 Hz, five-phase uncontrolled rectifier is connected to highly inductive load and takes 25 A. Calculate:*

a. *Average load voltage*

b. *The value of lowest order harmonic in DC side*

c. *Maximum and minimum DC side voltage and voltage ripple*

d. *The root mean square value of supply current*

e. *The root mean square value of the supply current fundamental*

f. *The third, fifth, and seventh harmonic factors for supply current*

Solution

Figure 8 shows the load voltage and supply voltage for the uncontrolled rectifier when the supply voltage has phase and rms value of 240 V.

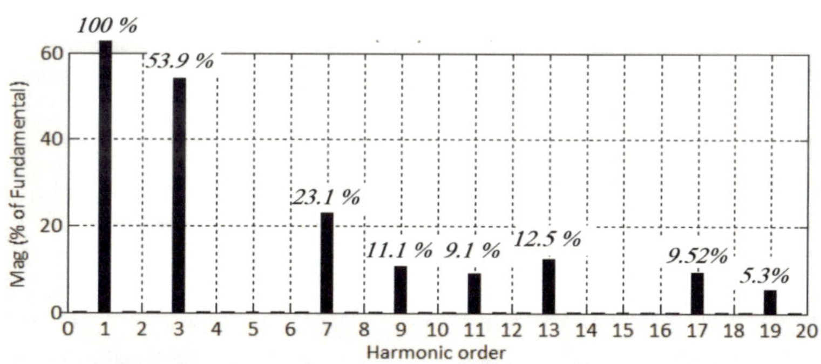

Figure 7.
Supply current spectrum normalized to the fundamental component.

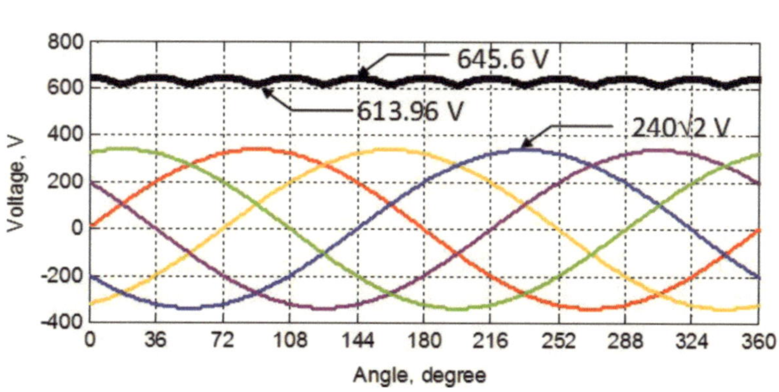

Figure 8.
Phase voltages and load voltage.

a. The average voltage is calculated using Eq. (3):

$$V_{mean} = 1.87\,V_{ph\_\max} = 1.87^{*}240^{*}\sqrt{2} = 634.7\ \ V$$

b. The value of lowest harmonic of DC-side voltage is 2.1% of average DC value and occurs at 10 times the supply voltage frequency, i.e. 500 Hz:

$$Lowest\ \ order\ \ harmonic = 0.021^{*}634.7 = 13.32\ \ V$$

c. The maximum voltage in DC side is the peak value of line voltage for nonadjacent phases:

$$V_{\max} = 1.902V_{ph\_\max} = 1.902^{*}240\sqrt{2} = 645.5\ \ V$$

The minimum voltage in DC side is the value of line voltage for nonadjacent phases at angle 18°:

$$V_{\max} = 1.902V_{ph\_\max}\cos 18 = 613.96\ \ V$$

The percentage of ripple voltage can be calculated using Eq. (4):

$$\%Voltage\ \ ripple = \frac{Maximum\ \ value - Minimum\ \ value}{V_{mean}}{}^{*}100 = \frac{645.6 - 613.96}{634.7}{}^{*}100 = 4.98\%$$

d. The rms of supply current is calculated using Eq. (9):

$$I_s = 0.6324\,I_{dc} = 15.81A$$

e. The root mean square value of supply current fundamental is calculated using Eq. (8):

$$I_{s1} = 0.529\,I_{dc} = 0.529^{*}25 = 13.225A$$

f. The third harmonic of supply current represents 53.9% of the fundamental = 7.13 A:

The fifth harmonic of supply current = 0.

The seventh harmonic of supply current represents 23.1% of the fundamental = 3.05 A.

Example 3: *A highly inductive DC load requires 100 A at 475 V. Give design details for the diode requirement using uncontrolled five-phase full-wave rectifier.*

Solution

The required load voltage is 475 V; hence the supply voltage can be calculated using Eq. (3):

$$V_{mean} = 1.87\,V_{ph\_\max} = 475 = 1.87\,V_{ph\_\max}$$
$$\therefore\ V_{ph\_\max} = 254\ \ V$$

The required rms value of phase voltage is 179.6 ≅ 180 V.

The line voltage of nonadjacent phases is

$$V_L = 1.902V_{ph\_\max} = 1.902^{*}179.6^{*}\sqrt{2} = 483\ \ V$$

Hence, the peak reverse voltage is 483 V. Hence, each diode should withstand maximum voltage of 483 V and 100 A.

4. Fully controlled five-phase rectifier

The five-phase fully controlled rectifier is similar to the uncontrolled rectifier, but it has 10 thyristors instead of diodes as shown in **Figure 9**. The challenge with five-phase controlled rectifier is gating signal generation where the development and implementation of gating signals algorithm use nonadjacent line voltages. Delaying the commutation of thyristors for a certain firing delay angle 'α' controls the mean load voltage.

Figure 1 illustrates how to get DC-load voltage or mean voltage from five-phase source using five-phase controlled rectifier. **Figure 10a** shows the phase voltages, positive voltage with respect to neutral point, and negative voltage with respect to neutral point. **Figure 10b** shows the gating signals for the small firing angle α. The load voltage or output voltage can be calculated by the addition of the two five-phase half-wave voltages, relative to the supply neutral point n, appearing at the positive and negative sides of the load, respectively. As shown in **Figure 10c**, the voltage is 10 pulses in nature, labeled from '1' to '10' like uncontrolled rectifier. When v_a is the most positive phase that the thyristor, *T1* conducts (*T1–T5* are connected to the positive terminals, and *T6–T10* are connected to negative terminals), and during this period, v_c is the most negative with thyristor *T8*, conducting until v_d becomes more negative, when the current in *T8* commutates to *T9*. The next firing pulse will be to thyristor *T9*, and *T1* is still conducting until phase b has the most positive value and is fired by $g2$, where *T9* is still conducting, and so forth. The load voltage follows, in turn, with 10 sinusoidal voltages during one cycle, those being v_{ac}, v_{ad}, v_{bd}, v_{be}, v_{ce}, v_{ca}, v_{da}, v_{db}, v_{eb}, and v_{ec}. For the period labeled '1' in **Figure 9c**, phase a is the most positive and conducting, when $g1$ is applied to *T1*, while at the same time, phase c is conducted when $g8$ is applied to *T8*. Consequently, the load voltage ($v_{ac} = v_a - v_c$) is plotted with respect to the reference of the intersection point between phase a and phase c. Phase c is commutated with phase d for the period labeled '2', and thus, the load voltage is ($v_{ad} = v_a - v_d$) and plotted with respect to the reference of the intersection point between phase a and phase d, where the load voltage, in turn, follows the phase voltage. **Figure 10d** and **e** shows currents in *T1* and *T6*, respectively. **Figure 10f** shows the supply current of phase 'a', which is the addition of *T1* and *T6* currents. The supply current is a quasi-square wave in nature. The supply current, the rms value of supply current, and rms value of fundamental supply current are identical to the uncontrolled rectifier case as given by Eqs. (7)–(10). **Figure 10g** shows the voltage across *T1* which can be explained using **Figure 11**. The period is labeled by '2' (**Figure 10g**) and shows that

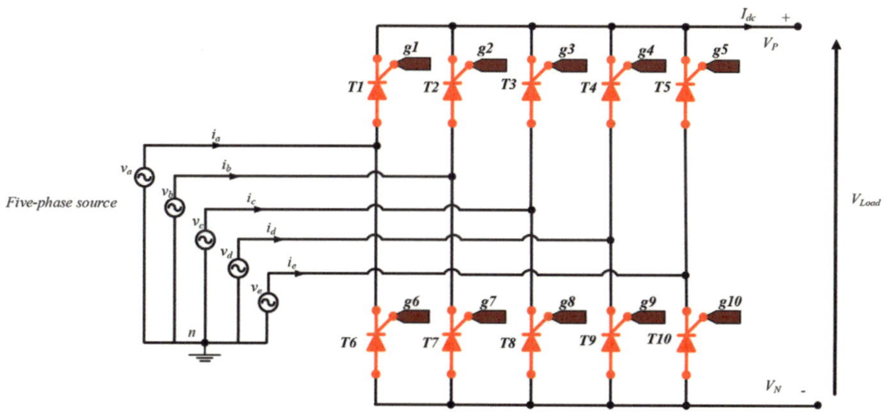

Figure 9.
Five-phase controlled line commutated rectifier.

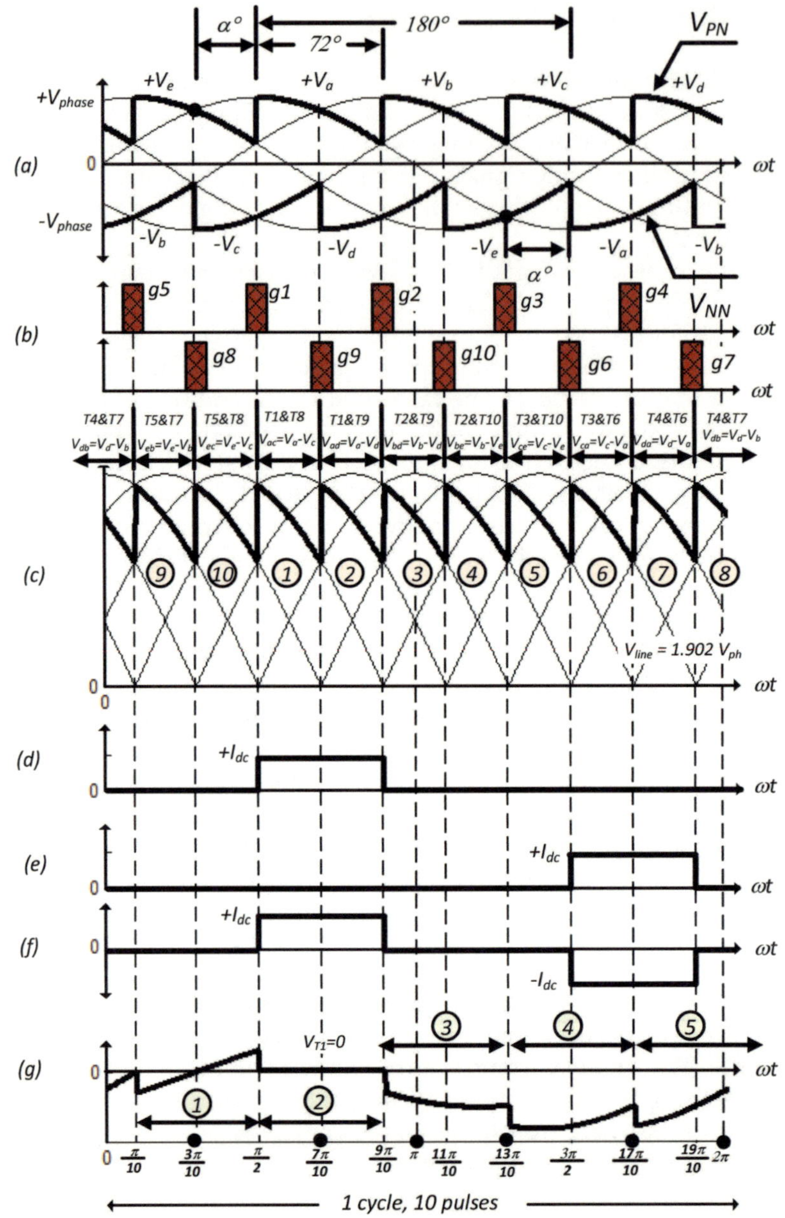

Figure 10.
Five-phase rectifier waveforms: (a) supply voltage, positive voltage with respect to neutral VPN, and negative voltage with respect to neutral VNN, (b) gating signals for firing angle α°, (c) output voltage waveform, (d) thyristor 1 current, (e) thyristor 6 current, (f) supply current for phase 'a', and (g) thyristor 1 voltage.

the voltage across the thyristor is near zero, owing to decrease in the thyristor voltage, while the period labeled by '3' (**Figure 10g**) shows that the voltage across *T1* can be calculated as v_{ba}, which has a maximum value of $1.1756V_{ph\_max}$, and the waveform reference point is the intersection between phases a and b. This period is explained using **Figure 11a**. The period label in 4 (**Figure 10**) is explained with the aid of **Figure 11b**.The switch has a constraint design that it should withstand during the off periods' peak reverse voltage (*PRV*) of $1.902V_{ph\_max}$ (line voltage for nonadjacent phases).

The mean load voltage can be evaluated with the aid of **Figures 10c** and **12** where it shows one pulse of the output load voltage:

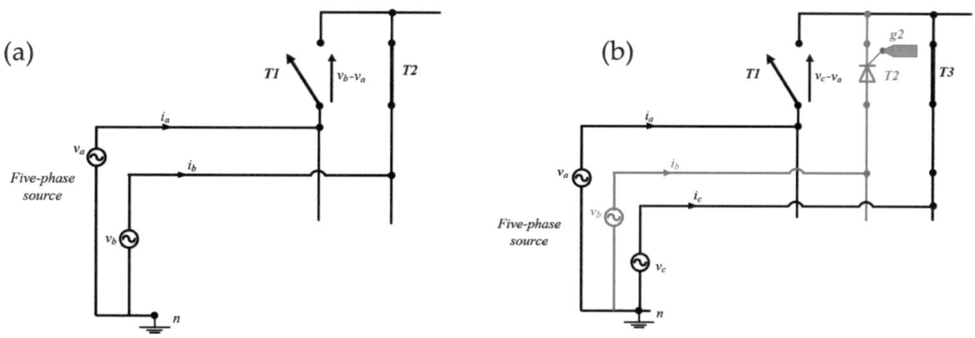

Figure 11.
Circuits calculates voltage across switch T1: (a) equivalent circuit for the period labeled 3 in ***Figure 10g*** *and (b) equivalent circuit for the period labeled 4 in* ***Figure 10g****.*

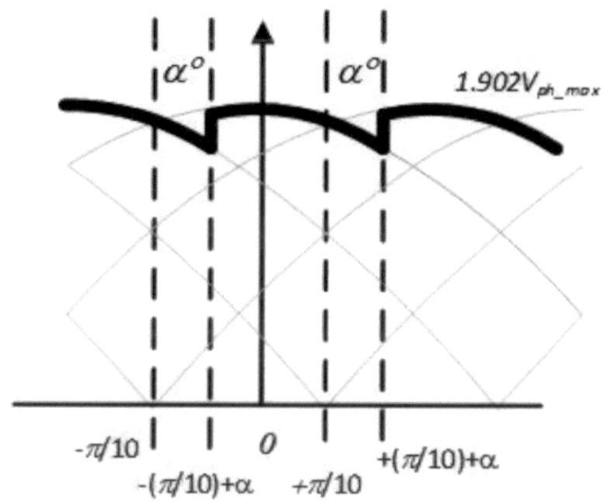

Figure 12.
One pulse of the load voltage for uncontrolled five-phase rectifier.

$$V_{mean} = \frac{1}{2\pi/_{10}} \int_{\frac{-\pi}{10}+\alpha}^{\frac{\pi}{10}+\alpha} 1.902\,V_{ph\_\max} \cos(\omega t) d\omega t \tag{11}$$

$$V_{mean} = 1.87 V_{ph\_\max} \cos(\alpha) \tag{12}$$

The angle between the fundamental input current and fundamental input phase voltage, ϕ, is called the displacement angle. The displacement factor is defined as [3, 5, 16].

$$DPF = \cos \phi_1 = \cos(\alpha) \tag{13}$$

The supply power factor is defined as the ratio of the supply power delivered P to apparent supply power S [16]:

$$pf = \frac{P}{S} \tag{14}$$

In the case of non-sinusoidal current and voltage, the active input power to the converter delivered by the five-phase source (e.g. output of wind turbine generator) is

$$P_{in} = \frac{5}{2\pi}\int_{0}^{2\pi} v_{s1}(t)\ i_{s1}(t)\ d\omega t = 5\ V_{s1}\ I_{s1}\ \cos\ \phi_1 \tag{15}$$

where V_{s1} and I_{s1} are the rms values of the fundamental components of the input phase voltage and current, respectively. The apparent power is given by

$$S = 5\ V_{s-rms}\ I_{s-rms} \tag{16}$$

where $V_{s\text{-}rms}$ and $I_{s\text{-}rms}$ are the rms values of the input phase voltage and current, respectively. The displacement factor (*DPF*) of the fundamental current was obtained in Eq. (13). By substituting (13), (15), and (16) into (14), the power factor can be expressed as

$$pf = \frac{V_{s1}I_{s1}\cos\ \phi_1}{V_{s-rms}I_{s-rms}} \tag{17}$$

if the source is pure sinusoidal supply $V_{s1} = V_{s\text{-}rms}$.

The output power (load power) can be calculated using

$$P_{out} = \frac{1}{T}\int_{0}^{T} v(t)\ i(t)\ dt = V_{mean}I_{mean} \tag{18}$$

The efficiency is

$$\eta = \frac{P_{out}}{P_{in}} \tag{19}$$

The total harmonic distortion, *THD*, is given by [3, 5, 15]

$$THD = \frac{\sqrt{I_s^2 - I_{s1}^2}}{I_{s1}} \tag{20}$$

For a five-phase rectifier, the *THD* is 65.5%.

Example 4: *A five-phase fully controlled rectifier is fed from 220 V (phase value), 50 Hz AC source. The DC load is highly inductive and requires 100 A at 475 V. Give design details for the thyristor requirement and the required firing angle. What is maximum DC load voltage that the converter can deliver to the load?*

Solution

The required load voltage is 475 V and the supply voltage is 220 V. The firing angle can be calculated using Eq. (12):

$$V_{mean} = 1.87\,V_{ph\_\max}\cos\alpha = 475 = 1.87^{*}200^{*}\sqrt{2}\cos\alpha$$
$$\therefore\ \cos\alpha = 0.8998$$
$$\therefore\ \alpha = 26.1^{\circ}$$

The peak line voltage of nonadjacent phases is

$$V_L = 1.902V_{ph\_\max} = 1.902^{*}220^{*}\sqrt{2} = 591.8\ \ V$$

Hence, the peak reverse voltage is ≅592 V. Hence, each thyristor should withstand maximum voltage of 592 V and 100 A.

The maximum possible DC load voltage is when α is adjusted to zero:

$$V_{mean} = 1.87\,V_{ph\_\max} = 1.87^{*}220^{*}\sqrt{2} = 581.8\ \ V$$

Example 5: *A five-phase fully controlled rectifier is fed from 230 V (phase value), 15 Hz AC source. The load is 45 Ω resistance and 1 H inductance (highly inductive load). Calculate for firing angle 36°, the converter power factor. Simulate your converter.*

Solution

The mean load voltage is

$$V_{mean} = 1.87\,V_{ph\_\max}\cos\alpha = 1.87^{*}230^{*}\sqrt{2}\cos 36 = 492\ \ V$$

The load current is

$$I_{mean} = \frac{V_{mean}}{R} = \frac{492}{25} = 19.68\ \ \Omega$$

From Eq. (8), the rms value for fundamental supply current is 10.41 A.

The angle of fundamental current is the firing angle.

From Eq. (9), the rms value of supply current is 12.44 A.

The power factor is calculated using Eq. (17):

$$pf = \frac{V_{s1}I_{s1}\cos\ \ \phi_1}{V_{s-rms}I_{s-rms}} = \frac{230^{*}10.41^{*}\cos\,(36)}{230^{*}12.44} = 0.677$$

The load voltage and supply current are shown in **Figure 13**. It should be noted that as the firing angle is 36°, the load voltage shown in **Figure 13a** starts for phase 'a' at 90° which equals 54°, the intersection point between two adjacent phases, and 36° which is the firing angle. The supply current for phase 'a' is shown in **Figure 13b**. Switch 1 accommodates 72° and starts at 90°. Switch 6 is shifted 180° for the same phase.

5. Gating pulse generation

A pulse generator synchronized on the source voltage provides the trigger pulses for the 10 thyristors that construct the 10-pulse rectifier. The pulses are generated by α degrees after increasing zero crossings of the commutation voltages which is indicated in **Figures 10a** and **12**. The input to the gating signal generator is 10 signals represent five-line voltages for nonadjacent phases and its inverted signals, and multiplexed, then determines the zero-crossing point and counts the delay angle α and hence applies the pulse for the specified switch through gate drive circuits as shown in **Figure 14**. A double-pulse technique is used to ensure that the specified gate signal fires its own thyristor [3, 5].

A sample of gating signal generation is shown in **Figure 15** where an example for gating signal generator output with delay angle 36° with double-pulse technique is indicated. The phase voltages of five-phase permanent magnet generator are measured as shown in part a where the rotational speed tends to have a frequency around 18 Hz and rms value of 56 V. The delay angle is calculated from the

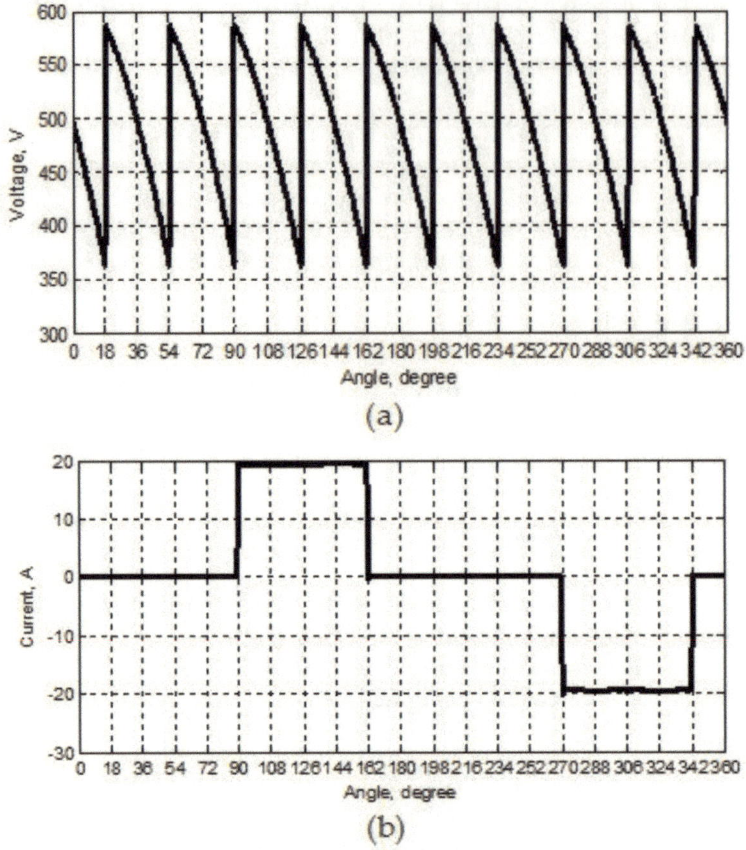

Figure 13.
Load voltage and supply current waveforms: (a) load voltage and (b) supply current.

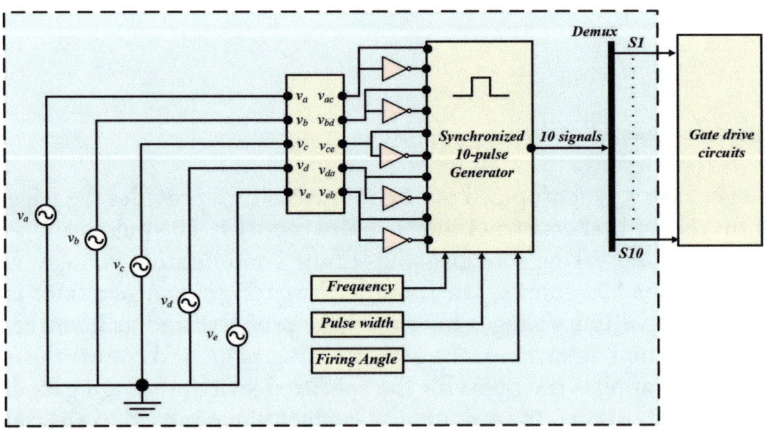

Figure 14.
Gating signal generation block.

zero-crossing point determined by the pulse generator as indicated in **Figure 15b** where gate signals 1, 2, and 6 confirm that the phase shift between gates 1 and 2 is 72° and between gates 1 and 6 is 180°. The delay angle is calculated after 54° from the start of the phase voltage [3, 5]. If the delay angle is increased to 36°, the results

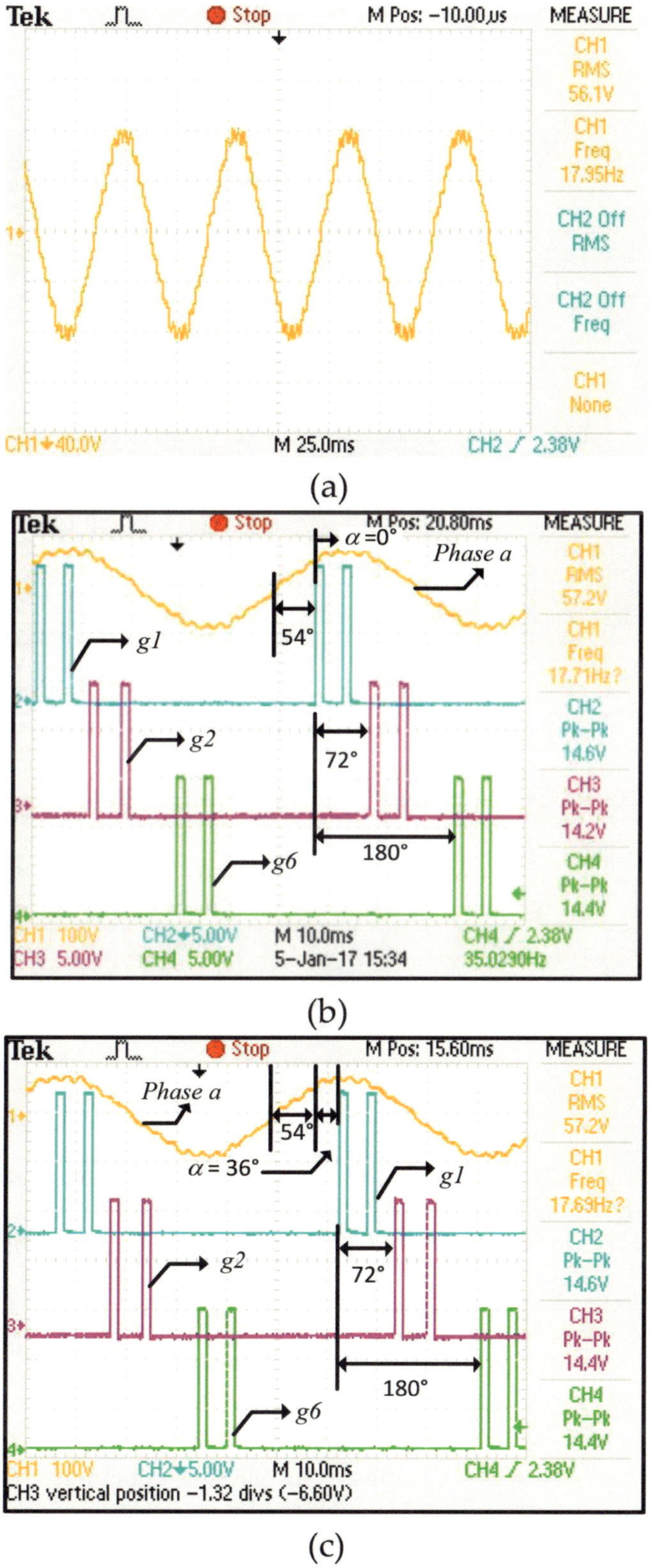

Figure 15.
Gating pulse generation: (a) Phase voltage for PM generator; (b) g1, g2, and g6 for firing angle zero; and (c) g1, g2, and g6 for firing angle 36°.

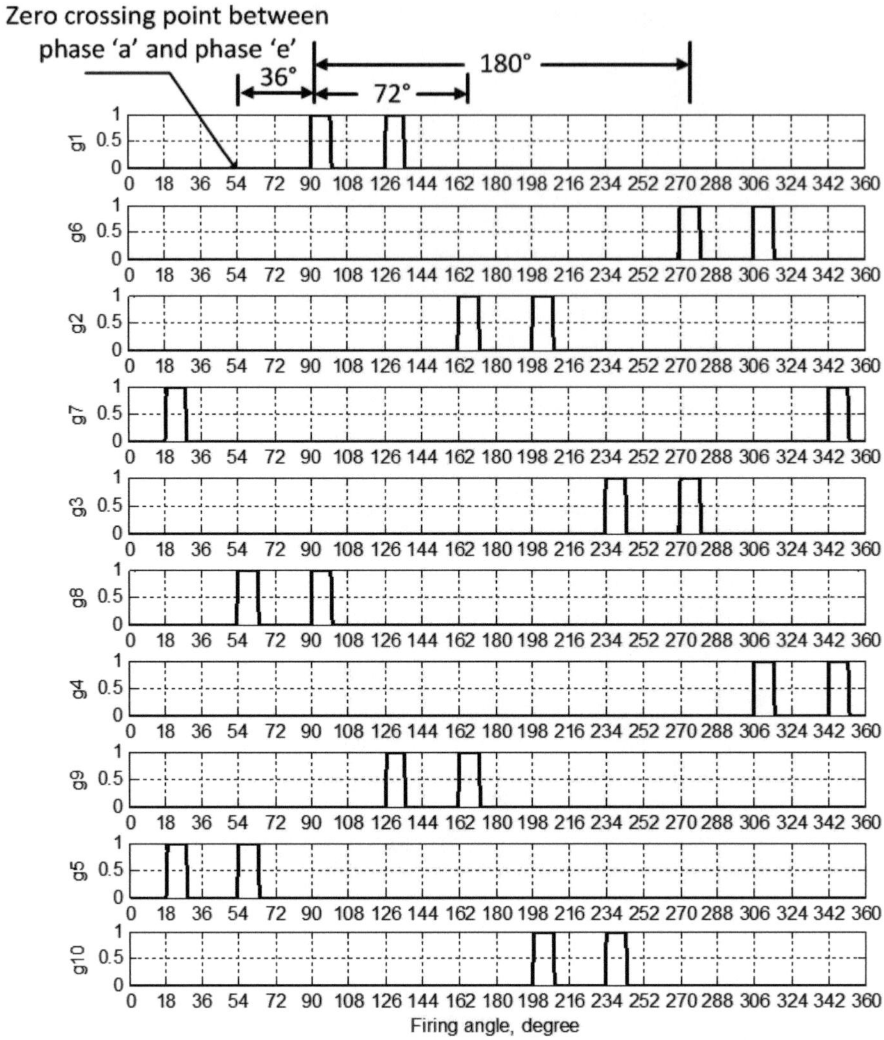

Figure 16.
Gating pulses for firing angle 36° (g1–g10).

show that the relation between $g1$, $g2$, and $g6$ will be the same but shifted to 90° from the starting of phase voltage as shown in **Figure 15c**.

Example 6: *For the converter given in '**Example 5**', plot the gating signals*

Solution

Figure 16 shows the gating signals for firing angle 36°.

6. Conclusion

This chapter introduces the five-phase, 10-pulse, line commutated rectifier. The relation between phase voltage and line voltage for five-phase source is introduced where line voltage equals $1.1756V_{ph\_max}$ and leads the phase voltage by 54° for adjacent phases, while it equals $1.902V_{ph\_max}$ and leads the phase voltage by 18° for nonadjacent phases. The converter performance and analysis for both AC side and DC side are detailed for both uncontrolled and controlled rectifiers. The average voltage of the DC side is 1.87 V_{ph_max} with ten pulses per one cycle, while it is 1.654

V_{ph_max} with six pulses for three-phase counterpart. Moreover, the DC-side voltage ripple is around 5%, while it is around 14% for full-wave three-phase converter. The switch (diode or thyristor) carries only 20% of the load current during one cycle, while it carries 33.3% in three-phase counterpart, and no-phase shifting transformer is required like 12-pulse converter. Also, the implementation of gating signal generation is introduced. The gating signal generation adopts double-pulse techniques and ensures the real-time synchronization between the five-phase source and gating signals. The phase shift pulse of adjacent phases is 72°, while the phase shift between pulses applied for switches on the same phase adopts 180°.

Author details

Mahmoud I. Masoud
Sultan Qaboos University, Muscat, Sultanate of Oman

*Address all correspondence to: mahmoudmasoud77@hotmail.com

References

[1] Renewables 2017 Global Status Report, REN21, Renewable Energy Policy Network. Available from: http://www.ren21.net/wp-content/uploads/2017/06/17-8399_GSR_2017_Full_Report_0621_Opt.pdf [Accessed: August 8, 2018]

[2] Levi E. Multiphase electric machines for variable-speed applications. IEEE Transactions on Industrial Electronics. 2008;**55**(5):1893-1909. DOI: 10.1109/TIE.2008.918488

[3] Masoud MI, Saleem A, Al-Abri R. Real-time gating signal generation and performance analysis for fully controlled five-phase, 10-pulse, line-commutated rectifier. IET Transaction on Power Electronics. 2018;**11**(4):744-754. DOI: 10.1049/iet-pel.2017.0587

[4] Abdel-Khalik AS, Ahmed S, Massoud AM, Elserouji A. An improved performance direct-drive Permanent Magnet wind generator using a novel single-layer winding layout. IEEE Transactions on Magnetics. 2013;**49**(9):5124-5134. DOI: 10.1109/TMAG.2013.2257823

[5] Masoud MI. Fully controlled 5-phase, 10-pulse, line commutated rectifier. Alexandria Engineering Journal, AEJ. 2015;**54**(4):1091-1104. DOI: 10.1016/j.aej.2015.07.004. Elsevier

[6] Abdel-Khalik AS, Masoud MI, Williams BW. Eleven phase induction machine: Steady-state analysis and performance evaluation with harmonic injection. The Institution of Engineering Technology, IET, Electric Power Applications, EPA. 2010;**4**(8):670-685. DOI: 10.1049/iet-epa.2009.0204

[7] Wu B. High Power Converters and AC Drives. Piscataway, NJ: IEEE Press; 2006. DOI: 10.13140/RG.2.1.1093.2967

[8] Lander CW. Power Electronics. 3rd ed. England: McGraw-Hill; 1993. ISBN: 0-07-707714-8

[9] Iqbal A, Payami S, Saleh M, Anad A, Behra RK. Analysis and control of a five-phase AC/DC/AC converter with active rectifier. In: The 22nd Australasian Universities Power Engineering Conference (AUPEC); Bali, Indonesia; 2012. pp. 1-6

[10] Blaabjerg F, Ma K. Future on power electronics for wind turbine systems. IEEE Journal of Emerging and Selected Topics in Power Electronics. 2013;**1**(3):139-152. DOI: 10.1109/JESTPE.2013.2275978

[11] Masoud MI, Al-Abri RS. Five-phase current source PWM rectifier. In: The 7th IEEE GCC International Conference and Exhibition; GCC, Doha, Qatar; 2013. pp. 148-152

[12] Rodriguez JR, Dixon LW, Espinoza JR, Pontt J, Lezana P. PWM regenerative rectifiers: State of the art. IEEE Transactions on Industrial Electronics. 2005;**52**(1):5-22. DOI: 10.1109/TIE.2004.841149

[13] Abdelsalam AK, Masoud MI, Finney SJ, Williams BW. Medium-voltage pulse width modulated current source rectifiers using different semiconductors: Loss and size comparison. International Journal of Engineering technology. IET, Transaction on Power Electronics. 2010;**3**(2):243-258. DOI: 10.1049/iet-pel.2008.0252

[14] Naseri F, Haidar S. A comparison study of high power IGBT-based and Thyristor-based AC to DC converters in medium power dc arc furnace plants. In: The 9th International Conference in Compatibility and Power Electronics; Lisbon, Portugal; 2015. pp. 14-19

[15] Rashid M. Power Electronics Circuits, Devices, and Applications. 3rd ed. USA: Prentice Hall; 2004. ISBN-10: 0131011405

[16] Tzeng YS. Harmonic analysis of parallel-connected 12-pulse uncontrolled rectifier without an interphase transformer. IEE Proceedings of Electric Power Applications. 1998; **145**(3):253-260. DOI: 10.1049/ip-epa: 19981850

Chapter 5

Power Device Loss Analysis of a High-Voltage High-Power Dual Active Bridge DC-DC Converter

Thaiyal Naayagi Ramasamy

Abstract

The insulated-gate bipolar transistor (IGBT) offers low conduction loss and improved performance and, hence, is a potential candidate for high-current and high-voltage power electronic applications. This chapter presents the power loss estimation of IGBTs as employed in a high-voltage high-power dual active bridge (DAB) DC-DC converter. The mathematical models of the device currents are derived, and the power loss prediction is clearly explained using the mathematical models. There are many parameters to consider when selecting an appropriate power device for a given application. This chapter highlights the step-by-step procedure for selecting suitable IGBTs for a 20 kW, 540/125 V, 20 kHz DAB converter designed for aerospace energy storage systems. Experimental results are given to demonstrate the device performance at 540 V, 80 A operation of high-voltage IGBTs and 125 V, 300 A operation of low-voltage IGBTs and thus validate the selection procedure presented.

Keywords: power loss prediction, HV-side IGBTs, LV-side IGBTs, zero voltage switching, zero current switching, dual active bridge, DC-DC converter

1. Introduction

The dual active bridge (DAB) [1, 2] is a bidirectional DC-DC converter that has a small transformer between the two active bridges. This topology has received significant attention from researchers for high-power applications for decades. Compared with a single-phase DAB, a three-phase DAB of similar rating offers the benefits of smaller passive components and improved magnetic utilization [3]. Acronyms used in this chapter are listed in **Table 1**. A high-voltage SiC IGBT is used to achieve optimum design of a DAB converter based on the consideration of the device and transformer characteristics in [4]. However, the high *dv/dt* of the SiC IGBT switches [5–7] causes a huge spike and ringing in the currents during hard switching at high voltage levels.

An optimization algorithm for non-active power loss minimization in the DAB DC-DC converter is discussed in [8] to improve the converter efficiency. Ref. [9] presents an advanced control strategy to reduce DAB converter losses. A survey of material-level developments in the key components of PWM power converters and the potential for improving system power density using advanced components is presented in [10]. Ref. [11, 12] indicate that a reduction of the switching loss by tenfold

Acronyms	Abbreviations
SiC	Silicon carbide
PWM	Pulse width modulation
HV	High voltage
LV	Low voltage
DAB	Dual active bridge
IGBT	Insulated-gate bipolar transistor
MOSFET	Metal-oxide-semiconductor field-effect transistor
DC	Direct current
ZVS	Zero voltage switching
ESR	Effective series resistance
ZCS	Zero current switching

Table 1.
List of acronyms and their abbreviations.

is achievable with the use of such advanced devices. Although wide bandgap devices have merits over Si devices, most have not been commercialized due to various reasons, such as the difficulty of mass producing large-diameter SiC wafers, the difficulty of controlling the impurity level of these devices, and high cost. Several parameters must be considered to select the correct power semiconductor device for any power electronic circuit. Device output characteristics, thermal characteristics, switching characteristics, blocking voltage, leakage current, and safe operating area (SOA) are all important factors with respect to reliability. It is important to note that at any point of time during device operation, the maximum ratings specified in the device datasheet should not be exceeded [13–15]. The gate charge characteristics and gate driver requirements of a power device and the associated driver circuit loss are also important factors to be considered when selecting an appropriate power device. The focus of this chapter is the estimation of the conduction power loss and the switching power loss. The key parameters and the device characteristics portrayed in the datasheets are discussed in [16].

The DAB DC-DC converter is shown in **Figure 1**. The converter has two H-bridge circuits interfaced via a high-frequency transformer. To correctly choose

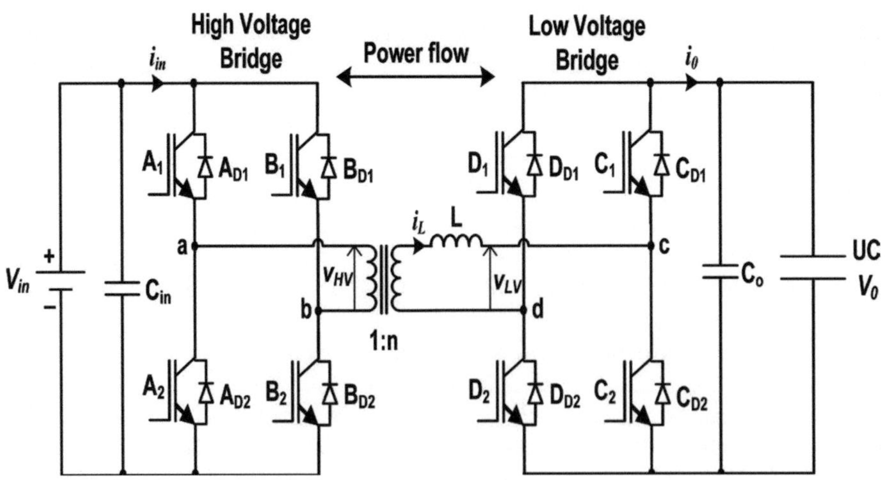

Figure 1.
Dual active bridge DC-DC converter.

the power device for the DAB, power loss calculation is performed for all power devices under the worst-case operating condition of the converter.

The performances of various metal-oxide-semiconductor field-effect transistors (MOSFETs) and IGBTs produced by commercial manufacturers are compared to determine the appropriate power semiconductor device for the high-voltage (HV) and low-voltage (LV) sides of the converter [17–21]. For mission-critical applications such as aerospace systems, voltage transients are inevitable and are taken into consideration when selecting the voltage ratings of devices: 1200 V IGBTs are chosen for the HV side, and 600 V IGBTs are chosen for the LV side [22]. In addition to the device voltage and current handling capabilities and the power density, the anti-parallel diode's on-state voltage and soft recovery characteristics should be taken into consideration. An analysis of MOSFET usage for the low-voltage side of the converter is presented in [23]. Due to the low voltage on the secondary side, the devices should be able to handle high currents. Hence, three or more MOSFETs should be connected in parallel, which makes the circuit very complex. Consequently, IGBT technology [24–26] has been chosen due to its high-current and high-voltage handling capabilities. The IGBT power modules have the benefit of reduced internal inductance and improved heat dissipation and are easy to connect. This chapter discusses the selection procedure of appropriate power devices for a high-power DAB DC-DC converter design. The mathematical model of the device current equations is presented. Researchers at an early stage of their research will find this information valuable as a reference for the design and development of converter prototypes.

This chapter significantly extends the experimental results given in [16]. Specifically, the experimental results for high-voltage IGBTs with the snubber capacitor and for low-voltage IGBTs with and without the snubber capacitors are included. In addition, the experimental measurements of power losses of the HV- and LV-side IGBTs are presented. The remainder of this chapter is organized as follows. The power loss estimation of HV- IGBTs of the converter from various manufacturers is presented in Section 2. Section 3 discusses the power loss calculation of LV-side IGBTs and the loss comparison of IGBTs with similar ratings from various leading manufacturers. The experimental results are presented in Section 4, and the conclusions are given in Section 5.

2. Power loss estimation of HV-side devices of the DAB DC-DC converter

To select the correct device, a power loss calculation should be performed for the worst-case scenario of the converter for the chosen application. In this work, we consider the design of a DAB converter for an aerospace energy storage system whose HV side is connected to the DC link of an aircraft and its LV side is connected to an ultracapacitor-based energy storage system. The working range for the ultracapacitor is assumed to be 2:1. The worst-case operating condition of the converter is V_{HV} = 540 V, V_{LV} = 62.5 V, inductor RMS current I_{RMS} = 427 A, peak inductor current I_P = 640 A, average ultracapacitor current I_0 = 320 A, power throughput P_0 = 20 kW, and d = 0.5. The main component values of the converter are determined by applying the worst-case operating condition to the equations derived based on the mathematical model of the DAB converter, which are presented in [16, 27].

The loss calculation is performed for the devices provided by various leading manufacturers for a 20 kHz switching frequency. The theoretical waveforms of the converter and the device currents are illustrated in **Figure 2**. The terms and symbols used in the analysis are listed in **Table 2**. Piecewise linearity is assumed for the

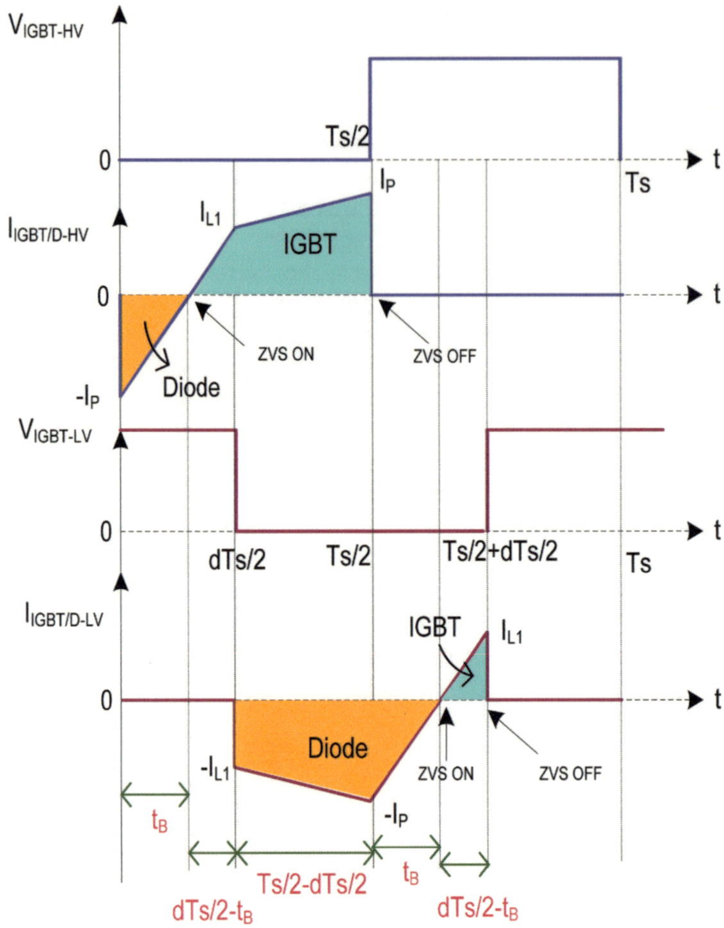

Figure 2.
Theoretical waveforms of devices on the HV side and LV side of the DAB DC-DC converter. (Device conduction intervals are marked in pink).

device current waveforms. Hence, the average current flow through the IGBT on the HV side of the converter during the on-time is given by

$$I_T = \frac{\frac{1}{2} \times I_{L1} \times \left(\frac{dT_S}{2} - t_B\right) + \frac{1}{2} \times (I_{L1} + I_P) \times \left(\frac{T_S}{2} - \frac{dT_S}{2}\right)}{\frac{T_S}{2} - t_B} \tag{1}$$

The IGBT has a constant voltage drop during the on-state. Hence, the conduction loss can be calculated using the average IGBT current and the duty cycle, which is given as

$$P_{CondT} = V_{CE(sat)} \times I_T \times d \tag{2}$$

The duty cycle is the ratio of the on-period of the transistor to the switching period. The datasheet specifies the forward voltage drop of the transistor ($V_{CE(sat)}$) and the anti-parallel diode (V_F) with respect to the main terminals of the modules, which includes the voltage drops across the terminals. Due to the high power densities of the devices, terminal losses cannot be neglected compared with the semiconductor losses. Hence, it is important to specify the voltage drop at the chip level and across the terminals (r_{CE}') separately. The voltage drop across the terminals is given by

$$V_{CE(sat)} = V_{CEO} + (r_{CE} \times I_T) \tag{3}$$

The average diode current on the HV side of the converter during the on-state, as shown in **Figure 2**, is given as

$$I_D = \frac{1}{2} \times I_P \tag{4}$$

In Eq. (4), the diode conduction interval and the base interval are the same on the HV side; hence, the diode current is computed regardless of time. Assuming a constant voltage drop, the conduction losses of the diode are estimated as

Symbols/terms	Description
d	Duty cycle
T_s	Switching time period
I_T	Transistor current
I_P	Peak inductor current
I_{L1}	First inductor current peak
t_B	Time interval from I_P to zero current
$V_{CE\ (sat)}$	Transistor forward voltage drop
P_{CondT}	Transistor conduction loss
r_{CE}	Collector to emitter resistance
V_{CEO}	Collector to emitter voltage with gate open
I_D	Diode current
P_{CondD}	Diode conduction loss
V_F	Diode forward voltage drop
P_C	Total conduction loss
P_{ON}	Transistor turn-on power loss
E_{ON}	Transistor turn-on energy loss
f	Switching frequency
P_{OFF}	Transistor turn-off power loss
E_{OFF}	Transistor turn-off energy loss
V_{CE}	Collector to emitter voltage
I_C	Collector current
t_r	Rise-time
t_f	Fall-time
P_{SW}	Total switching loss
$R_{th(c\text{-}s)}$	Case to sink thermal resistance
$R_{th(j\text{-}c)}$	Junction to case thermal resistance
P_T	Total power loss
T_{jIGBT}	Junction temperature of IGBT
T_{jDiode}	Junction temperature of diode

Table 2.
List of symbols/terms used in analysis and their description.

$$P_{CondD} = V_F \times I_D \times d \tag{5}$$

The DAB converter has four transistors and four antiparallel diodes on the HV side. Hence, the total conduction loss P_C of the semiconductor devices on the HV side of the converter is given as

$$P_C = 4 \times (P_{CondT} + P_{CondD}) \tag{6}$$

Switching losses are calculated from the turn-on (E_{ON}) and turn-off (E_{OFF}) energy loss curves, which are usually given as a function of the collector current at the switching instants of an IGBT, as given in the manufacturer datasheet. In the DAB converter configuration under zero voltage switching (ZVS), the diode always turns off with zero current, thus eliminating the diode reverse recovery losses. Therefore, the power loss equations for the turn-on and turn-off instants are

$$P_{ON} = E_{ON} \times f \tag{7}$$

and

$$P_{OFF} = E_{OFF} \times f \tag{8}$$

P_{ON} is the power loss during the turn-on instant, P_{OFF} is the power loss during the turn-off instant, and f is the switching frequency. In some manufacturers' datasheets, the energy loss curves for turn-on and turn-off may not be available. For those cases, it is necessary to consider the simplified voltage and current waveforms during the switching process. Switching losses are predominant during the rise-time and fall-time periods. Hence, the loss equations are approximated during the turn-on and turn-off instants, which are given as

$$P_{ON} = \frac{1}{2} V_{CE} \times I_C \times t_r \times f \tag{9}$$

and

$$P_{OFF} = \frac{1}{2} V_{CE} \times I_C \times t_f \times f \tag{10}$$

In DAB converter operation during ZVS, the anti-parallel diode current is always transferred to the IGBTs. Hence, the turn-on switching losses can be neglected. Therefore, the total switching losses (P_{SW}) are approximated as

$$P_{SW} = 4 \times [E_{OFF} \times f] \tag{11}$$

The junction temperature of the devices is calculated as

$$T_{jIGBT} = T_s + (P_T \times R_{th(c-s)}) + (P_{IGBT} \times R_{th(j-c)}) \tag{12}$$

and

$$T_{jDiode} = T_s + (P_T \times R_{th(c-s)}) + (P_{Diode} \times R_{th(j-c)}) \tag{13}$$

where T_S is the heat sink temperature and P_T is the total power loss of the IGBT module. The case to sink $R_{th(c-s)}$ and junction to case $R_{th(j-c)}$ thermal resistances can

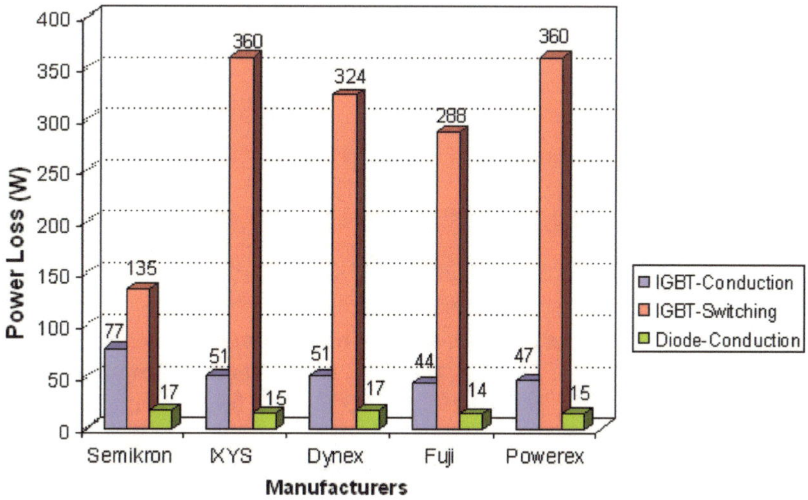

Figure 3.
HV-side IGBT and anti-parallel diode power loss comparison for various manufacturers.

be obtained from the device datasheet. The total power losses of the IGBTs and the antiparallel diode of the HV side of the DAB converter are compared for devices with ratings similar to 1200 V, 300 A from five different manufacturers:

- SKM300GB125D from Semikron, which has a rating of 1200 V, 300 A at T_{case} = 25°C;
- MII300-12A4 from IXYS, which has a rating of 1200 V, 330 A at T_{case} = 25°C;
- DIM200WHS12-A000 from Dynex Semiconductor, which has a rating of 1200 V, 200 A at T_{case} = 80°C;
- 2MB1300U4H-120 from Fuji Semiconductor, which has a rating of 1200 V, 400 A at T_{case} = 25°C; and
- CM300DY-24A from Powerex, which has a rating of 1200 V, 300 A at T_{case} = 25°C.

The power losses of IGBTs and anti-parallel diodes are predicted for the worst-case operating condition and are depicted in **Figure 3**. Based on a comparison of the losses of all devices, the ultra-fast SKM300GB125D 1200 V, 300 A phase leg IGBT modules from Semikron [28] are selected for the HV side of the converter due to their low loss. As shown in **Figure 3**, the switching power losses of the IGBTs are significant and are a major contributor to the overall power loss due to the high switching frequency and high turn-off current.

3. Power loss estimation of LV-side devices of the DAB DC-DC converter

The power loss of LV-side devices is predicted using a procedure similar to that discussed in Section 2. The main difference is that the shape of the LV-side device currents differs from those of the HV-side devices due to the phase shift introduced between the HV- and LV-side bridges as shown in **Figure 2**. The transistor and diode current equations during the on-state are

$$I_T = \frac{\frac{1}{2} \times I_{L1} \times \left(\frac{dT_S}{2} - t_B\right)}{\left(\frac{dT_S}{2} - t_B\right)} \tag{14}$$

and

$$I_D = \frac{\frac{1}{2} \times (I_P \times t_B) + \frac{1}{2} \times (I_{L1} + I_P) \times \left(\frac{T_S}{2} - \frac{dT_S}{2}\right)}{\left(\frac{T_S}{2} - \frac{dT_S}{2} + t_B\right)} \tag{15}$$

Similar to the HV side, the on-state and switching power losses of the IGBTs and the antiparallel diode of the LV side of the DAB converter are compared for devices with ratings similar to 600 V, 700 A at T_{case} = 25°C from four leading manufacturers:

- SKM600GB066D from Semikron, which has a rating of 600 V, 760 A;
- CM600DY-12NF from Mitsubishi, which has a rating of 600 V, 600 A;
- 2MB1600U2E-060 from Fuji Semiconductor, which as a rating of 650 V, 600 A; and
- CM600DY-12NF from Powerex, which has a rating of 600 V, 600 A.

Figure 4 portrays the power loss comparison of an LV IGBT and antiparallel diode for each module. As shown in **Figure 4**, all IGBT modules exhibit similar performance. During the forward conduction mode (charging mode), the diodes perform the rectification function. Hence, the conduction losses of the diodes are predominant. When the power flow reverses, the loss values are exchanged between the diode and IGBT. The phase leg modules from Semikron have lower losses than the other devices. In addition, the Semikron module has a maximum junction temperature of 175°C, whereas the other modules have a maximum junction temperature of 150°C. Hence, the Semikron SKM600GB066D high-temperature phase leg IGBT modules [29], which have a rating of 600 V, 760 A, were chosen for the LV side of the converter.

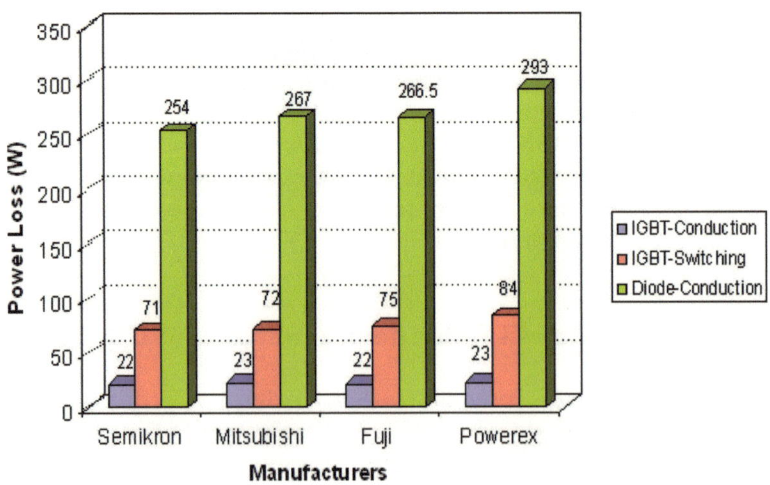

Figure 4.
LV-side IGBT and antiparallel diode power loss comparison from various manufacturers.

4. Experimental results

Experiments were performed for the HV- and LV-side devices of the converter using a suitable reactive load. A 39 μH air-core inductor was used for HV bridge testing by phase shifting the two half-bridge legs of the IGBTs, and the devices are subjected to maximum currents. Because the load was purely reactive, the current drawn from the source was used to meet the device conduction losses, switching losses, and the losses due to the passive components. A photograph of the DAB converter prototype is shown in **Figure 5**. Experimental results are shown in **Figures 6** and 7 for the HV-side IGBTs of the DAB converter for 540 V, 80 A operation with and without the snubber capacitor.

A 47 nF snubber capacitor was used on the HV side to minimize the turn-off losses and improve the switching transient of the HV-side IGBTs. A direct mount snubber capacitor type was used with a reduced effective series resistance (ESR) and with a low effective series inductance (ESL). As evident from the waveforms shown in **Figures 6** and 7, the diode begins conduction before the transistor, which ensures ZVS turn-on at peak current and ZVS/ZCS during turn-off. Similarly, the IGBT has ZVS/ZCS turn-on and ZVS turn-off at peak current.

The power loss measured during the experiments for the HV bridge converter at 16.6 kVA operation is plotted in **Figure 8**. Two cases are considered: in the first, a 47 nF

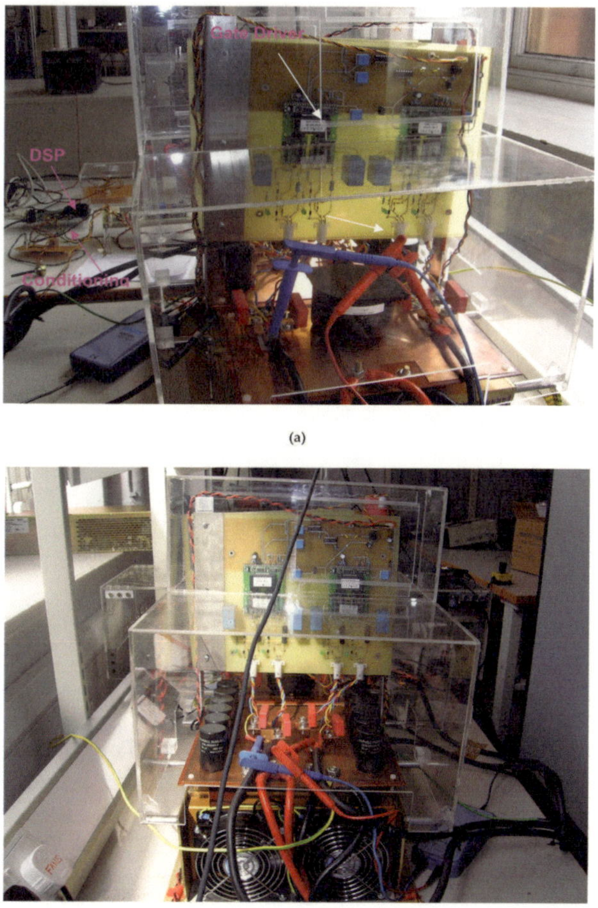

(a)

(b)

Figure 5.
Photographs of the DAB converter prototype: (a) front view and (b) back view.

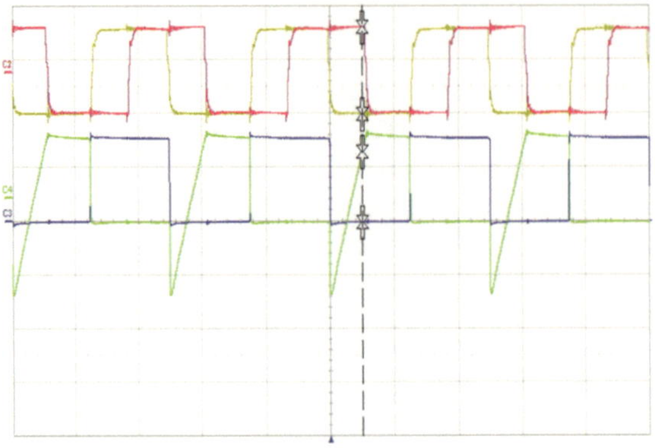

Figure 6.
Experimental results for the HV-side converter. V_{in} = 540 V, L = 39 μH, V_{Lrms} = 363 V, I_{Lrms} = 65.6 A, f_s = 20 kHz, I_{OFF} = 80 A. *Channel 1 (yellow)—gate pulse of transistor* A_2, *10 V/div. Channel 2 (pink)—gate pulse of transistor* B_2, *10 V/div. Channel 3 (blue)—IGBT* A_1 *voltage, 350 V/div. Channel 4 (green)—IGBT* A_1 *current, 50A/div. Time scale: 20 μs/div.*

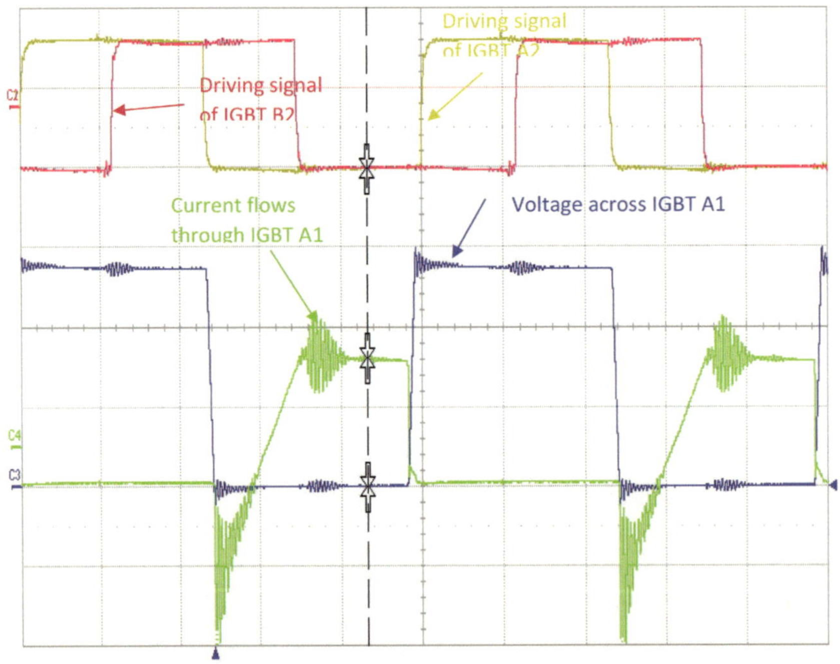

Figure 7.
Experimental results for the HV-side converter with a 47 nF snubber. V_{in} = 540 V, I_{in} = 1.318 A, L = 39 μH, f_s = 20 kHz, V_{Lrms} = 363 V, I_{Lrms} = 65.6 A, I_{OFF} = 80 A, R_G = 3 Ω, C_s = 47 nF. *Time scale: 10 μs/div. Channel 1 (yellow)—gate pulse of transistor A2, 10 V/div. Channel 2 (pink)—gate pulse of transistor B2, 10 V/div. Channel 3 (blue)—IGBT A1 voltage, 200 V/div. Channel 4 (green)—IGBT A1 current, 50A/div.*

snubber capacitor is connected across each of the four IGBTs of the HV-side bridge; in the second, no snubber capacitor is used. The power losses of the leading-phase and lagging-phase leg devices are displayed separately in **Figure 8** due to different conduction intervals of the devices. Power losses in the cables, busbar, and connections are neglected. Due to the ZCS turn-off of the diodes, their reverse recovery losses are neglected; due to ZVS/ZCS turn-on of the IGBTs, their turn-on losses are omitted. As shown in **Figure 8**, the air-core inductor copper losses are high although it has only 11 turns. Diode and IGBT conduction losses, however, remain approximately constant with or without the snubber capacitors. As evident in **Figure 8**, there is a considerable reduction in turn-off losses when the snubber capacitor is used.

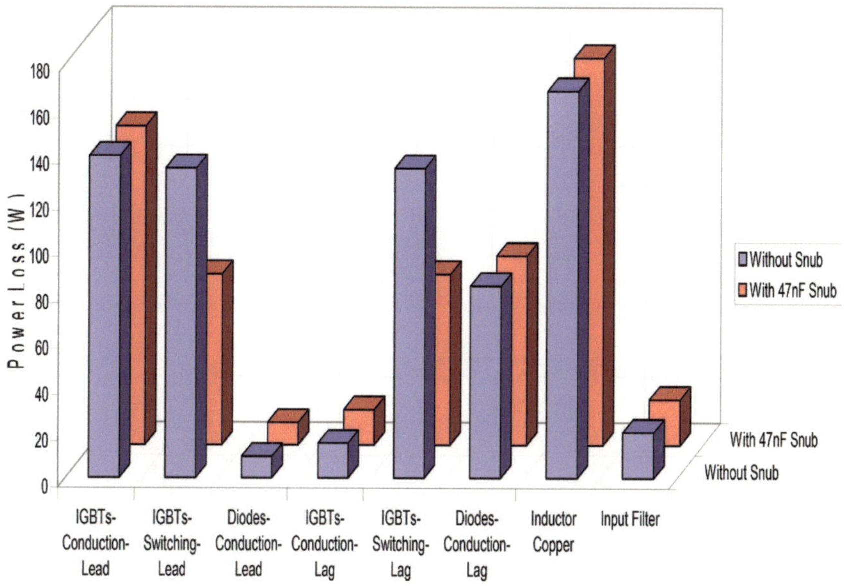

Figure 8.
Experimental power loss breakdown of the HV-side converter with and without the snubber capacitor. V_{in} = 540 V, L = 39 μH, f_s = 20 kHz, I_{OFF} = 80 A, V_{Lrms} = 363 V, I_{Lrms} = 65.6 A, R_G = 3 Ω.

The LV-side devices of the converter were tested, and the LV-side IGBTs were subjected to a peak current of 300 A at 125 V. The two legs of the LV-side bridge were connected through an air core inductor with a value of 4.17 μH. An external electrolytic capacitor bank of 19.8 mF capacitance (comprising three 6.6 mF capacitor banks) was added to the DC supply to smooth the input ripple current. **Figure 9** shows the current and voltage waveforms of the leading-leg IGBT and the driving signals of the IGBTs, D_2 and C_2. A CWT15 Rogowski current probe with a sensitivity of 2 mV/A was used

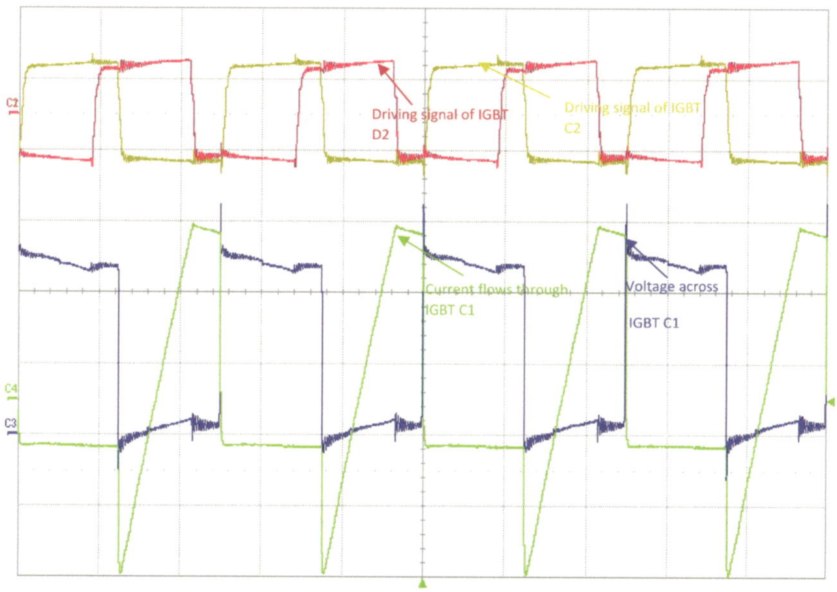

Figure 9.
Experimental results for the LV-side converter without the snubber. V_{in} = 125 V, I_{in} = 10.51 A, L = 4.17 μH, V_{Lrms} = 106 V, I_{Lrms} = 202 A, f_s = 20 kHz, I_{OFF} = 300 A, R_G = 2.5 Ω. *Time scale: 20 μs/div. Channel 1 (yellow)—gate pulse of transistor C2, 10 V/div. Channel 2 (pink)—gate pulse of transistor D2, 10 V/div. Channel 3 (blue)—IGBT C1 voltage, 100 V/div. Channel 4 (green)—IGBT C1 current, 100A/div.*

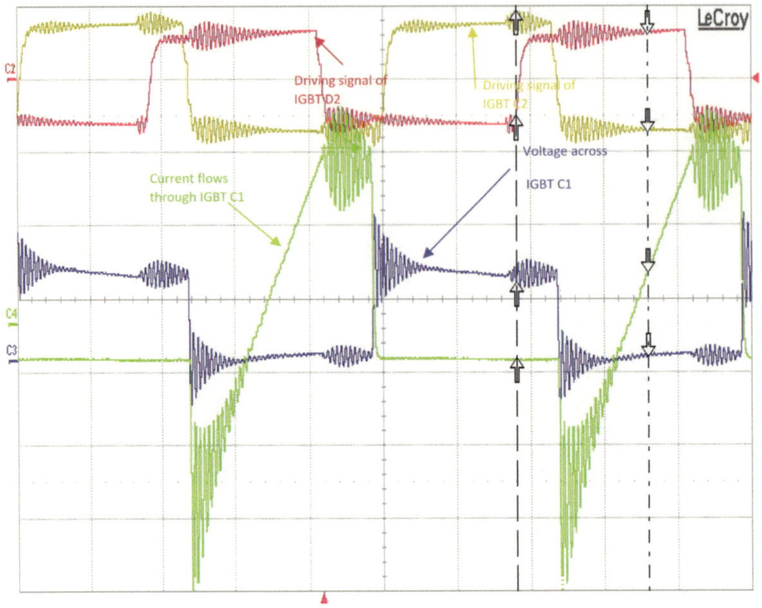

Figure 10.
Experimental waveforms of the LV-side converter with a 100 nF snubber. V_{in} = 125 V, I_{in} = 9.29 A, L = 4.17 μH, V_{Lrms} = 106 V, I_{Lrms} = 202 A, f_s = 20 kHz, I_{OFF} = 300 A, R_G = 2.5 Ω. Time scale: 10 μs/div. Channel 1 (yellow)—gate pulse of transistor C2, 10 V/div. Channel 2 (pink)—gate pulse of transistor D2, 10 V/div. Channel 3 (blue)—IGBT C1 voltage, 100 V/div. Channel 4 (green)—IGBT C1 current, 100A/div.

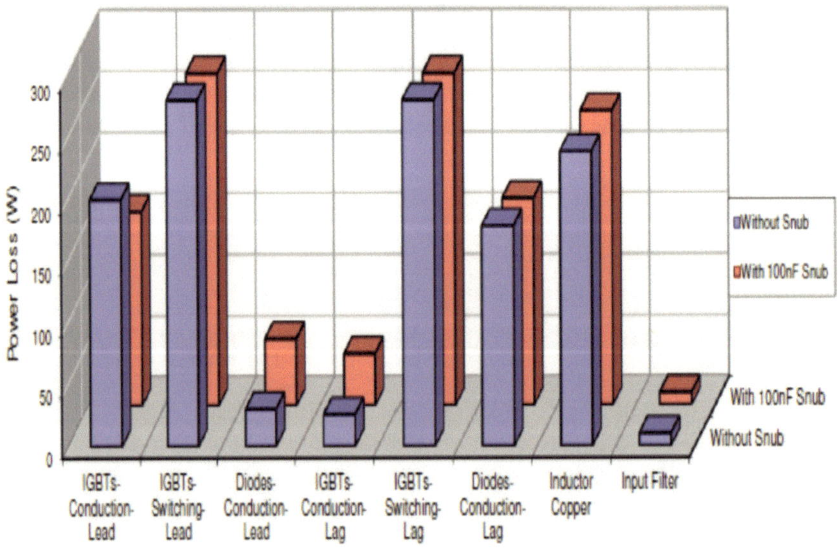

Figure 11.
Experimental measurement of power loss on the LV-side converter with and without the snubber. V_{in} = 125 V, L = 4.17 μH, f_s = 20 kHz, I_{OFF} = 300 A, V_{Lrms} = 106 V, I_{Lrms} = 202 A, R_G = 2.5 Ω.

to measure the device currents. The measured current fall-time is 205 ns for the initial current fall time and 620 ns for the tail current fall duration. High-frequency ringing is observed in the device waveforms when the snubbers are introduced across the devices. This may be resonance due to parasitic inductances of the module (with values of 15–20 nH), the busbars, and the snubber capacitor connections.

The input power drawn from the source using 100 nF snubber capacitors is lower than those observed for other snubbers. Moreover, a significant reduction in *dv/dt* during the switching transients is observed. Hence, 100 nF snubbers are chosen for the LV-side IGBTs. **Figure 10** shows the experimental waveforms for the LV-side IGBT C_1 with a 100 nF snubber capacitor. **Figure 11** illustrates the power loss breakdown for the LV-side bridge converter at 19.2 kVA operation with and without the snubber capacitors, as determined from the measurements. The IGBT switching losses are dominant. Power losses due to cables, busbars, and connections are not included. When 100 nF snubber capacitors are connected across the IGBTs, no significant reduction in switching losses occurs. The experimental power loss using the 100 nF snubber is 135 W, and *dv/dt* during the switching transient is reduced significantly, thereby minimizing the device stresses.

Comparisons of the theoretical and practical power losses of the HV- and LV-side devices are shown in **Figures 12** and **13**, respectively, for the leading leg devices to validate the mathematical models describing the device current equations. A close correlation between the calculated and experimental power loss values is observed, which demonstrates the effectiveness of the mathematical models presented.

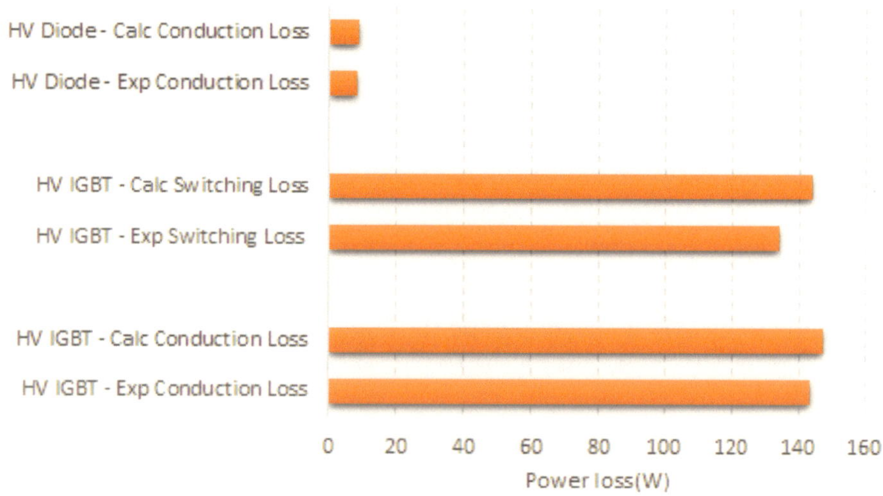

Figure 12.
Power loss comparison of HV-side devices—calculated (Calc) and experimental (Exp).

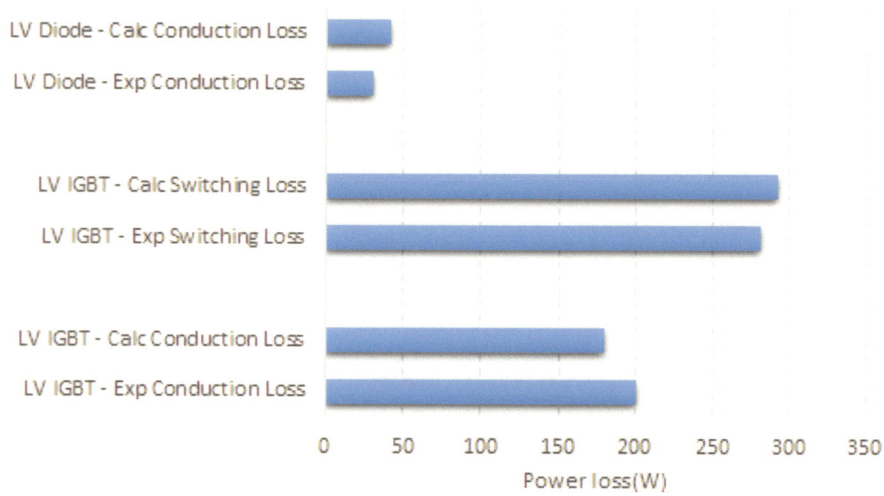

Figure 13.
Power loss comparison of LV-side devices—calculated (Calc) and experimental (Exp).

5. Conclusion

Power loss analysis of IGBTs in a high-voltage high-power DAB DC-DC converter intended for use in an aerospace energy storage system was presented. The guidelines for selecting the appropriate IGBTs for a 20 kW, 540/125 V, 20 kHz DAB DC-DC converter prototype suitable for aerospace applications were given. The important parameters provided in the device datasheet for calculating the device power losses were discussed. Power loss analysis was performed for HV IGBTs corresponding to five leading manufacturers and LV IGBTs corresponding to four leading manufacturers based on the DAB converter prototype design. Apart from the guidelines given in this chapter, some IGBT manufacturers offer customer support for device loss estimation through a software package. Such software packages may be used for preliminary studies to estimate the device losses and the junction temperature for various operating conditions. In this work, experimental results were presented for 540 V, 80 A peak current operation on the HV-side IGBT, and 125 V, 300 A peak current on the LV-side IGBT. Using snubber capacitors on the HV side of the converter resulted in a nearly 45% reduction of IGBT switching losses. Introducing snubber capacitors on the LV-side devices created parasitic ringing, with no significant reduction of switching power losses. However, snubbers on the LV side did reduce the device stresses by limiting *dv/dt*. The experimental results demonstrate that the use of snubber capacitors across IGBTs reduces switching losses and device stresses and thus improves converter performance.

Acknowledgements

The author thanks Rolls-Royce plc and the Engineering and Physical Sciences Research Council (EPSRC), UK, for the DHPA scholarship at the University of Manchester.

Author details

Thaiyal Naayagi Ramasamy
School of Electrical and Electronic Engineering, Newcastle University International Singapore, Singapore

*Address all correspondence to: naayagi.ramasamy@ncl.ac.uk

References

[1] DeDoncker RWRW, Divan DM, Kheraluwala MH. A three-phase soft-switched high-power-density dc/dc converter for high power applications. IEEE Transactions on Industry Applications. 1991;**27**(1):797-806. DOI: 10.1109/28.67533

[2] Kheraluwala MH, Gascoigne RW, Divan DM, Baumann ED. Performance characterization of a high-power dual active bridge dc-to dc converter. IEEE Transactions on Industry Applications. 1992;**28**(6):1294-1301. DOI: 10.1109/28.175280

[3] Segaran D, Holmes DG, McGrath BP. Comparative analysis of single and three-phase dual active bridge bidirectional dc-dc converters. In: Proc. Australian Universities Power Engineering Conference; 2008. pp. 1-6

[4] Tripathi AK, Mainali K, Patel DC, Kadavelugu A, Hazra S, Bhattacharya S, et al. Design considerations of a 15-kV SiC IGBT-based medium-voltage high-frequency isolated DC–DC converter. IEEE Transactions on Industry Applications. 2015;**51**(4):3284-3294. DOI: 10.1109/TIA.2015.2394294

[5] Kadavelugu A, Bhattacharya S, Ryu S-H, Brunt EV, Grider D, Agarwal A, Leslie S. Characterization of 15 kV SiC n-IGBT and its application considerations for high power converters. In: Proc. IEEE Energy Conversion Congress and Expo; 2013. pp. 2528-2535

[6] Kadavelugu A, Bhattacharya S. Design considerations and development of gate driver for 15 kV SiC IGBT. In: Proc. IEEE Applied Power Electronics Conference and Expo; 2014. pp. 1494-1501

[7] Hazra S, Madhusoodhanan S, Bhattacharya S, Moghaddam GK, Hatua K. Design considerations and performance evaluation of 1200 V, 100 a SiC MOSFET based converter for high power density application. In: Proc. IEEE Energy Conversion Congress and Expo; 2013. pp. 4278-4285

[8] Wen H, Xiao W, Bin S. Non active power loss minimization in a bidirectional isolated DC–DC converter for distributed power systems. IEEE Transactions on Industry Applications. 2014;**61**(12):6822-6831. DOI: 10.1109/TIE.2014.2316229

[9] Oggier GG, Garcia GO, Oliva AR. Switching control strategy to minimize dual active bridge converter losses. IEEE Transactions on Power Electronics. 2009;**24**(7):1826-1838. DOI: 10.1109/TPEL.2009.2020902

[10] Wang Y, de Haan SWH, Ferreira JA. Potential of improving PWM converter power density with advanced components. In: Proc. 13th European Conference on Power Electronics and Applications; 2009. pp. 1-10

[11] Gao Y, Huang AQ, Krishnaswami S, Richmond J, Agarwal AK. Comparison of Static and Switching Characteristics of 1200 V 4H-SiC BJT and 1200 V Si-IGBT. IEEE Transactions on Industry Applications. 2008;**44**(3):887-893. DOI: 10.1109/TIA.2008.921408

[12] Nomura T, Masuda M, Ikeda N, Yoshida S. Switching characteristics of GaN HFETs in a half bridge package for high temperature applications. IEEE Transactions on Power Electronics. 2008;**23**(2):692-697. DOI: 10.1109/TPEL.2007.915671

[13] Semikron. Operation Principle of Power Semiconductors-Application Manual; 2008

[14] Semikron. Calculation of the Junction Temperature-Hints for Application-3.2.22008. p. 146

[15] Fuji Electric. Fuji IGBT Modules Application Manual; 2004

[16] Naayagi RT. Selection of power semiconductor devices for the DAB DC-DC converter for aerospace applications. In: Proc. IEEE 11th International Conference on Power Electronics and Drive Systems; 2015. pp. 499-502

[17] Advanced Power Technology. IGBT Tutorial Application Note APT0201. Rev. B. 2002

[18] Microsemi. Advanced IGBT Driver Application Manual-Application Note 1903; 2006

[19] Powerex. General Considerations for IGBT and Intelligent Power Modules-Application Note A10; 2008

[20] International Rectifier. Application Characterization of IGBTs-Application Note AN-990, rev.2; 2008

[21] IXYS. Choosing the Appropriate Component from Data Sheet Ratings and Characteristics-Technical information IXAN0056; 2008

[22] Ramasamy TN. Bidirectional DC-DC converter for aircraft electric energy storage systems [PhD thesis]. School of Electrical and Electronic Engineering, University of Manchester; 2010

[23] International Rectifier. IGBT or MOSFET: Choose Wisely-Application note; 2008

[24] Semikron. Power Modules: Special Features of Multi-Chip Structures-Application Note-1.4, AN1404; 2014

[25] Semikron. New Low-Inductive IGBT Module Constructions for High Currents and Voltages-Application Note-1.5.4; 2014

[26] Semikron. New Developments in MOSFET and IGBT Technology-Application Note-1.2.4;2014

[27] Naayagi RT, Forsyth AJ, Shuttleworth R. High-power bidirectional DC-DC converter for aerospace applications. IEEE Transactions on Power Electronics. 2012;**27**(11):4366-4379. DOI: 10.1109/TPEL.2012.2184771

[28] Semikron. Ultrafast IGBT Module SKM300GB125D Datasheet; 2009

[29] Semikron. Trench IGBT Module SKM600GB066D Datasheet; 2009

Chapter 6

Resonant Power Converters

Mohammed Salem and Khalid Yahya

Abstract

Recently, DC/DC resonant converters have received much research interest as a result of the advancements in their applications. This increase in their industrial application has given rise to more efforts in enhancing the soft-switching, smooth waveforms, high-power density, and high efficiency features of the resonant converters. Their suitability to high frequency usage and capacity to minimize switching losses have endeared them to industrial applications compared to the hard-switching conventional converters. However, studies have continued to suggest improvements in certain areas of these converters, including high-power density, wide load variations, reliability, high efficiency, minimal number of components, and low cost. In this chapter, the resonant power converters (RPCs), their principles, and their classifications based on the DC-DC family of converters are presented. The recent advancements in the constructions, operational principles, advantages, and disadvantages were also reviewed. From the review of different topologies of the resonant DC-DC converters, it has been suggested that more studies are necessary to produce power circuits, which can address the drawbacks of the existing one.

Keywords: soft-switching, LLC resonant converter, control strategies, resonant power converters (RPCs),

1. Introduction

Several studies have been conducted on the switching mode DC-DC converters to ensure that they satisfy the most demanding criteria for power electronics application. The possibility of minimizing the switching and conduction losses in the switch-mode through increasing the switching frequency makes them more attractive. However, several switching topologies can attain a high-power transfer [1, 2] but the problem is the power switches (transistors or MOSFET), diodes, and energy storage passive elements (capacitors and inductors) contained in the structure of the power converters, which affects their efficiency. Efficient circuits have been developed for power converters along with the developments in materials, control systems, and devices technology. Such topologies can minimize the overall converter size and the switching losses, thereby, providing a higher efficiency [3–5]. The DC-DC converters can be categorized into three major groups (linear, hard-switching, and soft-switching mode resonant converters) based on their modes of operation as depicted in **Figure 1** [6–8].

The advantages of the linear regulator technologies include low noise, fast response, simplicity, and excellent regulation. However, they have some disadvantages such as power dissipation in any working condition, which can result in low efficiency. The switching-mode topologies can be categorized based on the isolation

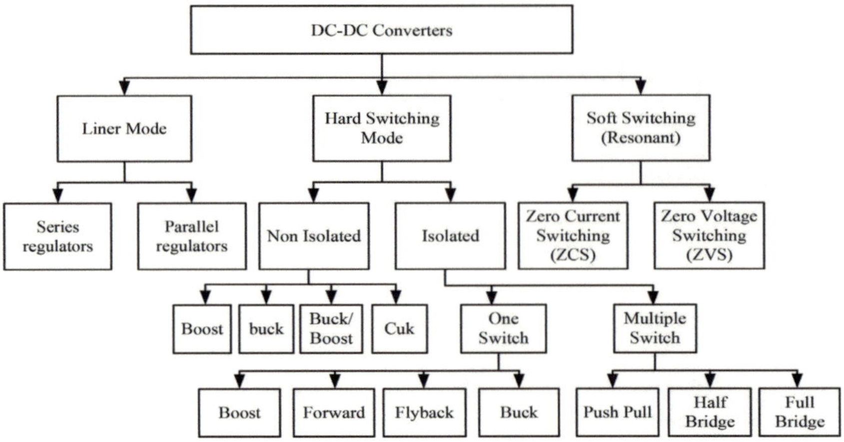

Figure 1.
The classification of the DC-DC converters.

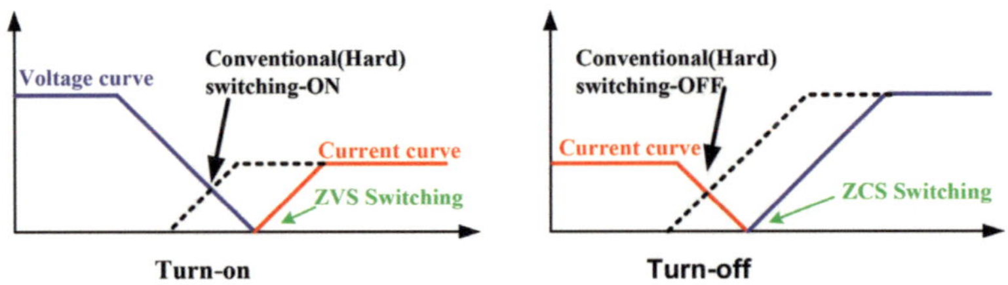

Figure 2.
Current waveform and voltage waveform of hard-switching and soft-switching at both turn-on and turn-off transitions.

feature to galvanic isolation converter, which also called chopper, and non-isolated converter. Galvanic isolation is required in most applications during the process of converting the power from the utility grid (voltages ranging from 100 to 600 V) for safety reason [9–11]. Since a higher power transformer is required in these converters, the single-switch converter may not be a proper solution for higher power applications, and as such, the other DC-DC isolated converters with more than one switch, such as the push-pull and full or half-bridge converters, are more appropriate for such high-power applications. Hard-switching transitions in devices operating with switched-mode converters will result in high power dissipation and a gradual reduction of the efficiency of the converter; it can damage the switching devices as well. Snubbers are used to reduce the stress of the switches and solve this problem of power dissipation. Furthermore, the hard-switching converters have other disadvantages such as limited frequency, high EMI, high switching losses, large size, and heavy weight. Two more problems are encountered during the control of the transferred power; the first one is the generated noise during the switching process while the second one is the energy lost in the switches.

Regardless of these drawbacks, this study implemented the hard-switching transitions [12]. The voltage and the active switches' current were modified to overcome or minimize their effects [13]. These modified methods work by either forcing the voltage across the switch or current through the switch to zero, and such a transition technique can only be achieved with a soft-switching technique (**Figure 2**). The current waveform is forced to reduce to 0 in the zero current switching (ZCS) circuit while the voltage waveform is treated as such by the zero voltage switching (ZVS) circuit. Power circuits with these types of transition techniques are called soft-switching converters (SSCs) [14]. Meanwhile, there are

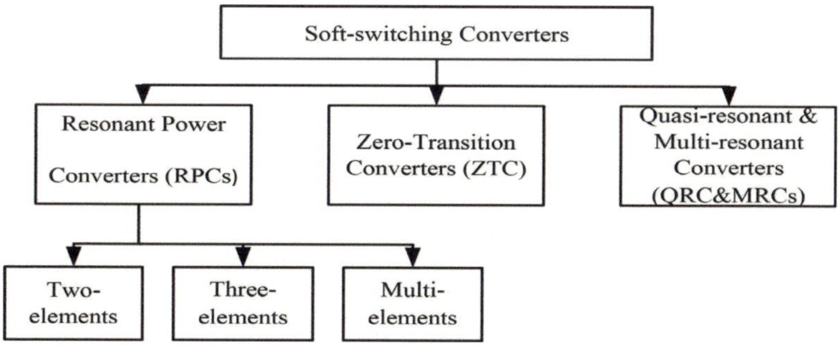

Figure 3.
Categories of the SSCs.

several advantages of the SSCs over the linear mode regulators. The advantages include (i) the possibility of using a small ferrite transformer core due to the high-switching frequency, making operation in a wider DC input voltage range possible compared to the linear regulators; and (ii) it offers a higher efficiency. However, the complexity of the control circuit is associated with several drawbacks compared to the linear control circuit; and the power switching technique may increase the supply noise.

The SSCs have been significantly improved in the areas of switching losses, the EMIs, and device stresses, allowing the converters to perfectly work even at higher frequencies and a consequent reduction in the magnetic components. Generally, the SSC families are categorized into resonant transition converters (RTCs), resonant power converters (RPCs), quasi-resonant converters (QRCs), and multi-resonant converters (MRCs) based on their modes of operation, as shown in **Figure 3**.

2. Resonant power converters (RPCs)

Figure 4 shows the structure of the RPCs and each stage represents a specific job to be carried out. The controlled switching network (CSN) is powered by the DC source; it rapidly switches on and off depending on the working frequency to generate the output voltage or current, which feeds the next stage. The sinusoidal voltage and current signals are generated at a stage of the high-frequency resonant tank network (RTN), where there are two or more reactive components. This is to ensure a reduced electromagnetic interference and harmonic distortion [15]. Being deployed as an energy cushioning stage between the load and the CSN, a frequency-selective network can identify this stage. The impedances of both capacitance and inductance at resonance condition are equivalent and will produce the resonant

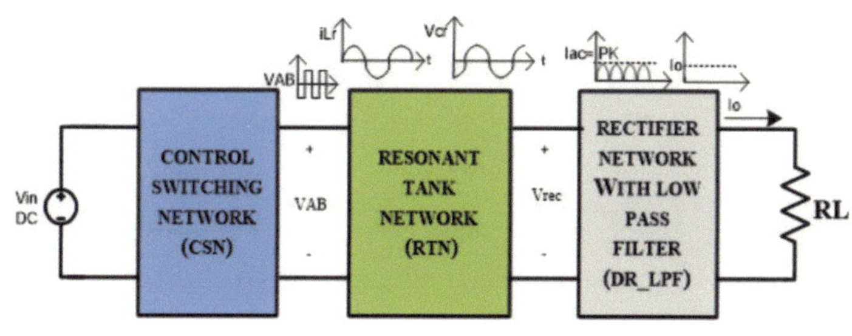

Figure 4.
The structure of the RPC.

frequency. A rectifier network and a pass filter are then used for rectifying and filtering the output signal to generate the anticipated DC output voltage [16].

2.1 Control switching network (CSN)

The full and half bridge are the commonest switching networks, whose usage is power-dependent. For high-power applications, the full bridge inverter is often used as opposed to the single ended or half bridge inverters, which can only supply the active switch with half of the input voltage. This indicates that full bridge inverters have a low rate of voltage switch, making them ideal for application in high input voltage conditions [17]. The RPCs and either a half or full-bridge inverter are often deployed together along with each of center-tapped or full-bridge rectifiers [18].

The CSN depicted in **Figure 5** generates a square waveform voltage Vs(t) of the switching frequency fs ($\omega_s = 2\pi$ fs) as represented in Eq. 1 by the Fourier series. Considering the response of the resonant tank, which is dominant to the basic constituent fs of the voltage waveform Vs(t), then, the infinitesimal response clearly demonstrated the harmonic frequencies nfs, n = 3, 5, 7, ... As a result, the power that correlates to the basic voltage waveform constituent Vs(t) is moved to the resonant tank as represented in Eq. (2). The basic constituent is a sinusoidal waveform with a peak amplitude of ($4/\pi$) times the DC source voltage. This basic constituent and the original waveform are in the same phase. However, when the S1 is on, there is a positive output sinusoidal switched current (t) but negative when S2 is off. This is due to the alternate working principle of the two switches, and its peak amplitude Is1 with phase equal to φs. The input current (DC) to the CSN can be computed by dividing the sinusoidal switched current with half the switching period [6, 16].

$$V_s(t) = \frac{4V_g}{\pi} \sum_{n=1,3,5,\dots} \frac{1}{n} \sin(n\omega_s t) \tag{1}$$

$$V_{s1}(t) = \frac{4V_g}{\pi} \sin(n\omega_s t) \tag{2}$$

$$i_s(t) = I_{s1} \sin(\omega_s t - \varphi_s) \tag{3}$$

$$I_{in} = \frac{2}{T_s} \int_0^{\frac{T_s}{2}} i_s(t) d_t = \frac{2}{\pi} I_{s1} \cos(\varphi_s) \tag{4}$$

2.2 Resonant tank network (RTN)

Resonant tank networks (RTNs) comprise of LC circuit (reactive elements) that stock oscillating energy with the frequency of circuit resonant. The LC circuit's resonance h attains the electromagnetic frequency useful in several applications, including the telecommunication technology. The tank can possibly be charged to a

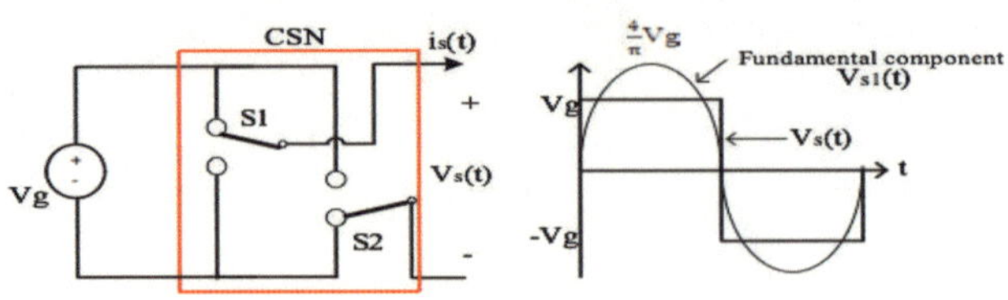

Figure 5.
The equivalent circuit of CSN.

certain resonant frequency through the adjustment of reactive element data. In addition, the essential phase of the resonant power converter is the resonant tank network. There are different kinds of RTN; it is mainly categorized into three parameters. Firstly, the resonant power converter can be sectioned through the connection technique used in tank element. The main common three resonant circuits include a series-parallel resonant converter (SPRC), a series resonant converter (SRC), and parallel resonant converter (PRC) [19]. The second factor lies in a quantity of the reactive elements (amount of transfer function order). However, the third one depends on the elements and multi-elements resonant tank [16]. Topographies of the three elements RTN (third order resonant tank) are controlled in overpowering the inadequacies in the two elements RTN. Most especially, the third element is put in the two elements RTN with a certain rumination to generate the three elements RTN. Thus, these can be taken as an integration of the advantages of mostly used two elements resonant converters PRC and SRC and enhance their inadequacies. The third order RTN contains 36 various tanks, the most common used tanks are LCC, LCL, and CLL [14, 20]. Multi-element resonant converters are RTN that contains four and beyond the number of elements. It should be noted that an increase in the number of reactive elements cause the network to be more complicated based on its size, analysis, and cost. For instance, **Figure 6** illustrates the fourth order RTN, which is referred to as the LCLC tank system [21]; this topology includes the characteristics of two main famous three-element systems such as LLC and LCC, and then reflects on their setbacks.

2.3 Diode rectifier network with low pass filter (DR-LPF)

The RTN generates voltage waveforms and sinusoidal current based on the output voltage and resonant frequency, which are taken to be the input to the last phase DR-LPF is the pulse waveform. This implies that the purpose of utilizing DR-LPF is to filter and correct the AC waveform to achieve the entailed DC output waveform. In the previous studies on resonant power converter, the DR-LPS had been illustrated as a full-bridge or center-tapped rectifiers. While the center-tapped rectifier is inappropriate due to an elevated voltage stress from the diodes, the low pass filter had been investigated for the entire occurrences of inductance or capacitive [16, 22].

2.3.1 Diode rectifier with capacitive low pass filter (DR-CLPF)

In the DR-CLPF, the input voltage VR(t) is appraised as the square wave of a resonant frequency as illustrated in **Figure** 7. The input voltage VR(t) can be estimated based on the resonant tank filtering with its basic component VR1(t) as shown in Eqs. (6) and (7), respectively. In addition, the basic component and current are in the same phase, any drop in current to zero alters the basic compartment because of the variations in the conducting diodes.

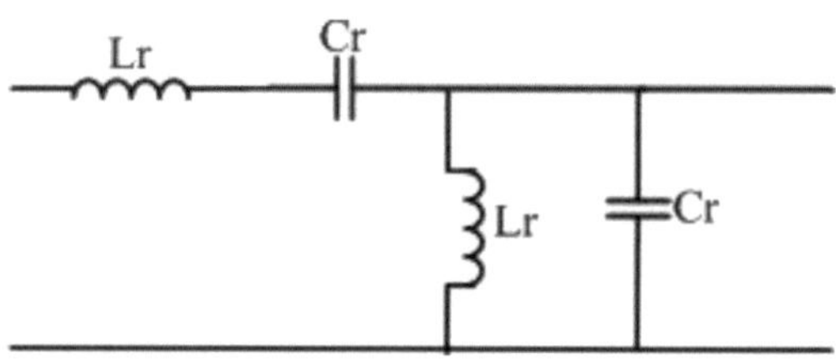

Figure 6.
An illustration of four-element RTN (LCLC).

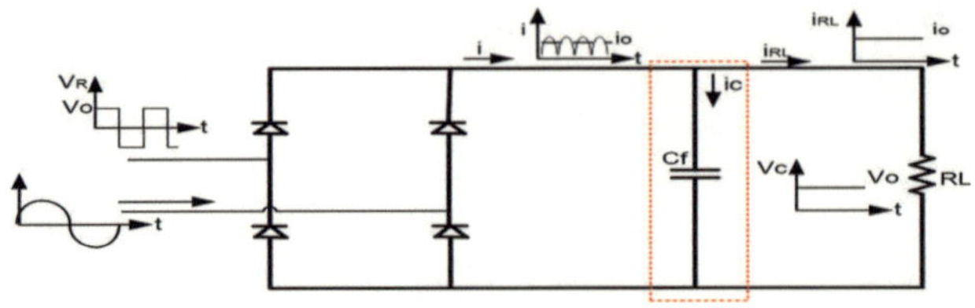

Figure 7.
DR-CLPF containing a capacitive pass filter and its variables waveforms.

$$i_R(t) = I_p \sin(\omega_s t - \varphi_s) \tag{5}$$

$$V_R(t) = \frac{4V_o}{\pi} \sum_{n=1,3,5,\ldots} \frac{1}{n} \sin(n\omega_s t - \varphi_s) \tag{6}$$

$$V_{R1}(t) = \frac{4V_o}{\pi} \sin(\omega_s t - \varphi_s) \tag{7}$$

$$I_o = \frac{2}{T_s} \int_0^{\frac{T_s}{2}} i_R(t) d_t = \frac{2}{\pi} I_R \tag{8}$$

2.3.2 Diode rectifier coupled with inductive low pass filter (DR-LLPF)

The diode rectifier is coupled with the input voltage VR(t) (sinusoidal waveform) and an inductive filter jacket. The current inputted is appraised as the square waveform iR(t) as illustrated in **Figure 8**.

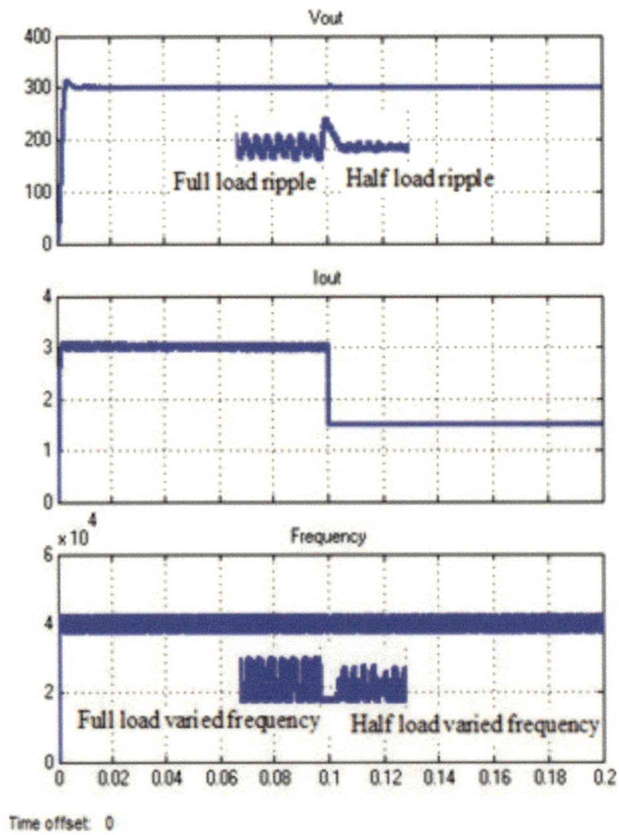

Figure 8.
DR-LLPF coupled with an inductive pass filter and its variables waveforms.

$$V_R(t) = V_p \sin(\omega_s t - \varphi_s) \tag{9}$$

$$i_R(t) = \frac{4I_o}{\pi} \sum_{n=1,3,5,\dots} \frac{1}{n} \sin(n\omega_s t - \varphi_s) \tag{10}$$

$$i_{R1}(t) = \frac{4I_o}{\pi} \sin(\omega_s t - \varphi_s) \tag{11}$$

$$V_o = \frac{2}{T_s} \int_0^{\frac{T_s}{2}} V_R(t) d_t = \frac{2}{\pi} V_R \tag{12}$$

3. Properties of resonant power converters

3.1 Parallel resonant converter (PRC)

The PRC is categorized into two elements tank converter. The resonant capacitor Cr needs to be parallel to the diode rectifier network DR and load. For the effective load resistance, Rac is much more enormous relative to resonant capacitor reactance Cr; this implies that the resonant current is unproportionable to the load. Moreover, in addition, the voltage over the resonant capacitor and parallel resistance Rac can be improved by declining the load [14]. **Figure 9** shows the relationship between the load quality factor and PRC voltage gain depending on switching frequency based on Eq. (13). It can be observed that the high voltage gain is attained from the switching frequencies and light load conditions, which are nearly equivalent to resonant frequency fs = fr. Therefore, PRC can either step the output voltage down or up depending on the variation in the control switching system frequency. The voltage output can be adjusted with load states, whereas, the resonant current is restricted to resonant inductor data, this causes the PRC to be appropriate for open and short circuit applications [23].

$$M_V = \frac{1}{\sqrt{\left[1 - \left(\frac{f_s}{f_r}\right)^2\right]^2 + \left[\frac{f_s}{f_r}\left(\frac{1}{Q}\right)\right]^2}} \tag{13}$$

3.2 LLC resonant converter

LLC resonant converter in the RTN comprises three reactive elements, whereby it is appraised as a conventional SRC and addition of inductor L_r equidistant to the

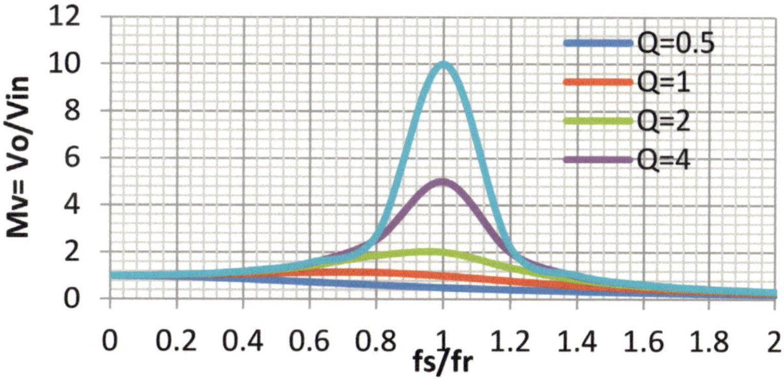

Figure 9.
Voltage gain characteristic of PRC.

load that is referred to as parallel inductor L_{rp}. The parallel inductor can be replaced by using the magnetizing inductance when using a transformer [14, 16]. The topology of LLC generates two resonant frequencies: firstly, the series resonant frequency fr1 depending on the series elements L_r C_r and secondly, the parallel resonant frequency fr2 depending on the entire three tank elements (L_{rp}, C_r, and L_r) as illustrated in Eqs. (14) and (15), where $f_{r1} > f_{r2}$.

$$f_{r1} = \frac{1}{2\pi\sqrt{(L_r + L_{rp})C_r}} \tag{14}$$

$$f_{r2} = \frac{1}{2\pi\sqrt{L_r C_r}} \tag{15}$$

Figure 10 illustrates the LLC voltage gain depending on the unity inductance ration (AL = 1), switching frequency, and load quality factor as shown in the Eq. (16).

$$M_V = \frac{1}{\sqrt{\left[1 - \left(\frac{f_s}{f_r}\right)^2\right]^2 + \left[\frac{f_s}{f_r}\left(\frac{1}{Q}\right)\right]^2}} \tag{16}$$

3.3 LCC resonant converter

The topology of LCC in the RTN comprises three reactive elements, the capacitor Crp that is connected to the load in parallel represents the third element. Thus, the topology contains two resonant frequencies: firstly, the series resonant frequency fr1 depending on the series elements Lr Cr and secondly, the parallel resonant frequency fr2 depending on the entire three tank elements (Lrp, Cr, and Lr) as illustrated in as they are shown in Eqs. (17) and (18), where fr1 < fr2.

$$f_{r1} = \frac{1}{2\pi\sqrt{L_r C_r}} \tag{17}$$

$$f_{r2} = \frac{1}{2\pi\sqrt{L_r\left(\frac{C_r C_{rp}}{C_r + C_{rp}}\right)}} \tag{18}$$

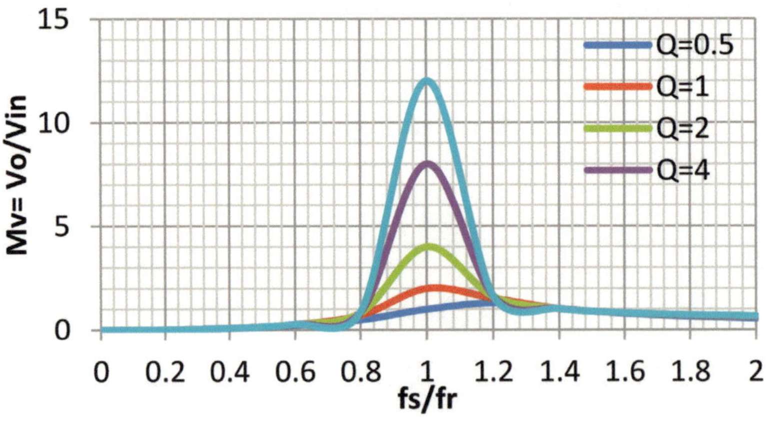

Figure 10.
Voltage gain property of LLC converter.

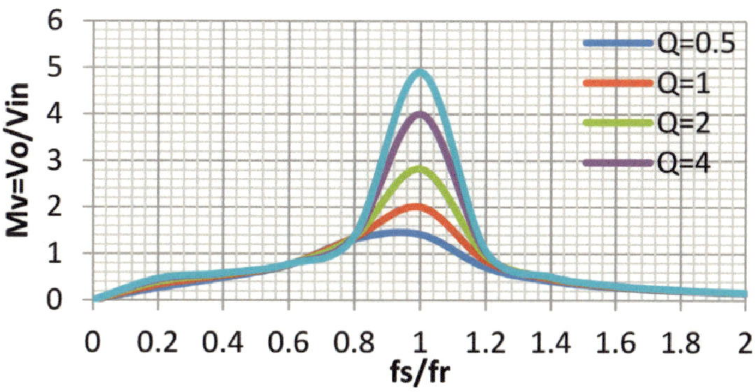

Figure 11.
Voltage gain property of LCC converter.

In the LCC converters, the proportion of resonant capacitors AC should be chosen prudently to be equal to the targeted peak gain. **Figure 11** explains the LCC voltage gain as depending on the load quality parameter and switching frequency with single inductance ratio (AC = 1) as described in Eq. (19). It is seen that the light load voltage gain advances across the converter properties and parallel resonant frequency fr2 acts as a parallel resonant converter PRC.

$$M_V = \frac{1}{\sqrt{(1+A_C)^2\left[1-\left(\frac{f_s}{f_{r2}}\right)^2\right]^2+\left[\frac{1}{Q}\left(\left(\frac{f_s}{f_{r1}}\right)-\frac{A_C}{A_C+1}\frac{f_{r1}}{f_s}\right)\right]^2}} \tag{19}$$

3.4 LCLC resonant converter

Similar to LCC, the topology of LCLC in the RTN comprises four reactive elements as illustrated in **Figure 6**, where this topology homogenizes the characteristics of LLC and LCC. The topology structure comprises parallel resonant capacitor Crp, parallel resonant inductance Lrp, and a series elements Lr Cr, which implies that the topology contains two proportions whereby the inductance AL and capacitance AC must be appraised in the design. Moreover, the topology comprises three frequencies: two parallel frequencies frp1, frp2, and one series resonant frequency frs. **Figure 12** reflected the relationship between the load quality factor, AC = AL = 1 and LCLC voltage gain depending on the switching frequency as shown in Eq. (20).

$$M_V = \frac{1}{\sqrt{\left[1+A_C+A_L-A_C\left(\frac{f_s}{f_r}\right)^2-A_L\left(\frac{f_r}{f_s}\right)^2\right]^2+\left[\frac{1}{Q}\left(\frac{f_r}{f_s}-\frac{f_s}{f_r}\right)\right]^2}} \tag{20}$$

4. Controlled strategies of resonant converters

The controlled strategies of the resonant converter is a bit disparate from the pulse width modulated (PWM) converters. A lot of parameters should be considered to attain a soft switching at a certain segment to fabricate the precise controller

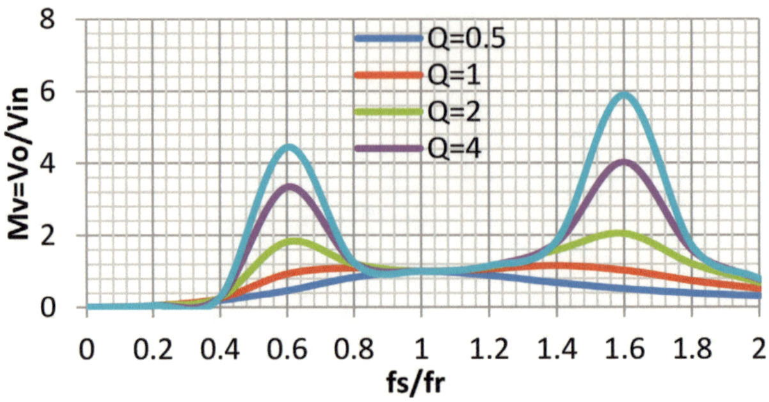

Figure 12.
Voltage gain property of LCLC converter.

that can achieve the desired results like load condition, energy storage elements, frequency range, and among others. There were several controlled topologies, which had been applied in the previous studies to manage the series resonant converters. For example, the pulse density modulation, voltage and current control, diode conduction control, and frequency control [24]. The full bridge resonant converter voltage had been regulated through the phase shift control; this phenomenon can be referred to as the switching signal primary control. In addition, to enhance the outcome of a control system, several improved techniques through adaptive controls had been reported [25], which include the passivity-based control and auto disturbance-rejection control (ADRC). A phase shift control had been used to control the current of resonant [26]. Because of this, the control outcome was increased in relative to the traditional PSRC control system. From the previous studies, the controlled techniques can be categorized into their implementation technique either through analog or digital. The digital controls are used because of their flexibility features in compact, programming, and light in comparison to analog controllers, they are also more resistant to inferences and noise. A three element DC-DC resonant converter type LLC has been discussed in this part, in order to compare the performance of frequency (duty-cycle) with variable frequency control, in terms of wide load variation.

Moreover, to ensure the expansion range of ZVS for entire inverter switches (S1–S4), and to improve the converter voltage gain. Then, the magnetizing inductance and resonant tank are used to generate a second resonant frequency $f_{r1} = \frac{1}{\sqrt{(L_r+L_m).C_r}}$.

From **Figure 14**, the LLC resonant converter gain M is evaluated using voltage divider law by considering the load quality factor Q and transformer step-up ratio as shown in Eq. (21).

$$M = \frac{V_o}{V_{in}} = \frac{Z_i}{Z_o} = \frac{F^2(A_L - 1)}{n\sqrt{\left(\frac{f_s^2}{f_{r2}^2} - 1\right)^2 + \left(Q(F^3 - F)^2(A_L - 1)\right)^2}} \qquad (21)$$

4.1 Fixed frequency control

The purpose of using this technique is that the voltage output is being influenced by changing the duty-cycle to attain a targeted voltage output. However, the error that exists between the constant desired voltage (reference voltage) $V_{err} = V_{ref} - V_{meas}$ and measured voltage is used for the PI controller. Thus, the controlled signal pertained to the PWM generator through the utilization of the switching frequency in generating a gate signal for the entire switches. From Eq. (21), the switching frequency that is equal to 42.5 kHz is the highest gain that can be obtained if the switching frequency and resonant frequency are closer. Therefore, a duty cycle is the only factor that measures the output voltage. Because the voltage ripple can vary by changing the load, then, the duty cycle will react to the variation of load variation relative to the estimated output voltage ripple.

4.2 Variable-frequency control

This control varies the switching frequency in adjusting the output voltage so as to reach the targeted load stage and save the output voltage of converter stable in any situation as illustrated in **Figure 5**. Moreover, the error that exists between the stable targeted voltage (reference voltage) $V_{err} = V_{ref} - V_{meas}$ and measured is used in the PI controller. Therefore, the controller increases or decreases the switching frequency based on the targeted output voltage if any variation occurs from the required output or error sign. Thus, a signal will be sent to the entire device switches. Depending on the magnetizing inductance and resonant impedance, the tank response needs to be saved inductively to appraise the attainment of the ZVS in the entire switches. However, the ranges of controlled switching frequency need to be restrained by higher and lower frequencies that affirm the fulfillment of ZVS (45 − 37.5 kHz).

4.3 Simulation results

The model simulation of resonant converter is carried out and implemented by utilizing the MATLAB/SIMULINK software using the factors enlisted in **Table 1**; this is to verify the LLC series resonant DC-DC converter design circuit as shown in **Figure 13** and control techniques analysis.

In this frequency control, the controlled signal is used for the PWM generator in generating the gate signals for the entire switches. Then, the switches S_1 and S_4 are enhanced concurrently and replace the S_2 and S_3 switches to generate the input voltage for the resonant tank V_{AB}. Therefore, the resonant elements generate the voltage V_{Cr} and sinusoidal current i_{Lr} as illustrated in **Figure 15**.

Figure 16 illustrates an output voltage of the duty-cycle controller dynamic response, the load is stepped up and down within half load of (200 Ω) and a full

V_{in}	100 (V)
L_r	200 (μH)
C_r	70 (nF)
C_f	4.7 (μF)
V_o	300 (V)
R_L	100 (Ω)
L_m	300 (μH)
Kp, Ki Duty-cycle control	2.7, 0.5
f_s ranges for VF control	37.5 − 45 (kHz)

Table 1.
The parameters of the LLC converter.

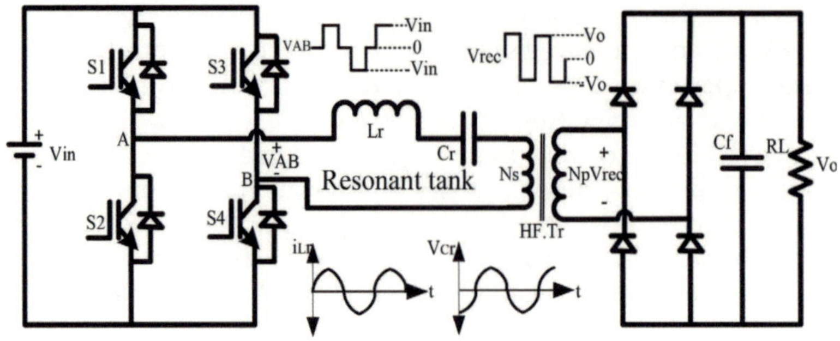

Figure 13.
A simplified illustration of full-bridge LLC resonant converter.

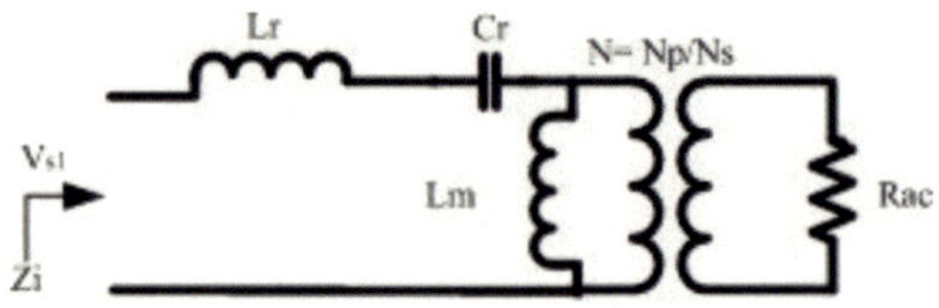

Figure 14.
The AC equivalent circuit between the rectifier and inverter.

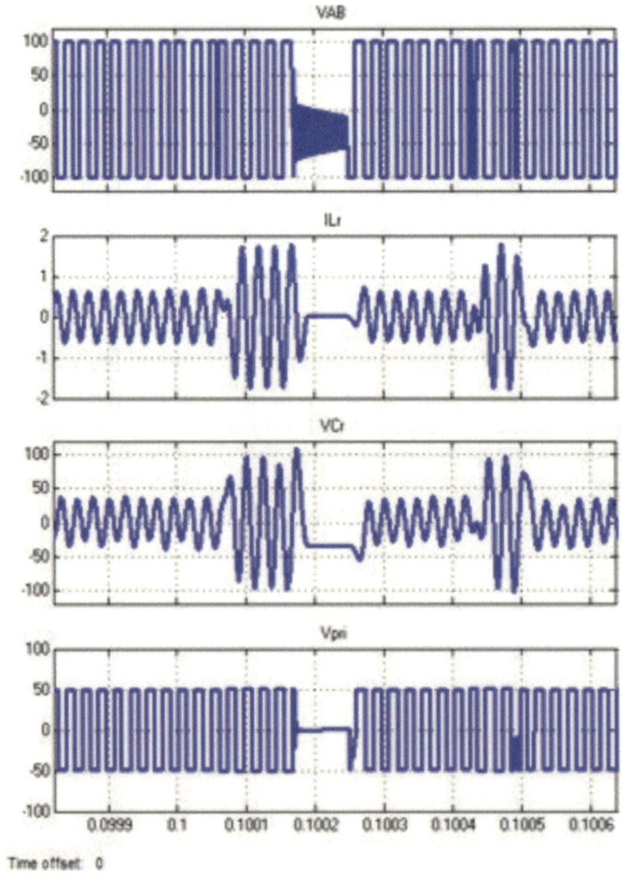

Figure 15.
Simulation waveforms of the resonant tank input voltage V_{AB}, resonant inductor current i_L, resonant capacitor voltage V_{cr}, and transformer primary voltage V_{pri}.

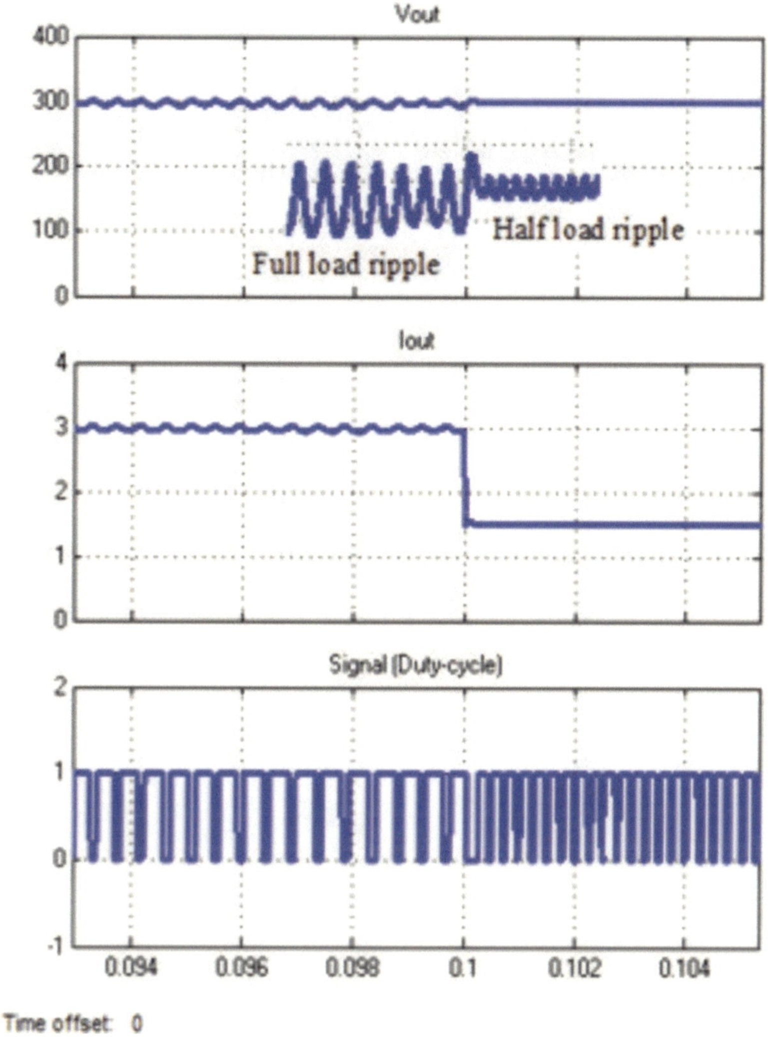

Figure 16.
The dynamic response of the output voltage, output current, and controlled duty-cycle signal with respect to the load changes.

load of (100 Ω). It was observed that the output voltage ripple is enormous at full load condition in relative to the half load state that reflects a direct duty cycle changes within the entire load conditions. Although, the system generates a favorable outcome by controlling the output voltage equivalent to 300 V, nevertheless, the resonant tank parameters mislaying the resonant concept during the changes in load changes as illustrated in **Figure 15**. At simulation time $t = 0.1$ s, the load varies the AC tank parameters hold-up by the $1.5e-3$ s, thereafter, the resonant parameters reproduce the targeted AC parameters depending on a value of the load, which can be appraised as a disadvantage of the duty-cycle control technique.

In the variable frequency control technique, the measured output voltage is used to detect frequency. Thereafter, the controlled signal is implemented on PWM to produce gate signals to the entire switches by considering the switching duration depending on the converter nature to generate enormous output voltage gain. Moreover, the variation in load is being tested and applied to affirm the controller dynamic responses. **Figure 18** illustrates the parameter frequency controller dynamic response of the output voltage, and the load is stepped up and down in the

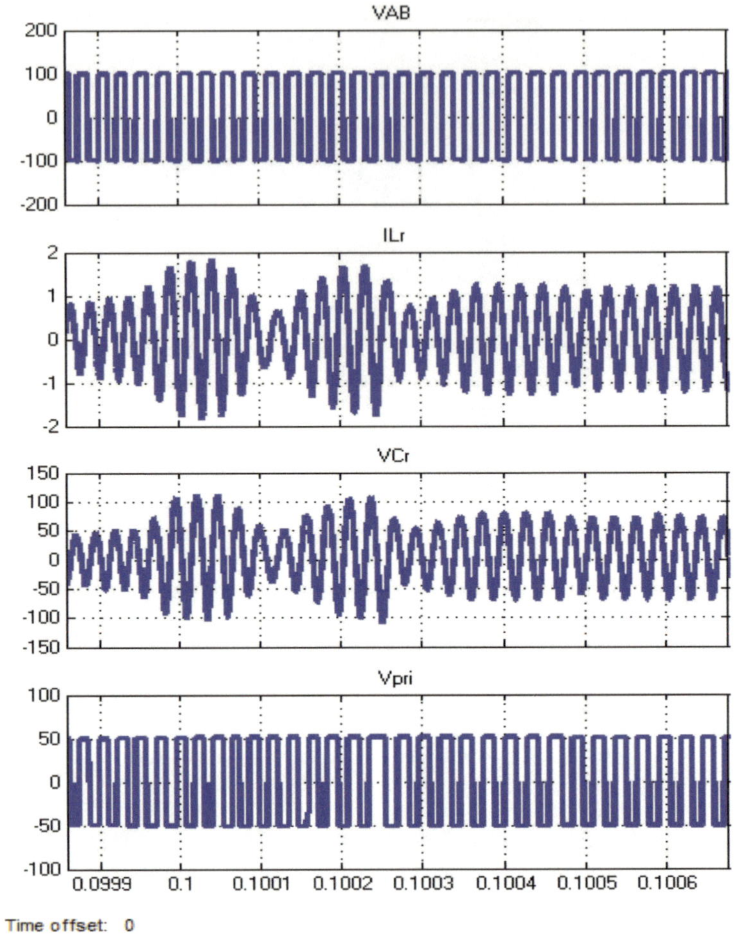

Figure 17.
Simulation waveforms of the resonant tank input voltage V_{AB}, resonant inductor current i_L, resonant capacitor voltage V_{cr}, and transformer primary voltage V_{pri}.

similar way of duty-cycle control. It can also be observed that in this control technique, the full load condition results in enormous output voltage ripple as compared to half load state, and this affirms the significant variation in frequency with the entire load condition. Moreover, the parameter frequency control gives a significant response to the tank AC parameters as illustrated in **Figure 17**. As the load varies, the AC parameters temporarily react by increasing or decreasing the voltage and resonant current values depending on the varied frequency and keeping a shape of the sinusoidal waveform.

5. Applications of resonant power converters

Based on the sufficient demerit of RPCs as earlier stated in the above segments, they have uncommon application in modern industries. The summary of the noticeable implementations is discussed in this segment. The main areas of RPCs application are household applications like induction cookers, portable power supplies, network connection of renewable energy mains, and hybrid and electric vehicles. In a case of the portable power supply, requirements of the converter

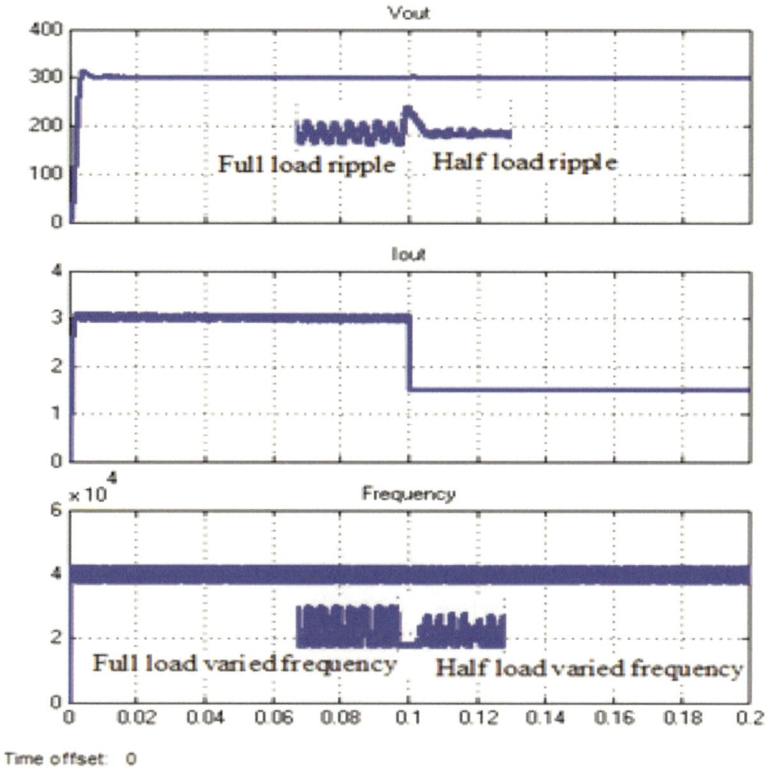

Figure 18.
The dynamic response of output voltage, output current, and controlled frequency signal as the function of load change from full to half load.

include a low price tag, light-weight and small size, high efficiency, high reliability, and low electromagnetic interference (EMI). Soft switching is the way of ensuring higher efficiency; it can be implemented by utilizing RPCs. Based on the area of application, the topology can be chosen to ensure maximum efficiency, ideal cost, and size. For example, the supply of power to an electron beam welding compartment uses a full bridge LLC resonant converter [16]. The soft switching technique and topology solves the problem associated with power utilization within the filament supply by staying away from the inverter heating challenge and ensures higher efficiencies. RPCs are used in the electrostatic precipitator. This is a high-power appliance industrially utilized for removing smoke and dust from a flowing gas. The series-parallel RPC coupled with phase control suggested by [27] is negligible in size; it gives a faster temporary response and possesses a higher efficiency as compared to the traditional line frequency power supplies. RPCs are known for charging hybrid vehicles whereby the batteries need to be charged either by wireless or wired. Due to being smaller and demonstrating higher effectiveness, the charging of EVs through wire [28] and plug-in EVs [29] RPCs are used. For example, the high-performance LLC converters are suggested in the two-stage smart battery chargers. The converter evacuates the low and frequencies ripples from the current output and increases the life of a battery without the size increment of the charger. Previous studies had proposed several wired charging topologies. Apart from wired chargers, wireless power transfer (WPT) has been a modern charging process used in the hybrid and electric vehicles. The accessible WPT technologies include electromagnetic, magnetic, and electric power transfer. Among all of these, the magnetic coupling method utilizes RPCs, which portray higher power transmission and higher efficiency within a closer distance. Application of RPC in WPT for hybrid and electric vehicles had been reported [30].

The grid integration of renewable energy mains such as fuel cell, wind, and solar PV needs converters of minimal current ripple and higher efficiency. DC-DC converters are the main prerequisite in processing power from renewable energy mains. Out of several options, RPCs can be the main competitor because of its low EMI, higher efficiency, robustness, and low output ripple current. RPCs are used in FC networks [31, 32], PVs [33], grid connection, and electrolyser interfaces [34]. Moreover, RPCs is applied to home induction cookers. The resonant inverter is the induction cooker key element; it produces an AC current that heats up the inductor-vessel compartment. The resonant inverters utilize in induction cooker are multilevel, half bridge, single-switch, and full bridge inverters [35].

6. Conclusion

This chapter has vividly explained the resonant power converters, which include the effective generation of RPCs as a high-switching converter that can serve as a solution to electromagnetic interference (EMI) and switching losses difficulties that surface by using the PWM converters. Furthermore, the resonant converter classifications depending on several points of views as well as controlled techniques (either varied or fixed frequency techniques) were elucidated. The controlled techniques and several application areas of resonant converters have been explained. The variable and fixed frequency controls are used to cross-check the LLC converter output voltage. The load is changed when using 50% of the full load in the entire control techniques; the obtained results affirmed the significant stable response of both controllers with little overshoot voltage. Nevertheless, the variable frequency control gives a significant outcome based on the resonant tank waveforms in relative to the stable frequency control when changing the load.

Author details

Mohammed Salem[1*] and Khalid Yahya[2]

1 School of Electrical and Electronic Engineering, Universiti Sains Malaysia, Nibong Tebal, Malaysia

2 Kocaeli University, İzmit, Kocaeli, Turkey

*Address all correspondence to: salemm@usm.my

References

[1] Outeiro MT, Visintini R, Buja G. Considerations in designing power supplies for particle accelerators. In: Industrial Electronics Society, IECON 2013-39th Annual Conference of the IEEE 2013 Nov 10. IEEE; 2013. pp. 7076-7081. DOI: 10.1109/IECON.2013.6700307

[2] Salem M, Jusoh A, Idris NR. Implementing buck converter for battery charger using soft switching techniques. In: Power Engineering and Optimization Conference (PEOCO), 2013 IEEE 7th International 2013 Jun 3. IEEE; 2013. pp. 188-192. DOI: 10.1109/PEOCO.2013.6564540

[3] Mostaghimi O, Wright N, Horsfall A. Design and performance evaluation of SiC based DC-DC converters for PV applications. In: Energy Conversion Congress and Exposition (ECCE), 2012 IEEE 2012 Sep 15. IEEE; 2012. pp. 3956-3963. DOI: 10.1109/ECCE.2012.6342163

[4] Ching TW, Chan KU. Review of soft-switching techniques for high-frequency switched-mode power converters. In: Vehicle Power and Propulsion Conference, 2008. VPPC'08. IEEE 2008 Sep 3. IEEE; 2008. pp. 1-6. DOI: 10.1109/VPPC.2008.4677473

[5] Shafiei N, Pahlevaninezhad M, Farzanehfard H, Bakhshai A, Jain P. Analysis of a fifth-order resonant converter for high-voltage DC power supplies. IEEE Transactions on Power Electronics. 2013;**28**(1):85-100. DOI: 10.1109/TPEL.2012.2200301

[6] Outeiro MT, Buja G, Carvalho A. Resonant converters for electric equipment power supply. In: Industrial Electronics Society, IECON 2014-40th Annual Conference of the IEEE 2014 Oct 29. IEEE; 2014. pp. 5065-5071. DOI: 10.1109/IECON.2014.7049270

[7] Salem M, Jusoh A, Idris NR, Sutikno T, Buswig YM. Phase-shifted series resonant converter with zero voltage switching turn-on and variable frequency control. International Journal of Power Electronics and Drive Systems (IJPEDS). 2017;**8**(3):1184-1192. DOI: 10.11591

[8] Jin K, Ruan X. Hybrid full-bridge three-level LLC resonant converter-a novel DC-DC converter suitable for fuel cell power system. In: Power Electronics Specialists Conference, 2005. PESC'05. IEEE 36th 2005 Jun 16. IEEE; 2015. pp. 361-367. DOI: 10.1109/PESC.2005.1581649

[9] Salem M, Jusoh A, Rumzi N, Idris N, Alhamrouni I. Steady state and generalized state space averaging analysis of the series resonant converter. In: 3rd IET International Conference on Clean Energy and Technology (CEAT). 2014. pp. 1-5. DOI: 10.1049/cp.2014.1488

[10] Alhamroun I, Salem M, Jusoh A, Idris NR, Ismail B, Albatsh FM. Comparison of two and four switches inverter feeding series resonant converter. In: Energy Conversion (CENCON), 2017 IEEE Conference on 2017 Oct 30. IEEE; 2017. pp. 334-338. DOI: 10.1109/CENCON.2017.8262508

[11] Rathore AK, Bhat AK, Oruganti R. A comparison of soft-switched DC-DC converters for fuel cell to utility interface application. IEEJ Transactions on Industry Applications. 2008;**128**(4): 450-458. DOI: 10.1541/ieejias.128.450

[12] Salem M, Jusoh A, Idris NR, Alhamrouni I. Comparison of LCL resonant converter with fixed frequency, and variable frequency controllers. In: Energy Conversion (CENCON), 2017 IEEE Conference on

2017 Oct 30. IEEE; 2017. pp. 84-89. DOI: 10.1109/CENCON.2017.8262463

[13] Yang Y, Huang D, Lee FC, Li Q. Analysis and reduction of common mode EMI noise for resonant converters. In: Applied Power Electronics Conference and Exposition (APEC), 2014 Twenty-Ninth Annual IEEE 2014 Mar 16. IEEE; 2014. pp. 566-571. DOI: 10.1109/APEC.2014.6803365

[14] Outeiro MT, Buja G, Czarkowski D. Resonant power converters: An overview with multiple elements in the resonant tank network. IEEE Industrial Electronics Magazine. 2016;**10**(2):21-45. DOI: 10.1109/MIE.2016.2549981

[15] Polleri A, Anwari M. Modeling and simulation of paralleled series-loaded-resonant converter. In: Second Asia International Conference on Modelling & Simulation. 2008. p. 974. DOI: 10.1109/AMS.2008.86

[16] Salem M, Jusoh A, Idris NR, Das HS, Alhamrouni I. Resonant power converters with respect to passive storage (LC) elements and control techniques–An overview. Renewable and Sustainable Energy Reviews. 2018; **91**:504-520. DOI: 10.1016/j.rser.2018.04.020

[17] Chuang YC, Ke YL, Chuang HS, Chen HK. Implementation and analysis of an improved series-loaded resonant DC-DC converter operating above resonance for battery chargers. IEEE Transactions on Industry Applications. 2009;**45**(3):1052-1059. DOI: 10.1109/TIA.2009.2018946

[18] Ivensky G, Bronshtein S, Abramovitz A. Approximate analysis of resonant LLC DC-DC converter. IEEE Transactions on Power Electronics. 2011;**26**(11):3274-3284. DOI: 10.1109/TPEL.2011.2142009

[19] Salem M, Jusoh A, Idris NR, Alhamrouni I. A review of an inductive power transfer system for EV battery charger. European Journal of Scientific Research. 2015;**134**:41-56

[20] Tan X, Ruan X. Equivalence relations of resonant tanks: A new perspective for selection and design of resonant converters. IEEE Transactions on Industrial Electronics. 2016;**63**(4): 2111-2123. DOI: 10.1109/TIE.2015.2506151

[21] Shafiei N, Farzanehfard H. Steady state analysis of LCLC resonant converter with capacitive output filter. In: Power and Energy (PECon), 2010 IEEE International Conference on 2010 Nov 29. IEEE; 2010. pp. 807-812. DOI: 10.1109/PECON.2010.5697690

[22] Lin BR, Hou BR. Analysis and implementation of a zero-voltage switching pulse-width modulation resonant converter. IET Power Electronics. 2014;7(1, 1):148-156. DOI: 10.1049/iet-pel.2013.0134

[23] Salem M, Jusoh A, Idris NR, Sutikno T, Abid I. ZVS full bridge series resonant boost converter with series-connected transformer. International Journal of Power Electronics and Drive Systems (IJPEDS). 2017;**8**(2):812-825. DOI: 10.11591

[24] Salem M, Jusoh A, Idris NR, Tan CW, Alhamrouni I. Phase-shifted series resonant DC-DC converter for wide load variations using variable frequency control. In: Energy Conversion (CENCON), 2017 IEEE Conference on 2017 Oct 30. IEEE; 2017. pp. 329-333. DOI: 10.1109/CENCON.2017.8262507

[25] Lu Y, Cheng KW, Ho SL, Pan JF. Passivity-based control of a phase-shifted resonant converter. IEE Proceedings-Electric Power Applications. 2005;**152**(6):2005, 1509-2015. DOI: 10.1049/ip-epa:20045261

[26] Lu Y, Cheng KE, Ho SL. Quasi current mode control for the phase-shifted series resonant converter. IEEE Transactions on Power Electronics. 2008;**23**(1):353-358. DOI: 10.1109/TPEL.2007.911846

[27] Salem M, Jusoh A, Idris NR, Alhamrouni I. Extension of zero voltage switching range for series resonant converter. In: Energy Conversion (CENCON), 2015 IEEE Conference on 2015 Oct 19. IEEE; 2015. pp. 171-175. DOI: 10.1109/CENCON.2015.7409534

[28] Musavi F, Craciun M, Gautam DS, Eberle W, Dunford WG. An LLC resonant DC-DC converter for wide output voltage range battery charging applications. IEEE Transactions on Power Electronics. 2013;**28**(12): 5437-5445. DOI: 10.1109/TPEL.2013.2241792

[29] Mousavi SM, Beiranvand R, Goodarzi S, Mohamadian M. Designing A 48 V to 24 V DC-DC converter for vehicle application using a resonant switched capacitor converter topology. In: Power Electronics, Drives Systems & Technologies Conference (PEDSTC), 2015 6th 2015 Feb 3. IEEE; 2015. pp. 263-268. DOI: 10.1109/PEDSTC.2015.7093285

[30] Bojarski M, Asa E, Outeiro MT, Czarkowski D. Control and analysis of multi-level type multi-phase resonant converter for wireless EV charging. In: Industrial Electronics Society, IECON 2015-41st Annual Conference of the IEEE 2015 Nov 9. IEEE; 2015. pp. 5008-5013. DOI: 10.1109/IECON.2015.7392886

[31] Outeiro MT, Carvalho A. Methodology of designing power converters for fuel cell based systems: A resonant approach. In: New Developments in Renewable Energy. Rijeka: In Tech; 2013. DOI: 10.5772/54674

[32] Outeiro MT, Carvalho A. Design, implementation and experimental validation of a DC-DC resonant converter for PEM fuel cell applications. In: Industrial Electronics Society, IECON 2013-39th Annual Conference of the IEEE 2013 Nov 10. IEEE; 2013. pp. 619-624. DOI: 10.1109/IECON.2013.6699206

[33] Rezaei MA, Lee KJ, Huang AQ. A high-efficiency flyback micro-inverter with a new adaptive snubber for photovoltaic applications. IEEE Transactions on Power Electronics. 2016;**31**(1, 1):318-327. DOI: 10.1109/TPEL.2015.2407405

[34] Beres RN, Wang X, Blaabjerg F, Liserre M, Bak CL. Optimal design of high-order passive-damped filters for grid-connected applications. IEEE Transactions on Power Electronics. 2016;**31**(3):2083-2098. DOI: 10.1109/TPEL.2015.2441299

[35] Mishima T, Nakagawa Y, Nakaoka M. A bridgeless BHB ZVS-PWM AC-AC converter for high-frequency induction heating applications. IEEE Transactions on Industry Applications. 2015;**51**(4): 3304-3315. DOI: 10.1109/TIA.2015.2409177

Section 4

Electric Power Conversion Applications

Chapter 7

Perspectives on Dual-Purpose Smart Water Power Infrastructures for Households in Arid Regions

Dana Alghool, Noora Al-Khalfan, Stabrag Attiya and Farayi Musharavati

Abstract

In hot arid climates, freshwater and power are produced simultaneously through seawater desalination since these regions receive little rainfall. This results in a unique urban water/power cycle that often faces sustainability and resilience challenges. Elsewhere, such challenges have been addressed through smart grid technologies. This chapter explores opportunities and initiatives for implementing smart grid technologies at household level for a case study in Qatar. A functional dual-purpose smart water/power nanogrid is developed. The nanogrid includes multiloop systems for on-site water recycling and on-site power generation based on sustainability concepts. A prototype dual-purpose GSM-based smart water/power nanogrid is assembled and tested in a laboratory. Results of case study implementation show that the proposed nanogrid can reduce energy and water consumptions at household level by 25 and 20%, respectively. Economic analysis shows that implementing the nanogrid at household level has a payback period of 10 years. Hence, larger-scale projects may improve investment paybacks. Extension of the nanogrid into a resilient communal microgrid and/or mesogrid is discussed based on the concept of energy semantics. The modularity of the nanogrid allows the design to be adapted for different scale applications. Perspectives on how the nanogrid can be expanded for large scale applications are outlined.

Keywords: water conservation, energy efficiency, smart water, smart grids, renewable energy, nanogrid, energy semantics

1. Introduction

Water and energy are among the most important commodities in life. They support growth, development, and human survival on earth. Consequently, sustainable water and energy supply have become critical issues of consideration in most parts of the world [1, 2]. Moreover, the water and energy nexus has been a great subject of debate for decades [3, 4]. For example, the United Nations has predicted a 40% global shortfall of water availability by 2030 and a 50% global short fall on energy [5, 6]. In spite of these observations, the demand for energy has been on the rise as various national economies become more and more advanced. In addition, climate change studies have projected unique changes in urban water cycles, thus making it more difficult for national economies to balance water supply and distribution now and in future plans. These difficulties add more strain on energy supply and freshwater access [7]. Freshwater concerns are even more critical in hot arid regions characterized by low rainfall and harsh climate. This chapter discusses potential solutions to sustainability challenges with reference to water and energy conservation. The underlying theme lies in that implementation of smart water and energy technologies has a significant impact on water and energy conservation practices in arid regions.

In most arid regions, water and energy supply networks are implemented as separate single-loop systems. This means that the stages of the current water/power cycle do not intersect and yet these commodities are produced simultaneously in dual-purpose water/power production plants. In practice, the urban water/power cycles often face challenges that require further investment by local authorities in a bid to mitigate the effects of sustainability challenges. In addition, population growths and rapid economic developments often strain water/power supply and distribution networks, thus making the urban water/power cycle less sustainable. It is therefore important to rethink the urban water/power cycle in a bid to develop water/power infrastructures that can improve the water/power use efficiencies at the household level by incorporating smart technologies.

Smart technologies have been implemented with benefits that support sustainability goals [8–10]. In the public literature, smart energy grids have been discussed thoroughly by many authors [11–13]. Smart energy grids have also been successfully implemented in various parts of the world [14–16]. A number of benefits associated with smart energy grids have been identified including economic [17, 18], environmental [18, 19], reliability [20, 21], and customer choice [22, 23]. Such benefits significantly contribute to both resilience and sustainability. While the literature is overwhelmed with smart energy grids, relatively little is known about smart water grids [24–26]. Of late, smart water grids have been found to hold a lot of potential for unlocking the requirements for a sustainable, stable, reliable, high-quality, resilient, and secure water supply system. Another prominent gap in the public literature lies in that water and energy smart grids are usually discussed separately. While this may be appropriate for other regions of the world, the unique connection between water and energy in arid regions requires special considerations and technologies that are more appropriate. In the Gulf Region, for example, water and energy are produced simultaneously [27–29]. It has been shown that there is an inherent link between energy and water [30–32]. It is, therefore, necessary to investigate opportunities and initiatives for developing dual-purpose water/power smart grids. Perspectives on the design and operations of such a smart grid will be discussed with reference to a case study in Qatar.

The common practice in Qatar is that once water and power are produced, they are distributed separately to residential areas. At the household level, water is pumped into a tank positioned at the roof of villas. After use at different end points, this water is directed into a sewer line where it is mixed with black water and further directed to wastewater reservoirs. Separating and reusing this water at the source (household) may prove beneficial. On the other hand, solar energy in Qatar is currently found in isolated areas that are far from grid connections. Most of the energy generated from solar is used as supplements to the main grid power. There is a need to increase the fraction of renewable energy in Qatar since the insolation is relatively high. This can position Qatar toward achieving sustainability goals as stipulated in the Qatar National Vision 2030.

One way of addressing sustainability issues is to closely examine the 6Rs of sustainability, i.e., reduce, reuse, rethink, recycle, refuse, and repair [33]. The purpose in implementing the 6Rs is to obtain the most practical benefits from products, processes, and systems and to generate the least amount of wastes. This approach also activates other external positive issues such as pollution reduction, resource saving, and avoidance of greenhouse gas emissions. The discussions in this chapter derive inspiration from four of these 6Rs, i.e., reduce, reuse, recycle, and rethink. *Rethink* is about trying to think (in a different way) how to generate electricity and provide useful water in order to minimize the consumption of the main grid electricity and freshwater. For example, generating electricity from the velocity of clean or wastewater in water pipes (in-pipe hydropower generation) and treating the wasted water instead of disposing it to the main sewage directly after use are noble initiatives that can help in conserving both energy and water. In addition, current system designs for water are single-loop system from utilities. The idea in this work is to investigate the usefulness of multi-loops of water (freshwater, gray water, and black water) and energy (main grid supply, renewable energy micro-generation, and in-pipe hydroelectricity) at the household level in a bid to rethink and reuse available resources to the maximum possible. Design, development, and implementations of such multiple loops of energy and water deviate from the common single-loop systems and thus constitute an initiative for rethinking the energy and water networks at end-use locations.

Reduce is about reducing and minimizing the wasted water "produced" in the household as well as reducing the consumption and electricity from the main grid supply by implementing renewable energy and smart technologies in the existing infrastructure. It is also about behavioral changes due to the conscious realization, by residents, of "wasteful" consumption of water and energy. *Reuse* is about reusing gray water produced in the house after treating it. The treated water can be "reused" for watering the gardens, car washing, as well as toilet flushing instead of "throwing the water down the drain." *Recycle* is about collecting the gray water that is produced at different end points in the house, such as sinks, showers, and washing machines for the purpose of treating and reusing the gray water at the source instead of sending it to the main sewage line where it is further contaminated by black water. Based on the concepts discussed in the previous paragraphs, this chapter discusses the development of a smart dual-purpose water/power nanogrid under the climatic conditions in Qatar.

According to the Qatar National Vision 2030, Qatar aspires to be an advanced society capable of sustaining its development and providing a high standard of living for its residents [34]. However, with the current population explosion and numerous construction projects, the utility companies in Qatar may face a number of water and electricity consumption challenges. For example, the residents in Qatar consume nearly twice the average consumption of water and electricity in other parts of the world, the EU being a specific example [35]. Statistical projections show that these consumption rates are expected to double in the near future, thus further straining the balance between water/energy supply and demand. In a bid to provide solutions for these challenges, the effects of implementing smart technologies are discussed in this chapter. A combination of smart water and smart energy technologies are discussed, and perspectives on how to integrate them into a functional nanogrid for a single household are outlined. The motivation emanates from the observation that residential water and energy infrastructures often waste substantial quantities of freshwater and energy. Therefore, there is a need to reduce water and energy consumptions at the household level.

2. Background

Water and energy resources are communally and reciprocally linked since meeting energy needs requires water and vice versa [3, 4]. The consensus is that saving water saves energy and energy efficiency opportunities are often linked to water savings. Albeit, both initiatives result in less carbon emissions. In hot arid climates, this relationship is intertwined since water and energy are produced simultaneously in dual-purpose water/power production plants. Therefore, addressing water and energy issues in tandem can result in significant benefits for utility companies.

Improving efficiency of energy and water in the supply and demand sides can allow national economies to reduce resource consumptions as well as maximize benefits for utilities, consumers, businesses, and communities. National economies need to increase water and energy security while reducing the environmental impacts of water and energy use. This means that available water and energy must be used more efficiently. Energy consumption in water reticulation systems can be reduced by using energy recovered from household water systems and wastewater at nanoscale to produce power on-site. Power consumption can be reduced at the household level by, for example, giving residents detailed energy consumption information that can be used by residents to decide on how best to use energy in their homes.

Assessment of end-use energy and water efficiencies provides information that can be used to find ways of reducing the strain on the main power grid and water distribution network. However, a number of barriers and challenges may exist. In the Gulf countries, for example, there is currently an overall trend toward larger homes and a greater variety of appliances and electronics in each home. This trend further strains the water and energy resources at the national level and hence contributes to the imbalance on water and electricity supply and demand. Options for increasing end-use energy efficiencies include renewable on-site power generation, implementing well-designed energy codes and standards, improving end-use appliance energy efficiency, using efficient plumbing fixtures, and educating homeowners about behavioral changes that will result in significant reductions in energy consumptions. Since water and power are produced in expensive seawater desalination plants, water conservation and water recycling are important initiatives that can be used to leverage end-use efficiencies. Furthermore, such initiatives support sustainable developments.

Energy use in residential buildings account for about 17% of US greenhouse gas (GHG) emissions [36, 38]. Unlike the Gulf countries, the large share of residential building energy consumption is attributable to space heating and cooling, which varies with climate conditions. In the Gulf countries, cooling accounts for about 70% of energy used in residential buildings [37]. Other energy uses are related to providing power to various household appliances that are used randomly. Reducing energy consumptions of these end uses is difficult since it requires different technological improvements for each appliance as well as behavioral changes that aim at increasing energy efficiency and conservation . This represents significant challenges to sustainability goals.

While many options are available for providing clean water, seawater desalination has taken the center stage in the Gulf countries. Common technologies for seawater desalination include multistage flash distillation, multi-effect distillation, and reverse osmosis. Since environmental concerns are on the rise, renewable energy technologies are becoming more important and attractive partners for powering water desalination projects in arid regions [39], while desirable, renewable energy cannot cope with the quick, discontinuous, and uncontrollable falls and peaks in electricity demand. Since renewable energy technologies depend on the season and the time of the day, their integration poses challenges to the traditional grid systems. Generating electricity from renewable energy, mainly photovoltaics (PV), wave, and wind power depend extremely on the unpredictable nature of weather conditions and status [40]. If new electric devices are employed in the renewable energy-based electricity systems, great achievements can be realized. Examples of electric devices and components that can support renewable energy electricity integration include advanced batteries, inverters, advanced controllers, and smart technologies [40].

In the case of clean water, drivers that support water security are water conservation, water recycling, and efficient water use. A number of mechanisms are available for conserving water. Typically, groundwater aquifers collect less than 40 million m^3 annually as natural recharge. This imbalance makes the need for changes and rethinking the water cycle obvious. Minor changes such as changing a showerhead to a more efficient one can save small amounts of water at the household level. Hence, the impact of such changes is limited if one household is considered. This impact can be significant if large communities and neighborhoods are the bases of the analysis. Other opportunities include industries and commercial sectors taking the initiative to recycle and reuse both gray and black water on-site. A collective support of this kind from residential areas, industries, and commercial sectors can significantly impact the strain on the main grid freshwater supply.

Water use patterns are critical to any water conservation solutions. For example, in the urban areas in Portugal, the residential sectors were observed to have the highest water demand when compared with the industrial, commercial, and institutional sectors [41]. Reducing the domestic water consumption rises important benefits like the postponement of investments in the water supply system expansion and pump nanogrid upgrade. It also reduces peak and average effluent loading to the wastewater system [42]. A significant reduction on energy requirement is caused by a lower water demand in the household (e.g., for water heating). In addition, the water end-use sector of a distribution system (i.e., activities that use water in buildings and homes) has been found to be the highest energy intensive part in the urban water supply systems [43]. Such analysis, data, and information gathering can provide useful insight into practical water end-use efficiency programs that can be used by utility companies for the benefit of national economies.

Many studies have been conducted, for example, by Loh and Coghlan [44], Willis et al. [45], Beal et al. [46], Matos et al. [47], Cole and Stewart [48], and Omaghomi and Buchberger [49], to describe and characterize the types of water uses. These studies show that water end-use characteristics generally differ from place to place. Hence, it is important to analyze water end-use within local context in order to develop tools, mechanisms, and techniques for improving water end-use efficiencies. Studies by Willis et al. [50], Matos et al. [51], and Hunt and Rogers [52] demonstrated the relationships between consumer sociodemographic characteristics, end uses and consumer attitudes, and water end-use efficiencies. Willis et al. [53], Lee et al. [54], and Carragher et al. [55] reviewed the effect and the influence of the residential water use efficiency measures on water demand.

Improving the collection of gray water might significantly decrease the amount of clean water that is being used in landscaping, gardening, and toilet flushing at the household level. In Qatar, for example, gardening consumes around 5% of the total freshwater at household level. Albeit, gray water produced at houses is usually sent down the drain in sewage pipelines. Although the amount at one household may seem small, it is the collective actions of all residential areas that will affect the main grid strain on freshwater supply for a national economy. In addition, a lot of gray water is produced in other places such as mosques, air-conditioning units, shopping malls, and corporate buildings. By rethinking this practice, gray water collected for recycling from different places can be treated using simple processes to make this water suitable for gardening, landscaping, agriculture, construction works, and district cooling services. In spite of these potential reuses and recycling possibilities, gray water and black water in the case study villa is currently being channeled into a shared sewage system, which makes the gray water highly unusable. Although Qatar has a huge and a broad network system for collecting and treating the domestic wastewater, separation of the wastewater at the source can be more beneficial and more cost-effective in the long run than the central collection and treatment practices in the case study.

Due to the problems and challenges faced by the water sector, a number of water intelligence tools have been developed worldwide to alleviate global water issues. Information and communications technology (ICT) offers valuable chances to improve the efficiency and the productivity within the water sector, with the purpose of contributing to the sustainability of the resource. The increasing availability of more intelligent, ICT-enabled means to manage and protect the water resources of various national economies has led to the development of smart water management (SWM). The SWM approach promotes the sustainable consumption of water resources through coordinated water management, by integrating ICT products, solutions, and systems, targeted at maximizing the socioeconomic welfare of a society without compromising environment [56]. In Qatar, the potential use of ICT-enabled technologies has been initiated by Ooredoo, the telecommunication company in a pioneer project that aims to make Qatar's Lusail City a smart city.

The concept of smart water involves gradual convergence and integration of ICT solutions applied within the water domain. The smart water concept seeks to promote a sustainable, well-coordinated development and management of water resources by the integration of ICT products, tools, and solutions, thus providing the basis for a sustainable method to water management and consumption. An alternative way for more efficient water management could be offered by an approach that is fully linked to the quality of the vision developed for the Water Business Information System [57]. The more advanced ICTs used in the water system, the smarter the water becomes. For example, a water system with smart meters and smart pumps and valves is smarter than a system with smart meters only.

The concept of smart water and the level of water smartness depend on the number and the advancement of ICTs successfully implemented in the system. The implementation of the smart water concept has enabled significant improvement in water distribution, has helped to enhance wastewater and storm water management, and has helped to decrease losses due to nonrevenue water. The advantages of applying the smart water concept include increasing water quality and reliability, decreasing water loss due to leakage, reducing operational costs, ensuring proper management of green systems, and improving customer control and choice. At the household level, these advantages increase water end-use efficiencies, while at the national level, they increase the efficiency of the water sector and hence play a significant role in conserving water and thus reducing the strain on the main grid water supply and distribution [56]. A number of countries and communities have embraced smart water technologies [58–60]. Data obtained from implementation of smart technologies can help utilities in discovering problems on the consumer end of the water system. Consumption rates of water and power in Qatar are relatively high. This puts a large strain on the utility company. While the utility company has successfully implemented a number of projects to conserve water and power along the supply and distribution network, relatively few projects have been done in Qatar to reduce water and power consumptions at the household level. Most of the successful attempts at the household level have been through plumbing fixtures and conservation programs aimed at making residence aware of the need to conserve both water and power through Tarsheed, a proponent of the local water and power company.

3. Methodology

Both qualitative and quantitative approaches were used to synthesize the smart water/power nanogrid for households. Data was collected from a case study villa in Qatar. Analysis of the current situation revealed that at the household level, a number of factors influence water and power consumptions in Qatar. One of the factors is the water and power technologies implemented at the household level. Although a number of water and power saving tips have been provided by the utility company, no strict rules, regulations, or policies that directly influence water and power consumptions at the household level are available, although there are plans underway. Typically, most houses in Qatar are water and power metered, but the water and power rates are heavily subsidized by the government. In addition, no smart metering was available in households at the time when this project was carried out, although plans for smart metering were in place in the developing smart city of Lusail. Moreover, no information devices and except readings from conventional meters were available at the household level. Among the various types of residential unit villas were selected for case study since they composed the vast majority of residential preference of Qatar's residents. With various customer sectors, the focus on residential units also stems from its consumption contribution, with almost 90% of the national water consumption concentrated in residential areas [61]. A description of the case study villa is given in the following section.

3.1 Case study description

A typical three-storey villa was chosen as a case study in the development of the proposed smart water/power system. At the villa, consumption points include the bathroom and kitchen sinks, bathtubs, toilet, washing appliances, and garden watering. Wastewater from the houses is classified into two categories, i.e., black water (water produced from toilets and bidets) and gray water (water produced from all sinks, showers, and bathtubs). Black water and gray water are not separated in the current household water network outlet piping but are collected into one sewer pipe before disposal into the main sewage network. It is important to note that most villas in Qatar have a flat roof and there is no provision for harvesting rainfall since there is very little rainfall. In addition, vertical roof-mounted tanks are a common site in most villas in Qatar. Usually, a camouflage or protection structure is provided to make the roof tanks less visible. It is also important to note that typical families in Qatar are relatively large, with an average of 10 incumbents. A pictorial, schematic, and plan view of the case study villa is shown in **Figure 1**.

3.2 Assumptions

Based on survey results, a number of assumptions incorporated into the analysis were made as follows:

- A typical house in Qatar is designed for a family of 10 people.
- Consumption of water is the same in all households, as they are averaged per residential unit.
- All household pipes can be customized to meet the functional requirements of smart water and power technologies. This allows retrofitting.

Figure 1.
Case study villa: (a) pictorial view; (b) schematic side view; and (c) schematic plan view of the main water and electricity consumption points in a typical villa.

- All smart water and power technologies will not cause any interference or degradation of the water and power quality or services provided by the main grid components.

3.3 Data collection

Sources of data include data logging of water and power consumption data, in-home interviews, as well as water and electricity meter billing data. Smart meters and a logger were installed in the case study to collect flow observations from each water consumption point in the house. The collected data was used to determine the various water consumption events in the household such as volume, average flow, and maximum flow. Water and power consumptions are the amounts of water and power that reach consumers or end users and are usually estimated by water or electricity meters at the consumer and end-user points.

3.3.1 Water and energy consumptions

Pattern of water consumption on a typical day for the case study villa is shown in **Figure 2**. Typical monthly water consumption for the case study villa is also shown in **Figure 3**. **Figure 2** shows that water consumption per day varies from hour to hour depending on the needs of the people in the household. For example, **Figure 2** shows peaks at certain times (e.g., 7 am and 5 pm–7 pm) of the day corresponding to the times when water is required by most people in the household.

Figure 3 shows that the daily water consumption in the case study varies greatly from day to day. For example, it can be observed that there are peaks of water consumption at regular intervals throughout the month corresponding to high water consumptions. Interviews with residents in the case study villa revealed that more water is required on these respective days of the month for other uses such as various types of cleaning activities. Although variations are inevitable, the analysis in this work is based on the fact that there is a consistency in the flow patterns of residential water uses [62]. Pattern of power consumption on a typical day for the case study villa is shown in **Figure 4**. Typical monthly power consumption for the case study villa is also shown in **Figure 5**. **Figure 4** shows that power consumption per day varies from hour to hour depending on the needs of the residents. For example, **Figure 4** shows peaks at certain times (e.g., 7 am–9 am and 12 noon–8 pm) of the day corresponding to the times when power is required most in the household. **Figure 5** shows that power consumption per month varies from day to day depending on the needs of the residents. For example, **Figure 5** shows peaks on certain days of the month corresponding to days when power is required most in the household.

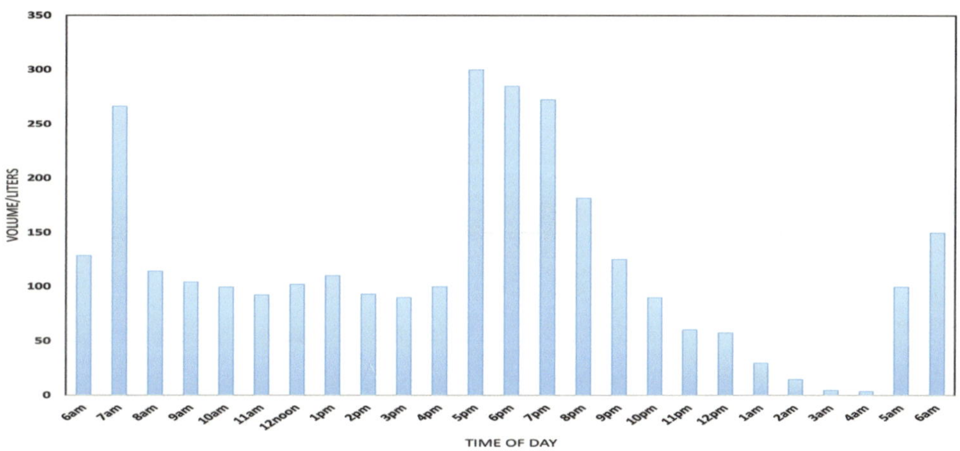

Figure 2.
Water consumption for a typical day at the case study villa.

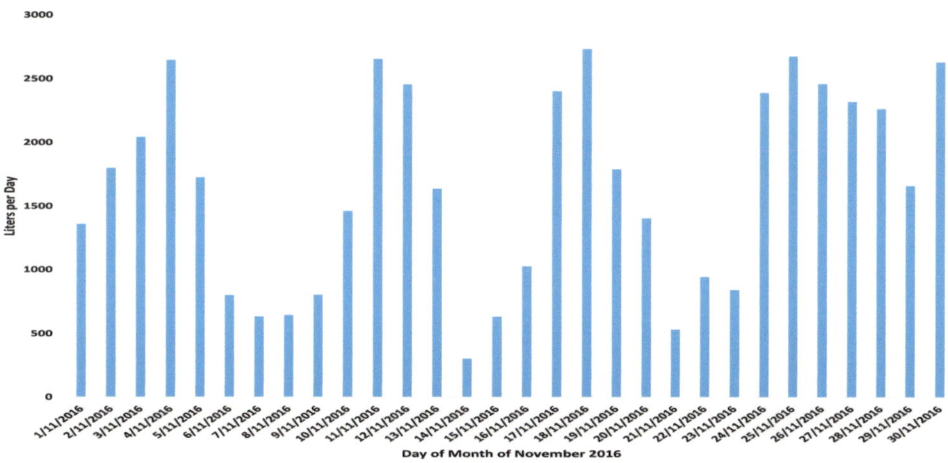

Figure 3.
Water consumption for a typical month at the case study villa.

3.3.2 Water and power end-use fractions

It has been observed that the majority of Qatar's water consumption is centered on residential areas [35]. Hence, more has to be done to conserve water in residential areas. The main sources of leakages in households are the faucets, toilet seats, bidets, showerheads, tubs, and junction points between pipes. Some of the reasons cited for these leakages include different types of materials used in pipping, changes in different pipes sizes, high water pressure at junctions' points between pipes, and the materials' corrosions of pipes. Besides losses at these leakage points, a lot of water is used by residents for various reasons. **Figure 6** shows the daily water and power fraction end use for the case study villa.

From **Figure 6**, it can be observed that the main points of potable water consumptions in the house are bathing, personal washing, and toilets. Bathing contributes 43% of the total daily potable water consumed in a typical house, while air conditioning contributes 60% of power consumed in a typical house in Qatar. Since air conditioners are used most of the day during summer, a lot of condensate is drained and redirected into the sewer line as gray water. The proposed household nanogrid collects gray water from various consumption points and redirects it for reuse.

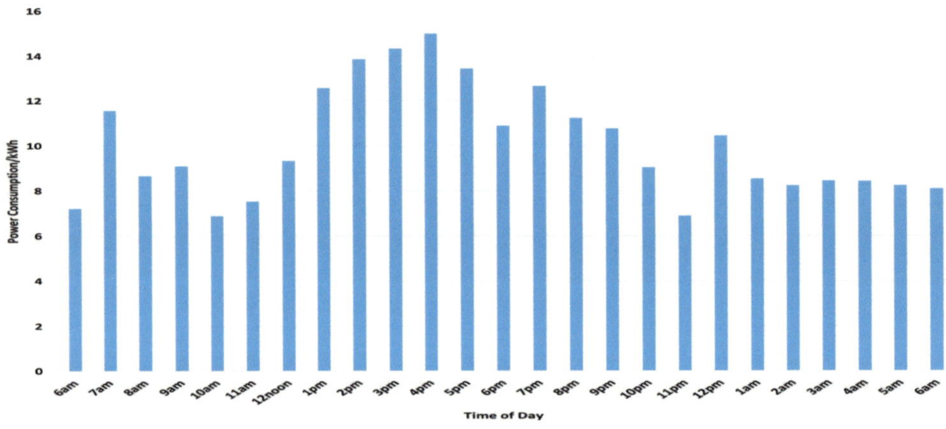

Figure 4.
Power consumption from the main grid for a typical day at the case study villa.

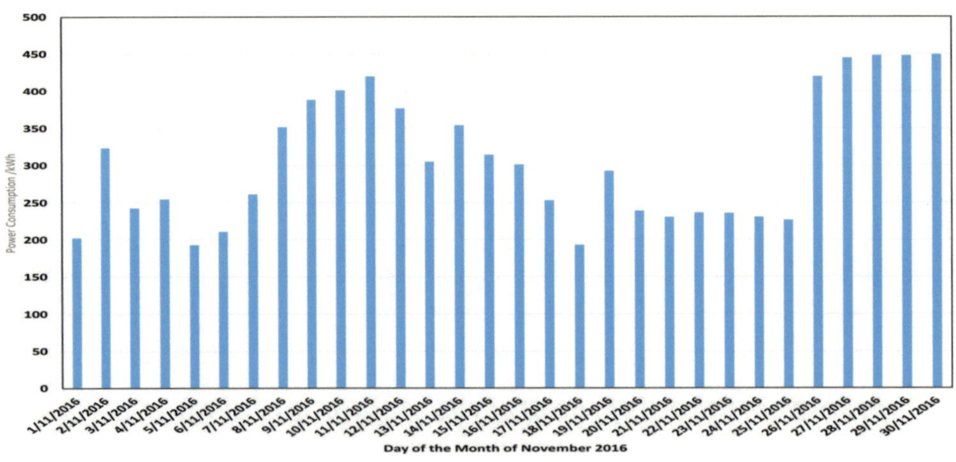

Figure 5.
Power consumption for a typical month at the case study villa.

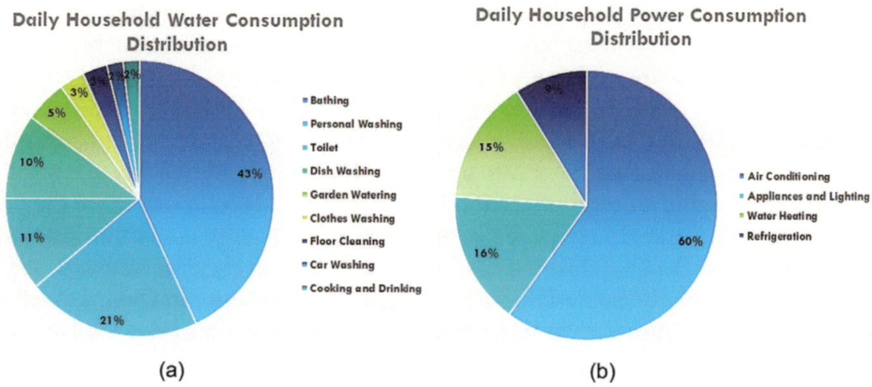

Figure 6.
Typical household freshwater and power use.

4. Design and analysis

The design analysis presented in this section is based on the information obtained from the case study villa. Proposed design perspectives are based on rethinking the urban water and energy cycles. The theme is devised based on methods, techniques, and technologies for transforming the current water and energy infrastructures into a completely redesigned setup based on the following concepts: multi-utility loops, smart water and energy, integrated gray water infrastructures at the household level, and separation of water resources at the household level.

4.1 Multi-utility loops

Multi-utility design features can be used to collect data from all types of smart meter installations. This requires implementations of multi-utility metering and multi-utility controllers that ensures security in data communication. For the case study, the proposed design implies a system that enables multiple loops for multiple alternative water sources, i.e., water from the main utility grid, water collected from various consumption points in the house, and water collected from air-conditioning units. Water from the household consumption points and water from the air-conditioning systems are gray water that can be treated on-site. The aim in implementing the multi-loop system is to ensure maximum use of water available at a household. This requires additional piping as well as wastewater treatment systems. The objective is to design and implement a water nanogrid that minimizes the carbon footprint of water use and reduces water leakages. The electricity network is envisaged to have multiple loops, each loop representing the source of on-site energy generation. Available power loops that have been included in the analysis are main power grid, solar PV, and in-pipe hydroelectricity. In such a design, a multi-utility controller will enable communication among the various consumption meters installed in the house in order to determine how much water or power is at disposal. Several costs are involved when upgrading the current water and energy infrastructures in households. Such costs include cost of smart meters, cost of installation and maintenance, as well as costs of data communication tools.

4.2 Design requirements

The dual-purpose smart water/power nanogrid is envisaged to be made up of water and energy smart technologies integrated into a nanogrid for household use. The purpose of such a nanogrid is to help in conserving water and energy. Such a nanogrid includes on-site power generation, on-site water recycling and reuse, as well as communication interfaces that will provide real-time information to household users about water and energy consumption levels. On-site power generation will reduce dependence on the main power grid and hence alleviate the strain on balancing power supply and demand. On-site water recycling will reduce consumption of freshwater, thus relieving the strain on freshwater supply networks. Information provided through the communication interface is expected to influence the behavior of household users in terms of sensible water and energy consumptions at the household level. Design parameters were collected from the case study. Based on the results of a survey carried out in a residential community in Qatar as well as the survey from the utility company, a number of design requirements for a smart water/power system at the household level were identified and are summarized as follows:

1. Minimize water and power consumptions at the household level in order to (a) reduce water and electricity bills for household owners and (b) reduce the strain on the main power and grid for utility company

2. Implementation of renewable energy technologies to alleviate the strains on the main power grid

3. Minimum breakdown of the system in order to ensure reliability in water and energy supply

4. Remote control for managing the water/power system with respect to household owner choice and preferences

5. Real-time notification about water and power consumption status for household owner decision-making

6. Ease of retrofitting the multi-loop nanogrid

7. Having an "acceptable" cost of procurement and installation of the system

8. Minimum safety threats to household users of the system

The conceptual design of the smart water/power system consists of different types of components. System design parameters for visualizing the architecture of the proposed smart water/power system were derived from the general nanogrid concept, i.e., nanogrids are autonomous renewable energy systems that do not interfere with the main grid. This consideration was important since currently the utility company in Qatar does not allow transfer or sharing of power across the main grid. The conceptual extension of nanogrids relates to an integral nanogrid composed of both smart water and energy technologies. The combined inclusion of smart water and power technologies in one nanogrid constitutes an important nanogrid design worth pursuing. With such nanogrids in place, it will be easier to translate existing nanogrid into a functional smart microgrid. Technologies selected for the nanogrid include solar PV, reverse osmosis, pumped storage, in-pipe hydropower generation, as well as energy storage components such as batteries. Target design specifications for the smart water/power system were derived as follows:

1. On-site generated power must be able to supply at least 20% of total household energy requirements (based on the Qatar National Vision 2030 aspirations).

2. Solar PV panels (monocrystalline type—as per utility company preference) are to be roof mounted, and the total area of these panels must not exceed the roof area ≈ 400 m^2 for a typical villa in Qatar.

3. Since the AC load will be provided by the main power grid, there must be two controllers in the system: one for the DC load and another one for energy storage.

4. In-pipe generator unit has to be installed within 25 mm of water supply pipes in the other floors of the house based on the current water reticulation pipe network system in the case study.

5. In-series generator units must have a distance of 4× pipe diameter apart.

6. Reverse osmosis unit should be able to process gray water with the following parameters: temperature ≈25°C and pressure ≈1.5–7 bar. The treated gray water must be suitable for flushing toilets, gardening, and on-site landscaping.

7. The system must include the following components to enable a certain level of smartness: smart meters, smart valves, pumps, and pH sensors.

8. Selected system components must be able to communicate with components in the existing household nanogrid.

9. The water/power system must be able to send information to household users for decision-making.

10. System components should be able to work based on the following specifications: 240 V, electrical frequency of about 50 Hz, and 900 GSM frequency (as per Qatar specifications).

The smart water/power system consists of three main units: (i) on-site power generation, (ii) on-site gray water recycling and reuse, and (iii) communication unit that will provide users with information about water and energy consumption as well as quality of the recycled gray water. Technologies used to assemble the smart water/power system include in-pipe electricity generators, pumped storage, solar photovoltaics, reverse osmosis, and a control system. The in-pipe electricity generator will be used to produce electricity by utilizing the water pressure as the water moves through the water supply pipe network as well as the gravity from the pumped storage. The roof-mounted tanks will facilitate pumped storage that will be used to maximize the use and reuse of recirculated water in the household. Solar photovoltaic panels will be used to generate solar electricity. In cases when the power generated on-site is not sufficient, the main grid power will be used instead. A reverse osmosis unit was used to treat gray water to sufficient quality for use in watering gardens, landscaping, car washing, or flushing the toilets. A control system was used for managing the operation of components and devices in the system as well as to provide household users the information on the system status. Smart meters were used to digitally send meter readings to household users so that they know their water and power consumptions will be added. Smart shut-off valves were used to facilitate remote control of water in the household. pH sensors were incorporated to facilitate the effective control and communication of the water quality in the system. A plumbing network, additional to the existing infrastructure, was used for water circulation in the system. **Figure 7** shows a schematic representation of the proposed system components.

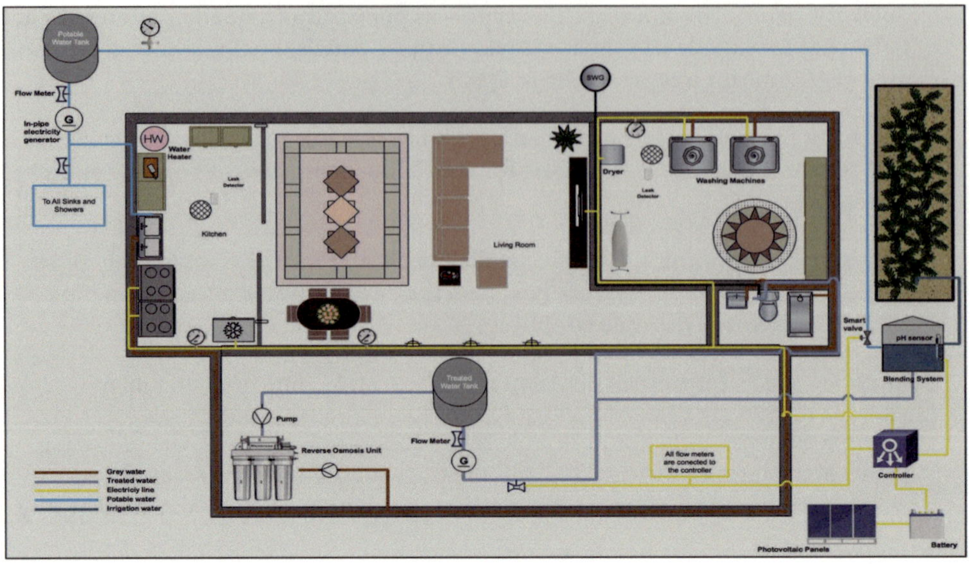

Figure 7.
Schematic diagram showing position of the various technologies implemented in the project.

Most houses in Qatar have little space available outside the building area. Therefore, the best place for the solar panels and storage tanks is on the roof. For practical implementation in the case study, these components must meet the minimum standards stipulated by the local utility company. The standards for the pipe network type and materials are controlled by the available local construction standards, codes, and regulations. As further requirement, additional components of the communication network must not degrade performance of currently existing infrastructure.

4.3 On-site power generation

The major components for on-site power generation are a solar PV system and in-pipe hydropower generation. The solar PV components include an array of PV modules, a charge controller, inverter, and battery bank. In pipe hydropower, generation was considered for both the existing network and the auxiliary network meant for recycling gray water at the villa. Available in the existing network is a pump that pumps water to the roof tank. The movement, flow, and velocity of this pumped water are captured to form the first type of in-pipe hydropower generated on-site. When the gray water is recycled, through the reverse osmosis unit, the water is pumped to the rooftop so that it can be conveniently used by taking advantage of the gained potential energy.

4.4 Design of system elements

4.1.1 Solar PV system sizing

In order to size the solar PV system, the following steps were followed:

1. Calculate the total power for all loads that will use solar PV electricity by adding the total watt-hours for each appliance used and finding the total watt-hours per day needed from the PV modules.

2. Sizing the PV modules by calculating the total watt-peak rating needed for PV modules and finding the number of required solar PV panels.

3. Sizing the inverter.

4. Sizing the battery.

Eqs. (1)–(6) were used for sizing the solar PV system.

$$\text{Total watt - hours per day for appliances used} = \text{Sum of watt - hours for all household appliances (excludes microwave ovens, cooker and any electrical machines, AC units, and any e lectric tools with power rating more than 1500 W)} \quad (1)$$

$$\text{Total watt - hours per day provided by PV modules} = 1.3 \times \text{sum of watt - hours for all appliances (to cater for energy lost in the system)} \quad (2)$$

$$\text{Total watt - peak rating needed for PV modules} = \text{total watt - hours per day provided by PV modules/panel generation factor} \quad (3)$$

$$\text{Minimum number of PV panels required} = \text{total watt - peak rating needed for PV modules/the rated output watt - peak of the PV modules} \quad (4)$$

$$\text{Inverter size} = 1.3 \times \text{total watts of household appliances} \quad (5)$$

$$\text{Battery capacity (Ah)} = \text{total watt - hours per day used by appliances} \times \text{days of autonomy } (0.85 \times 0.6 \times \text{nominal battery voltage}) \quad (6)$$

The requirements for the solar PV system and parameters used for sizing the solar PV system are shown in **Table 1**.

4.4.2 Pico hydroelectricity

In-pipe hydropower (or pico hydroelectricity) represents a clean source of energy that focusses on recovering the energy used to supply water to households. The energy used to treat gray water can also be partially recovered by taking advantage of pumped storage. In pico hydroelectricity generation, a turbine is forced to rotate due to flow and pressure of water in a water pipe network. The rotating turbine is connected to a generator that generates electricity. This technology has been successfully implemented in various contexts [62–64]. The amount of power generated at the household level is relatively small [65–67]. However, this amount becomes significant to the utility provider if the technology is implemented in all houses as a national level project. Since in-pipe generators are preferred in the aboveground location with gravity-fed delivery pipelines, their position outside the villa's walls is ideal for maintenance and requires minimal changes to the system's operations when retrofitted. The in-pipe generators were designated to the main supply pipe based on the following criteria:

Parameter	Value	Solar PV system components	Specifications
Total watts for appliances	1500 W	**Household power consumption demands** Total watt-hours per day for appliances used Total watt-hours per day to be provided by PV modules for house appliances	55 kWh 71.5 kWh
Total watt-hours for lighting and appliances per day (excludes microwave ovens, cooker and any electrical machines, AC units, and any electric tools with power rating more than 1500 W)	55,000 Wh		
Panel generation factor	5.84		
Days of autonomy	2	**PV modules size** Total watt-peak rating needed for PV modules Minimum number of PV panels required	7.15kWh 200
Nominal battery voltage	12 V		
Nominal panel wattage	350 W		
Battery loss	0.85	**Size of inverter**	1800 W
Depth of discharge	0.6	**Size of battery**	11,000 Ah

Table 1.
Solar PV requirements and design parameters.

- Size: The main supply pipe is installed with the largest diameter among the gravity-fed plumbing pipes, thus allowing the installation of the largest possible generator to yield the highest possible power.

- Pressure: The pressure within the supply pipe is maintained by the roof pump at 20 psi, with a maximum pressure head at the ground floor before it reaches the first floor's outlet.

- Water quality: The water flowing through the main supply pipe is potable and does not carry the risk of containing debris, solid waste, or mixed fluids (e.g., oil from the kitchen) that may affect the operations of the turbine blades or cause damage to its physical structure.

Assumptions:

- The in-pipe generator's installation begins 0.25m above the ground level to avoid the interference with the distribution inlet of the first floor.

- The in-pipe generators operate with an efficiency of 65%, a practical estimation since in-pipe generation is a proven technology with high reliability and capacity, allowing it to maintain high efficiency even when facing variable flows.

- Cumulative water consumption occurs in the house for 8 hours a day.

The number of in-pipe generators to be installed is determined with consideration to:

- Spacing factor: In the current installation procedures, it is recommended by most suppliers to space the generators by a 4- diameter factor. The spacing is to prevent generator from affecting the functionality of the following one; such effect includes allowing the water turbulence to dissipate.

- The loss of power generated: When positioning the generators 4 diameters apart, the height of the water column causing the pressure head will decrease, resulting in a drop in the power produced between successive in-pipe generators.

Eqs. (7)–(10) were used to calculate hydropower generated at the household level. **Table 2** shows the design parameters for the hydropower generation.

$$\text{Reynolds number, Re} = (Q \times D)/(\upsilon \times A)\ (0.0017 \times 0.035)/\left((1.004 \times 10^{-6}) \times (9.62 \times 10^{-4})\right) \quad (7)$$

$$\text{Head loss across pipe, } hl = (16/\text{Re}) \times (L/D) \times \left(V^2/(2 \times g)\right) \quad (8)$$

$$\text{Head loss due to turbine, } ht = (z) - \left(V^2/(2 \times g)\right) - hl \quad (9)$$

$$\text{Power output per in-pipe generator} = Q \times ht \times W \times \eta \quad (10)$$

A typical multi-loop power network for the case study is shown in **Figure 8(a)**. **Figure 8(a)** shows multiple power flows from three different sources: main power grid, solar PV, and in-pipe hydropower. **Figure 8** also shows converters that facilitate the use of generated power in the household depending on whether the appliance requires AC or DC power. A typical multi-loop water network for the case study is shown in **Figure 8(b)**. **Figure 8(b)** shows multiple water flows from two different sources: main water supply from utility and flow of recycled gray water.

4.5 Gray water recycling

Gray water recycling was achieved by installing a reverse osmosis (RO) unit. Requirements for the feed water include the water pressure inside the pipes, the quantity of gray water to be treated, and the temperature of feed water. The size and quantity of membranes required to produce the desired volume of permeate were selected based on off-the-shelf units. **Table 3** shows a comparison of the properties of feed water data at the household level, reverse osmosis requirements, and local authority requirements for recycled gray water. From **Table 3**, it can be observed

Parameter	Value
Flow rate of the water (Q)	0.0017 m^3/s
Velocity of the water (V)	1.84 m/s
Diameter of the pipe (D)	0.35 m
Area of diameter (A)	9.62×10^{-4} m^2
Total length (L)	11.9 m
Dynamic viscosity (μ)	1.002×10^{-3} N s/m^2
Kinematic viscosity (υ)	1.004×10^{-6} m^2/s
Specific weight of water (W)	9790 N/m^3
Efficiency of in-pipe turbine (η)	0.65

Table 2.
Design parameters for the hydropower generation.

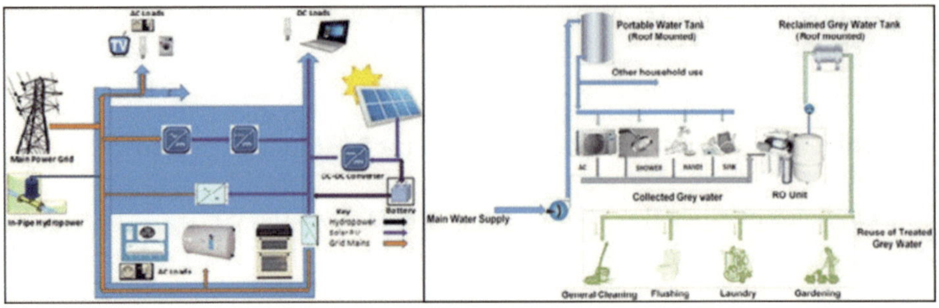

Figure 8.
(a) Multisource power loops and (b) multisource water loops for a typical household with solar energy and nano-hydropower.

that the feed water temperature is 28°C, i.e., 3° more than the required. Although this difference will have an effect on the quantity of treated water, the produced quantity is expected to be within the range of that stipulated by the local authority. The range of pressure (95–100 psi) is suitable for the reverse osmosis unit since it is high enough to allow all the solutes to be rejected from the solvent, thus creating treated gray water with the required specifications.

Since the total dissolved solids (TDS) of gray water is less than the TDS of what the feed water supposed to be, no filter was required for reducing the total dissolved solids. However, a sediment filter was required in order to remove dust, sand, suspended solids, particles, and rust, down to 5 μm. The hardness of the feed water is higher than that required. Hence, a hardness filter (water softener) was required in order to decrease the hardness of the gray water so that the reverse osmosis unit can function and produce treated gray water with the desired hardness limits. The concentration of chlorine in feed water is too high. Therefore, an activated carbon cartridge prefilter was required in order to minimize the level of chlorine in the gray water to an acceptable range before it goes to the granular activated carbon filter. The granular activated carbon filter was used to get rid of unpleasant chlorine, tastes, odors, cloudiness, colors, organic chemicals, sulfur, suspended particles, and dirt. This filter will also reduce the amount of the chlorine in the water to a desired value. Since the turbidity of feed water is high, a micro cartridge filter was used in order to reduce the turbidity of the gray water. This will help in achieving the required turbidity in the treated gray water.

4.6 Communication unit

Smart meter sensors allow water consumers to gain information on the water usages, on the water leaks, and on the quantity of water that is being drawn from the main grid and consumed in the house. This information is expected to allow users to control water leaks and abnormal usages. The implemented sensor is noncontact with water and makes use of the "pulse output" facility that is built in to most water meters. The smart meter sends information related to the water flow and the water quantity withdrawn from the grid to the control unit using a wireless connection.

The control unit sends this information to the user's phone at regular predetermined intervals to indicate real-time water consumption from the water meter as well as provide visual and sound alerts if there is an incident such as an abnormal water usage. Water leak detection is programmed to notify the users of the house when there is a water leak at a specific point in the house by producing a sound alert and an SMS for the user to take an action. If the user does not take any action

Constituent	Gray water	Reverse osmosis requirements	Requirements of local authority
Total solids	700 PPM	<1000 PPM	≤5 PPM
Chloride	50 PPM	<0.1 PPM	≥1 exiting treatment system, ≥0.2 at user end
Alkalinity (as $CaCO_3$)	100 PPM	<50 PPM	N/A
Biochemical oxygen demand (BOD5)	200 PPM	N/A	≤10 PPM
Turbidity	2–5 NTU	<1 NTU	≤5 NTU
Temperature	28°C	25°C	N/A
Pressure	95–100 psi	21.8–101.526 psi	N/A
pH	7.5	3–11	6–9
E. coli	0	0	Non-detectable
Dissolved oxygen in reclaimed water	N/A	N/A	≥2 PPM
Color	N/A	N/A	≤20 Hazen unit
Threshold odor number (TON)	N/A	N/A	≤100
Ammoniacal nitrogen	N/A	N/A	≤1 mg/l as N
Synthetic detergents	N/A	N/A	≤5 mg/l

Table 3.
Comparison of feed water data with reverse osmosis and local authority requirements.

after 10 min, a signal will be sent to the control unit via a wireless connection. In return, the control unit will send a message to the user's phone asking him/her to take an action as a soon as possible. The smart valves are smart due to their ability to open and close automatically based on specified conditions and commands from the control unit. As an example, pH sensors are used to measure the pH of the treated gray water. The pH sensor plays an important role in blending the treated gray water. Water blending was required to ensure that the pH of the treated gray water is suitable for garden watering and other applications. If the pH of the treated gray water is less or more than the required value, the pH sensor sends a signal to the control unit to open the smart valve in order to allow potable water to flow from the potable water tank to treated gray water tank. This flow is expected to neutralize the pH of the treated gray water so that the gray water is suitable for different purposes.

5. Construction and testing of the prototype

5.1 Prototype materials and assembly

In order to realize the functionality of the proposed design, a prototype was constructed. Components for the smart water/power system were assembled from standard components available off the shelf. The prototype construction included the physical structure and the control system. The physical structure consists of the positioning of devices such as in-pipe generators and flowmeters in addition to the plumbing network. The control system was assembled from Arduino Mega boards, and the control actions were programmed using the C language and the Arduino software. Prototype components included pumps, wastewater tank, treated water tank, a battery, storage tanks, reverse osmosis unit, in-pipe electricity generator, flowmeters, smart valves, photovoltaic panels, pipes, flexible hose, pipe fittings, sensors, pH meter, Arduino Mega board, and a GSM shield board. The selected pump has a voltage of 12 V, so it can be connected to a battery of 12 V, since this 12 V battery will supply the prototype with electricity. The battery was continuously charged by a 100 W solar panel. The suction lift of this pump is 1.2 m, which means that this pump will be able to pump the water to the storage tank at a height of 0.91 m and circulate water in the pipe network for simulated water use in the house. The voltage of the in-pipe electricity generator is 5 V, which is compatible with the battery and to the Arduino, which can take a maximum of 5 V. The in-pipe electricity generator has a water pressure of 0.05 MPa, which is suitable for the pipe dimensions used to construct the prototype. Reverse osmosis unit was chosen to have specifications that depend on the flow rate and the pressure of the water within the pipe networks. The reverse osmosis unit used in the prototype has a maximum capacity of 280 L/day, which is sufficient for prototype demonstrations. In addition to measuring the flow of water, flowmeters were used as devices to detect the leakages within the pipes. This was done by installing two flowmeters at a junction point or "leak hole" to simulate water leakages. Differences in the flowmeter readings would indicate a leakage. The function of smart shut-off valve was linked to that of flowmeters in such a way that when there is a difference between the values of flowmeters at leak points the smart shut-off valve will stop water flow in the pipe. **Figure 9** shows a pictorial view of the assembled prototype as well as a sample of main prototype elements.

Experiments with in-pipe hydropower demonstrated that electricity was produced and used to light a bulb in the prototype. Initial testing of the water section of the prototype included running potable water through the prototype with normal flows. It was observed that the reverse osmosis unit was taking too long to treat gray water, due to low pressure. The low efficiency and slow speed processing of the reverse osmosis unit was identified as a bottleneck in processing gray water. Gray water was supplied to the reverse osmosis unit to determine if the quality of the treated gray water was good enough for its intended purposes. Treated gray water parameters were found to be 4.3 ppm for total dissolved solids, 10 ppm biochemical oxygen demand, and a pH of 7. These values are close to the treated gray water requirements as stipulated by the local authority. Further testing of the prototype was done with the simulated pipe leak and running both portable and gray water in the prototype. The results of this test showed that the proposed method for identifying water leaks was suitable since differences in flowmeter reading were observed when a simulated leak occurs. The prototype was also able to send user notifications (to a smart phone) regarding the condition of the treated gray water and the presence of a leak.

6. Case study results

After experimenting with the prototype, the main components of the nanogrid were installed parallel to the existing infrastructure at the case study villa. The parallel installation was designed to replicate functionality of the prototype as well as for minimum interruption of normal household activities as per the requirements of the household owner. In addition, the parallel installation allowed easy removal of installed components after data collection. The following subsections summarize the analysis of data collected from the case study.

Figure 9.
Assembled prototype and a sample of the main components.

6.1 In-pipe hydropower generation

Table 4 shows the expected power to be generated from the in-pipe hydropower generators installed on the main water supply line.

From **Table 4** the insertion of four in-pipe generators provides the house with 2.59 kWh of electricity, for 8 h of water consumption per day. Taking in account assumptions and constraints, four in-pipe generators were positioned as follows:

- Supply in-pipe generator 1 at elevation 25 mm, at the lowest point considering the assumed 25 mm clearance.

- Supply in-pipe generator 2 at elevation 165 mm, 140 mm apart from generator 1 after spacing an amount of 4 diameters.

- Supply in-pipe generator 3 at elevation 305 mm, 140 mm apart from generator 2 after spacing an amount of 4 diameters.

- Supply in-pipe generator 4 at elevation 445 mm, 140 mm apart from generator 1 after spacing an amount of 4 diameters.

Since the water tank will be placed on the roof, power and placement of the in-pipe generators for the recycled gray water will be a replica of that shown in **Table 4**.

In-pipe generator	Elevation (mm)	Distance from the roof (m)	Losses across pipe (m)	Turbine losses (m)	Power output (kW)	Estimated daily output (kWh)
1	25	11.90	0.015	11.70	0.0819	0.655
2	165	11.76	0.015	11.60	0.0813	0.650
3	305	11.62	0.014	11.40	0.0806	0.645
4	445	11.48	0.014	11.29	0.0793	0.634
Total					0.3231	2.585

Table 4.
Expected power output from case study villa.

The total power generated from the recycled gray water's main water supply pipe is 2.59 kWh. Therefore, the total potential power generated from in-pipe hydropower generators at the case study villa is 5.17kWh (i.e., 155.06 kWh per month). These values agree with other research findings [68–72].

6.2 On-site solar PV power generation

The specifications for the solar PV power generated on-site are shown in **Table 5**. From **Table 5**, solar PV panels cover 40% of the total roof area. This leaves enough space for pumped hydro tank, AC units, and other equipment. From **Figure 6**, 60% of the energy at the household level is used for air conditioning, and this will be provided from the main grid power.

6.3 Water savings

Water saving calculations per month for the case study are shown in **Table 6**. From **Table 6**, total freshwater savings amounts to 25% per month due to reuse of on-site treated gray water. This calculation takes into account losses in gray water collection system, efficiency of equipment used, as well as water loses during treating and recycling of gray water on-site. Efficiency improvements in the gray water collection network and gray water treatment systems can increase the total amount of reusable gray water with additional benefits at the household level.

System component	Quantity	Comment
Number of panels	200	At 350 W
Performance ratio	0.62	Experimentally estimated loses {AC/DC conversion loses, shading effect, dust, temperature effects}
Efficiency of panels	20%	Manufacturer specification
Power output per month from PV system	2145 kWh	Since the basic appliances require 1650 kWh per month, the solar PV system is more than able to meet the power requirements for these appliances
Total power required for the case study villa per month	9251.48 kWh	Estimated from case study data (see **Figure 5**)
Power from pico hydro generation	155.06 kWh	Includes generation from the main water supply and treats gray water circulation at a household
Total on-site generated power per month	2300.06 kWh	Pico hydro + solar PV
Percentage (%) contribution of power generated from solar PV	22.86%	This percentage (%) is estimated for 1-month operation. Due to the intermittent nature of solar energy and its dependency on climate, on the long run, this percentage (%) is expected to decrease
Percentage (%) contribution of power generated from pico hydro	2%	
Percentage (%) contribution of power generated on-site	24.86%	This percentage (%) is estimated for 1-month operation. Due to the intermittent nature of solar energy and its dependency on climate, on the long run, this percentage (%) is expected to decrease

Table 5.
On-site power generation.

	Quantity (liters per month)	**Gray water generated** (liters per month)	**Recycled grey water** (liters per month)	**Freshwater savings** (liters per month)
Total freshwater use in the case study	**49755.49**			
Toilet flushing	5473.12			
Gardening	2487.77			
Floor cleaning	1492.66			
Car washing	995.11			
Cooking and drinking	995.11			
Total unrecoverable	**11443.77**			
Bathing	21394.86			
Personal washing (hands and face on sinks)	10448.65	6791.62		
Dish washing	4975.55	3234.11		
Clothe washing	1492.66	970.23		
Total air-conditioning requirements for 1200 m2 (three-storey villa at 400 m^2 per floor area) = 361.34 kW. For 3.5 kW, 0.125 liters per kWh of condensate is collected from the case study villa = 45.17 l Total condensate collected per day (considering 8 working hours) = 361.36 l AC condensate	10840.8	6504.48		
Total gray water generated (losses in gray water collection system)		**17500.44**		
Total recycled gray water (loses during treating and recycling)			**9975.25**	
Total monthly water savings in case study				**9975.25**
Percentage (%) freshwater savings				**20%**

Table 6.
Water saving calculations based on case study villa data.

6.4 Economic analysis

The economic analysis was done for the case study villa. The proposed nanogrid at the household level is viewed as the responsibility of the house owner. Therefore, investment costs are borne by the household owner for individual villas. While economic benefits are important to the household owner, the utility company is also interested in how the proposed nanogrid can be used to mitigate the effects of climate change through a reduction of the energy and water consumption. The following equation was used in the cost analysis.

$$\text{Simple Payback Period} = \text{Investment}/\text{Annual Savings} \tag{11}$$

$$\text{Discounted Payback Period} = \{ -\ln(1-(\text{investment amount} \times \text{discount rate})/(\text{annual savings})\}/\{\ln(1+\text{rate})\} \quad (12)$$

Monthly water bill as per local utility company tariff = (20) 4.4 + (49.76–20) 5.4 = QR 248.7

Monthly water bill savings = QR 248.7 – QR {(9.975) 4.4} = QR 204.81.

Monthly power bill as per local utility company tariff = QR (2000)0.08 + QR (4000–2000)0.09 + QR (6000–4000)0.1 QR+(9251.48–6000)0.12 QR = QR 930.18.

Monthly power bill savings = QR 930.18 – {(2000) 0.08 + QR (4000–2000)0.09 + QR (6000–4000)

1+ QR (6951.42–6000)0.12 = QR 276.01.

Total monthly saving (water and power bill) = QR 480.82.

Table 7 shows the cost saving parameters in Qatari Riyals for retrofitting system components to enable multi-utility loops at the household level.

The payback periods at a discounting rate of 10% from the case study household owner's perspective are shown in **Tables 8** and **9**. From **Tables 8** and **9**, it can be inferred that the payback period based on the simple payback calculation for smart water metering (7.25 years) is less than that of smart energy metering (9.59 years). As expected, the value of the discounted payback period is always higher than that from simple payback period. Further improvements to these investment paybacks can be realized by improving the efficiency of the gray water collection and treatment system.

7. Toward a smart water/power microgrid

The smart water/power concept discussed in this chapter is at the household level. In the proposed implementations, smart water/power nanogrid is the smallest unit of a dual-purpose smart water/power distribution network that is capable of independent operation to support the main grid water and power distribution and utilization at the household level. This essentially represents a smart water/power nanogrid composed of local small-scale generators of water (recycled gray water) and electricity (solar PV and in-pipe hydropower electricity). The proposed smart water/power nanogrid can be used to conserve water end use at the household level, thus relieving the strain on the main water grid as well as to supplement power supply at the household level, thus relieving the strain on the main power grid.

Water	**Annual cost saving QR**
Total cost for the multisource water circulation system and recycling units (cost of smart water system components + pipe network + pumped storage tank + RO unit and accessories)	19,661.76
Annual water savings due to reuse of recycled water and smart metering	2457.72
Power	
Total cost of the multisource power loop system (cost of smart meter, solar PV modules, inverters, pico hydro components, and battery bank)	46,369.68
Annual power savings due to on-site power generation	3312.12
Total annual savings (water and power bills)	5769.84

Table 7.
Cost saving parameters.

Payback period			**7.25 years**	
Discounted payback period			**12.71 years**	
Cash flow return rate			**6.79% per year**	
	Cash flow	**Net cash flow**	**Discounted cash flow**	**Net discounted cash flow**
Year 0	QR-19,661.76	QR-19,661.76	QR-19,661.76	QR-19,661.76
Year 1	QR2,457.72	QR-17,204.04	QR2,234.29	QR-17,427.47
Year 2	QR2,531.45	QR-14,672.59	QR2,092.11	QR-15,335.36
Year 3	QR2,607.40	QR-12,065.19	QR1,958.97	QR-13,376.39
Year 4	QR2,685.62	QR-9379.58	QR1,834.31	QR-11,542.07
Year 5	QR2,766.19	QR-6613.39	QR1,717.58	QR-9824.49
Year 6	QR2,849.17	QR-3764.22	QR1,608.28	QR-8216.21
Year 7	QR2,934.65	QR-829.57	QR1,505.94	QR-6710.27
Year 8	QR3,022.69	QR2,193.11	QR1,410.11	QR-5300.16
Year 9	QR3,113.37	QR5,306.48	QR1,320.37	QR-3979.79
Year 10	QR3,206.77	QR8,513.25	QR1,236.35	QR-2743.45

Table 8.
Payback period for smart water metering at the household level.

Payback period			**9.59 years**	
Discounted payback period			**23.18 years**	
Cash flow return rate			**0.76% per year**	
	Cash flow	**Net cash flow**	**Discounted cash flow**	**Net discounted cash flow**
Year 0	QR-36,369.68	QR-36,369.68	QR-36,369.68	QR-36,369.68
Year 1	QR3,312.12	QR-33,057.56	QR3,011.02	QR-33,358.66
Year 2	QR3,411.48	QR-29,646.08	QR2,819.41	QR-30,539.25
Year 3	QR3,513.83	QR-26,132.25	QR2,639.99	QR-27,899.26
Year 4	QR3,619.24	QR-22,513.01	QR2,471.99	QR-25,427.27
Year 5	QR3,727.82	QR-18,785.19	QR2,314.68	QR-23,112.59
Year 6	QR3,839.65	QR-14,945.53	QR2,167.39	QR-20,945.20
Year 7	QR3,954.84	QR-10,990.69	QR2,029.46	QR-18,915.74
Year 8	QR4,073.49	QR-6917.20	QR1,900.31	QR-17,015.43
Year 9	QR4,195.69	QR-2721.50	QR1,779.38	QR-15,236.05
Year 10	QR4,321.57	QR1,600.06	QR1,666.15	QR-13,569.89

Table 9.
Payback period for smart energy metering at the household level.

A network of smart nanogrids could be interconnected into a microgrid without any central entity. Thus, the proposed smart water/power nanogrid for a single household can be connected to another nanogrid of a neighboring house. A group of interconnected nanogrids can be configured into a microgrid, and a group of microgrids can be configured into a mesogrid as shown in **Figure 10**.

When extending the proposed nanogrid concept to microgrid and mesogrid, each smart residential unit is viewed as a single node with interconnectivity. Such interconnectivity provides the household users with water/power availability and

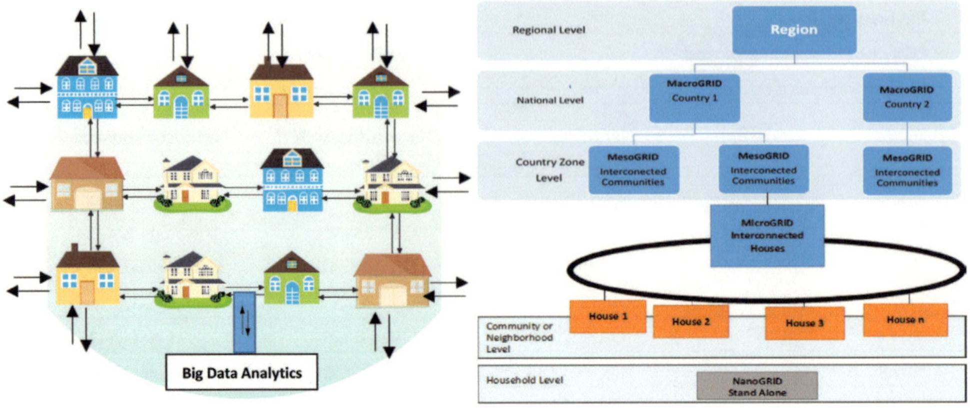

Figure 10.
Interconnected households.

consumption status in neighboring units through real-time user notifications. The information would include excess power and water generation in a neighboring node and neighbor's willingness to share such excess, along with the sharing conditions. The ability of the system to share provides the unit owner with the option of setting usage priority to or from the main grid or the mesogrid as preferred when the unit's power and water generation do not satisfy the user's demands. Such access promotes the status of the residential unit to that of a prosumer, a producer, and a consumer simultaneously. In the evolution process of the nanogrid toward the smart grid, the scales of water and power production are expected to increase. For example, simple water recycling and reuse at nanogrid can be expanded to freshwater production sources (small-scale water desalination, bigger water treatment structures, and water reservoirs) supported by zero water discharge policies. Power production scales can evolve from the rooftop solar PV panels to solar PV arrays. The increase in scales will help in stabilizing the water/power decentralization plans.

A node's sharing conditions are dependent on the individual prosumer's discretion in terms of selling cost, quantity, and the threshold of personal consumption. The exchange of information between various nodes in the microgrid or mesogrid and the level of access is to be governed through applications of energy semantic networking [73]. In the energy semantic concept, the system's "big data" enables it to function efficiently by operating with a high level of context awareness. The contextual awareness of the system will guarantee that the shared network between the nodes is capable of interpreting information and user commands as well as communicating them. In addition to data and command processing, the systems elevate user concern when determining the source of power and water by constantly indicting the optimum alternative based on the originator consumption and the varying nodes and main grid availability and pricing. Operating at the community level, the mesogrid enabler will be a semantically capable software, which receives data from the various sensors, devices, user preferences, and other data sources to allow user control over the system's hardware without clashing with the systems operations as well as water and energy consumption patterns. This evolution of the proposed smart water nanogrid provides the following advantages to the resultant smart water/power grid: operational excellence, environmental compliance, grid reliability, safety in operations, energy and water access, security of water and energy supply, consumer participation, grid resilience in normal operation, and disaster situations.

8. Concluding remarks

In this chapter, a smart water/power technological solution for residential areas has been analyzed based on case study specifications and operating conditions. The solution includes on-site power generation using PV modules, in-pipe hydropower generation from water supply and distribution networks, treatment of gray water via reverse osmosis technology, and reuse of treated gray water at the household level. Management and control of the water/power technological solution at the household level was done through a centralized controller. Coordination of the water/power components was achieved through networking and communication capabilities facilitated by the controller and GSM technology. This coordination provides the user with real-time data and information about water and power consumptions, flows, and water quality. The user can then make decisions and control actions based on the data and information provided. This allows the user to be in total control of water and power consumptions within their residential area. Although the analysis is based on one case study villa, the same concepts can be applied to other villas and large-scale residential units such as compounds and residential towers without loss of much generality. Experiments with the developed prototype showed that the proposed system is able to (1) generate, store, and provide information that can be used to control water/power consumptions at the household level, (2) allow two-way flows of data and information on the current state of power and water, and (3) treat and recycle treated gray water for use at the villa. The proposed system is expected to reduce freshwater consumption by 20% and power consumption by 25% in residential villas in Qatar. The research study has shown that the in-pipe hydro system can generate small amounts of electricity and contributes to 5% of on-site power generation based on the configurations discussed in this chapter. This contribution is expected to rise in large-scale applications. Payback analysis shows that the combined smart water power nanogrid is moderately attractive and yet environmentally friendly by nature. Prototype tests demonstrated that the proposed system could function properly when implemented in homes. Improvements in gray water collection and treatment processes could result in more benefits. A future improvement of the prototype is to devise the capability to identify the number of leaks as well as determine the exact location of the leaks. Results of such findings can shed light on the further contribution of nanogrids in reducing (a) water losses and (b) water and energy consumptions, thus making homes more energy efficient.

Author details

Dana Alghool, Noora Al-Khalfan, Stabrag Attiya and Farayi Musharavati*
Department of Mechanical and Industrial Engineering, College of Engineering, Qatar University, Doha, Qatar

*Address all correspondence to: farayi@qu.edu.qa

References

[1] Marlow DR, Moglia M, Cook S, Beale DJ. Towards sustainable urban water management: A critical reassessment. Water Research. 2013;**47**(20):7150-7161

[2] Asif M, Muneer T. Energy supply, its demand and security issues for developed and emerging economies. Renewable and Sustainable Energy Reviews. 2007;**11**(7):1388-1413

[3] Siddiqi A, Anadon LD. The water–energy nexus in Middle East and North Africa. Energy Policy. 2011;**39**(8):4529-4540

[4] Rao P, Kostecki R, Dale L, Gadgil A. Water-energy nexus: The role of technology and engineering. Annual Review of Environment and Resources. 2017;**42**(1)

[5] WWAP (United Nations World Water Assessment Programme). The United Nations World Water Development Report 2015: Water for a Sustainable World. Paris: UNESCO; 2015

[6] United Nations World Water Day 2014. UN Stresses Water and Energy Issues retrieved October 2017 from https://unu.edu/media-relations/releases/wwd2014-un-stresses-water-energy-issues.html

[7] DeNicola E, Aburizaiza OS, Siddique A, Khwaja H, Carpenter DO. Climate change and water scarcity: The case of Saudi Arabia. Annals of Global Health. 2015;**81**(3):342-353

[8] The European Electricity Grid Initiative (EEGI) Roadmap 2010-18 and Implementation Plan 2010-12. Presented at SET-PLAN conference; 2010

[9] Li F et al. Smart transmission grid: Vision and framework. IEEE Transactions on Smart Grid. 2010;**1**(2):168-177

[10] Grijalva S, Tariq MU. Prosumer-based smart grid architecture enables a flat, sustainable electricity industry. In: 2011 IEEE PES Innovative Smart Grid Technologies (ISGT). IEEE. 2011, January. pp. 1-6

[11] Mathiesen BV, Lund H, Connolly D, Wenzel H, Østergaard PA, Möller B, et al. Smart energy systems for coherent 100% renewable energy and transport solutions. Applied Energy. 2015;**145**:139-154

[12] Bahrami S, Sheikhi A. From demand response in smart grid toward integrated demand response in smart energy hub. IEEE Transactions on Smart Grid. 2016;**7**(2):650-658

[13] Dominković DF, Bačeković I, Sveinbjörnsson D, Pedersen AS, Krajačić G. On the way towards smart energy supply in cities: The impact of interconnecting geographically distributed district heating grids on the energy system. Energy. 2017

[14] Fadaeenejad M, Saberian AM, Fadaee M, Radzi MAM, Hizam H, AbKadir MZA. The present and future of smart power grid in developing countries. Renewable and Sustainable Energy Reviews. 2014;**29**:828-834

[15] Galo JJ, Macedo MN, Almeida LA, Lima AC. Criteria for smart grid deployment in Brazil by applying the Delphi method. Energy. 2014;**70**:605-611

[16] Cardenas JA, Gemoets L, Rosas JHA, Sarfi R. A literature survey on smart grid distribution: An analytical approach. Journal of Cleaner Production. 2014;**65**:202-216

[17] Siano P. Demand response and smart grids—A survey. Renewable and Sustainable Energy Reviews. 2014;**30**:461-478

[18] Aghaei J, Alizadeh MI. Demand response in smart electricity grids equipped with renewable energy sources: A review. Renewable and Sustainable Energy Reviews. 2013;**18**:64-72

[19] Verbong GP, Beemsterboer S, Sengers F. Smart grids or smart users? Involving users in developing a low carbon electricity economy. Energy Policy. 2013;**52**:117-125

[20] Boroojeni KG, Amini MH, Iyengar SS. Reliability in smart grids. In: Smart Grids: Security and Privacy Issues. Springer International Publishing; 2017. pp. 19-29

[21] Moslehi K, Kumar R. A reliability perspective of the smart grid. IEEE Transactions on Smart Grid. 2010;**1**(1):57-64

[22] Brown RE. Impact of smart grid on distribution system design. In: Power and Energy Society General Meeting-Conversion and Delivery of Electrical Energy in the 21st Century, 2008 IEEE. IEEE. 2008. pp. 1-4

[23] Shao S, Pipattanasomporn M, Rahman S. Grid integration of electric vehicles and demand response with customer choice. IEEE Transactions on Smart Grid. 2012;**3**(1):543-550

[24] Hajebi S, Song H, Barrett S, Clarke A, Clarke S. Towards a reference model for water smart grid. International Journal of Advances in Engineering Science and Technology. 2013;**2**(3):310-317

[25] Lee SW, Sarp S, Jeon DJ, Kim JH. Smart water grid: The future water management platform. Desalination and Water Treatment. 2015;**55**(2):339-346

[26] Spinsante S, Squartini S, Gabrielli L, Pizzichini M, Gambi E, Piazza F. Wireless m-bus sensor networks for smart water grids: Analysis and results. International Journal of Distributed Sensor Networks. 2014;**10**(6):579271

[27] Dawoud MA. The role of desalination in augmentation of water supply in GCC countries. Desalination. 2005;**186**(1-3):187-198

[28] Darwish MA, Al-Najem NM, Lior N. Towards sustainable seawater desalting in the Gulf area. Desalination. 2009;**235**(1):58-87

[29] Khan SUD, Khan SUD, Haider S, El-Leathy A, Rana UA, Danish SN, et al. Development and techno-economic analysis of small modular nuclear reactor and desalination system across Middle East and North Africa region. Desalination. 2017;**406**:51-59

[30] Duan C, Chen B. Energy–water nexus of international energy trade of China. Applied Energy. 2017;**194**:725-734

[31] Wang S, Cao T, Chen B. Urban energy–water nexus based on modified input–output analysis. Applied Energy. 2017;**196**:208-217

[32] Fang D, Chen B. Linkage analysis for the water–energy nexus of city. Applied Energy. 2017;**189**:770-779

[33] Ramachandra T, Ramachandra T, Karunasena G, Karunasena G. Emerging issues in the built environment sustainability agenda. Built Environment Project and Asset Management. 2017;7(4):350-352

[34] Rizzo A. Rapid urban development and national master planning in Arab Gulf countries. Qatar as a case study. Cities. 2014;**39**:50-57

[35] Qatar General Electricity & Water Corporation. Statistics Report 2012. Retrieved October 2017, from: https://www.km.com.qa/MediaCenter/Publications/Statisticesreport-2012.pdf

[36] U.S. Environmental Protection Agency. 2009 U.S. Greenhouse Gas Inventory Report. Retrieved October 2016 from: https://www.epa.gov/ghgemissions/us-greenhouse-gas-inventory-report-archive

[37] Karlsson P-O, Decker C, Moussall J. Energy efficiency in the UAE: Aiming for sustainability. Retrieved October 2017, from: https://www.strategyand.pwc.com/reports/energy-efficiency-in-uae

[38] Hodge BK. Alternative Energy Systems and Applications. John Wiley & Sons; 2017

[39] Al-Karaghouli A, Renne D, Kazmerski LL. Solar and wind opportunities for water desalination in the Arab regions. Renewable and Sustainable Energy Reviews. 2009;**13**(9):2397-2407. DOI: 10.1016/j.rser.2008.05.007

[40] Zhang Q, Mclellan BC, Utama NA, Tezuka T, Ishihara K. A Methodology for Designing Future Zero-Carbon Electricity Systems with Smart Grid and Its Application to Kansai Area, Japan. Design for Innovative Value Towards a Sustainable Society. 2012. pp. 50-54. DOI: 10.1007/978-94-007-3010-6_11

[41] Cobacho R, Arregui F, Gascó L, Cabrera E. Low-flow devices in Spain: How efficient are they in fact? An accurate way of calculation. Water Science and Technology: Water Supply. 2004;**4**(3):91-102

[42] Vieira P, Jorge C, Covas D. Assessment of household water use efficiency using performance indices. Resources, Conservation and Recycling. 2017;**116**:94-106. DOI: 10.1016/j.resconrec.2016.09.007

[43] Siddiqi A, de Weck OL. Quantifying end-use energy intensity of the urban water cycle. Journal of Nanogrid Systems. 2013;**19**(4):474-485

[44] Loh M, Coghlan P. Domestic Water Use Study: Perth. Perth, Western Australia: Water Corporation; 2003

[45] Willis RM, Stewart RA, Panuwatwanich K, Capati B, Giurco DP. Gold coast domestic water end use study. Water. 2009;**36**(6):79-85

[46] Beal C, Stewart RA, Huang TT, Rey E. SEQ residential end use study. Journal of Australian Water Association. 2011;**38**(1):80-84

[47] Matos C, Teixeira CA, Duarte AALS, Bentes I. Domestic water uses: Characterization of daily cycles in the north region of Portugal. Science of the Total Environment. 2013;**458**:444-450

[48] Cole G, Stewart RA. Smart meter enabled disaggregation of urban peak water demand: Precursor to effective urban water planning. Urban Water Journal. 2013;**10**(3):174-194. DOI: 10.1080/1573062x.2012.716446

[49] Omaghomi T, Buchberger S. Estimating water demands in buildings. Procedia Engineering. 2014;**89**:1013-1022

[50] Willis RM, Stewart RA, Giurco DP, Talebpour MR, Mousavinejad A. End use water consumption in households: Impact of socio-demographic factors and efficient devices. Journal of Cleaner Production. 2013;**60**:107-115

[51] Matos C, Pereira S, Amorim EV, Bentes I, Briga-Sá A. Wastewater and greywater reuse on irrigation in centralized and decentralized systems—An integrated approach on water quality, energy consumption and CO_2 emissions. Science of the Total Environment. 2014;**493**:463-471

[52] Hunt D, Rogers C. A benchmarking system for domestic water use. Sustainability. 2014;**6**(5):2993-3018. DOI: 10.3390/su6052993

[53] Willis RM, Stewart RA, Panuwatwanich K, Williams PR, Hollingsworth AL. Quantifying the influence of environmental and water conservation attitudes on household end use water consumption. Journal of Environmental Management. 2011;**92**(8):1996-2009

[54] Lee M, Tansel B, Balbin M. Influence of residential water use efficiency measures on household water demand: A four year longitudinal study. Resources, Conservation and Recycling. 2011;**56**(1):1-6. DOI: 10.1016/j.resconrec.2011.08.006

[55] Carragher BJ, Stewart RA, Beal CD. Quantifying the influence of residential water appliance efficiency on average day diurnal demand patterns at an end use level: A precursor to optimised water service nanogrid planning. Resources, Conservation and Recycling. 2012;**62**:81-90. DOI: 10.1016/j.resconrec.2012.02.008

[56] Carr G, Potter RB. Towards effective water reuse: Drivers, challenges and strategies shaping the organisational management of reclaimed water in Jordan. The Geographical Journal. 2013;**179**(1):61-73

[57] Gourbesville P. Why smart water journal? Smart Water. 2016;**1**(1):1-2

[58] Mutchek M, Williams E. Moving towards sustainable and resilient smart water grids. Challenges. 2014;**5**(1):123-137

[59] Harris R. WaterSmart Toolbox: Giving customers online access to real-time water consumption is the right tool for water efficiency success. In: Presented at the Smart Water Innovations Conference and Exposition, 6-8 October; Las Vegas, NV. Available from: http://watersmartinnovations.com/2010/PDFs/10-T-1001.pdf. 2010 [Accessed: 10 October 2016]

[60] Weng KT, Lim A. Pursuit of a smart water grid in Singapore's water supply network. In: Presented at the American Water Works Association Sustainable Water Management Conference, 18-21 March; Portland, OR. 2012

[61] Alghool D, Al-Khalfan N, Attiya S. Design of a smart water/power system for households. Project Report. Qatar University; 2016

[62] Sai Y, Cohen S, Vogel RM. The impacts of water conservation strategies on water use: Four case studies. Journal of the American Water Resources Association (JAWRA). 2011;**47**(4):687-701. DOI: 10.1111/j.1752-1688.2011.00534.x

[63] Gabbar HA. Engineering design of green hybrid energy production and supply chains. Environmental Modelling & Software. 2009;**24**(3):423-435

[64] Gabbar HA, Bondarenko D, Hussain S, Musharavati F, Pokharel S. Building thermal energy modeling with loss minimization. Simulation Modelling Practice and Theory. 2014;**49**:110-121

[65] Casini M. Harvesting energy from in-pipe hydro systems at urban and building scale. International Journal of Smart Grid and Clean Energy. 2015;**4**:316-327

[66] Porkumaran K, Tharu RP, Sukanya S, Elezabeth VV, Gowtham N. Micro in-pipe hydro power plant for rural electrification using LabVIEW. In: 2017 International Conference on Innovations in Green Energy and Healthcare Technologies (IGEHT); IEEE. 2017. pp. 1-5

[67] Rakesh C, Nallode C, Adhvaith M, Anwin TJ, Krishna AA, Rakesh C, et al. Theoretical study and performance test of lucid spherical turbine. International Journal. 2016;**3**:418-423

[68] Haidar AM, Senan MF, Noman A, Radman T. Utilization of pico hydro generation in domestic and commercial loads. Renewable and Sustainable Energy Reviews. 2012;**16**(1):518-524

[69] Williams AA, Simpson R. Pico hydro-reducing technical risks for rural electrification. Renewable Energy (Elsevier). 2009;**34**:1986-1991

[70] Smits M, Bush SR. A light left in the dark: The practice and politics of pico-hydropower in the Lao PDR. Energy Policy (Elsevier). 2010;**38**:116-127

[71] Abbasi T, Abbasi SA. Small hydro and the environmental implications of its extensive utilization. Renewable and Sustainable Energy Reviews (Elsevier). 2011;**15**:2134-2143

[72] Fodorean D, Member SL, Miraoui A. Generator solutions for stand alone pico-electric power plants. Proceedings of IEEE Conference on Electric Machines and Drives. 2009:434-438

[73] Gabbar HA, Eldessouky AS. Energy semantic network for building energy management. Intelligent Industrial Systems. 2015;**1**(3):213-231

Chapter 8

Comparative Study and Simulation of Different Maximum Power Point Tracking (MPPT) Techniques Using Fractional Control & Grey Wolf Optimizer for Grid Connected PV System with Battery

Mohamed Ahmed Ebrahim and R.G. Mohamed

Abstract

This chapter presents the comparative analysis between perturb & observe (P&O), incremental conductance (Inc Cond), and fractional open-circuit voltage (FOCV) algorithms using fractional order control & a new meta-heuristic called Grey Wolf optimizer (GWO) for extracting the maximum power from photovoltaic (PV) array. PV array systems are equipped with maximum power point tracking controllers (MPPTCs) to maximize the output power even in the case of rapid changes of the panel's temperature and irradiance. In this chapter, three cost effective MPPTCs are introduced: FOCV, P&O and Inc. Cond. The output voltage of the array is boosted up to a higher value so it can be interfaced to the local medium voltage distribution network.

Keywords: maximum power point tracking, grid connected photovoltaic, battery, grey wolf optimizer, boost converter, fractional order PI control

1. Introduction

Solar photovoltaic array system (SPVS) is one of the most prominent sources of electrical energy. SPVS is environmentally friendly and as a result there are no CO_2 emissions [1]. The energy dilemma represents in increasing the electricity production from the resources matching with the environment requires searching for new green, renewable, and sustainable ideas. SPVS along with wind turbines and fuel cells are possible innovative solutions for this dilemma [2, 3]. It was stated [4] that solar power capacity has expanded rapidly to 227 GW by the end of 2017 with a global growth rate of 26% which was higher than in 2016 (16.4%). Solar energy production was around 11% of the global renewable generation capacity, and increasing [4]. The installed capacity of SPVS in Egypt was about 1% of the total electricity production from renewable energy sources in March 2018 [5]. In SPVSs, the operation at the maximum power point (MPP) is necessity. As a result for this, various MPP tracking (MPPT) techniques are developed, investigated and implemented in the last decades [6]. One of the most powerful techniques is the fractional order PID (FOPID) based MPPT controller (MPPTC) [7]. These kinds of controllers have merged the merits of classical MPPTCs and PID controller [8]. However, FOPID based MPPTCs require efficient tuning methods to improve the dynamic response especially in the presence of system disturbances [9]. Thanks to Meta heuristic optimization techniques that can be employed to significantly tune MPPTCs.

In this chapter, design methodology for three different types of MPPTCs using fractional order PID (FOPID) is summarized.

This chapter also presents a new meta-heuristic called Grey Wolf Optimizer (GWO) inspired by grey wolves. The GWO algorithm mimics the leadership hierarchy and hunting mechanism of grey wolves in nature. Three main steps of hunting, searching for prey, encircling prey, and attacking victim, are implemented in this algorithm [10].

2. Practical case study

This study present PV solar power plant connected to the Egyptian national grid and installed in Kom Ombo, Aswan, Egypt. This power plant will have a total capacity of 20 MW which can be considered one of largest Egyptian PV project. The PV system is constructed using MATLAB/SIMULINK to mimic the actual system. Different scenarios are considered to test the effectiveness of the proposed MPPTCs. These scenarios include small as well as large environmental conditions changes.

3. Proposed system simulation

The simulation of grid-connected PV system with battery contains various simulation blocks such as PV array, battery, battery charge controller, three-phase voltage source inverter, the filter circuit, load, utility grid, and MPPT. **Figure 1** shows proposed system simulation. PV array is connected to the 11-kV network via a DC-DC boost converter and a three-phase three-level voltage source converter (VSC). In this paper, PV array generates a voltage of 666 V DC for a solar irradiance

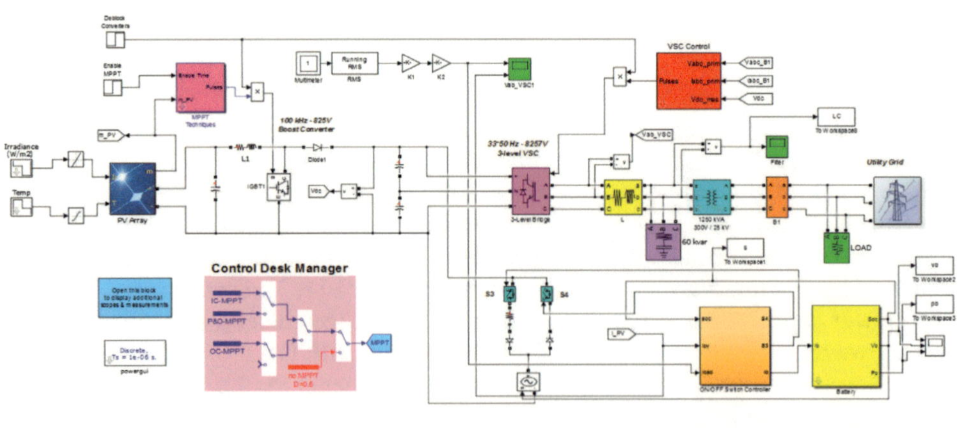

Figure 1.
Proposed system simulation.

of 1000 W/m^2. The 100-kHz DC-DC boost converter is increasing voltage from PV natural voltage (666 V DC at maximum power) to 825 V DC. Switching duty cycle is optimized by an MPPT controller that using different techniques such as 'Incremental Conductance, Hill Climbing/Perturb and Observe (P&O), and Fractional Open-Circuit Voltage (VOC) techniques. This MPPT technique automatically varies the duty cycle to generate the required voltage to extract maximum power 1980-Hz 3-level 3-phase VSC. The VSC converts the 825 V DC link voltage to 300 VAC and keeps unity power factor.

4. Problem overview

The most challenging problem considered by PV array system is how to automatically maintain the operation at maximum output power under environmental conditions continuous variation. In this chapter, a power converter that can vary the current coming from the PV array is employed as illustrated in **Figure 2** [6]. **Figure 2** shows pulse width modulation based boost converter. The philosophy of operation of the converter depends on the on and off states of the switch S [11, 12]. The power converter (boost converter) parameters can be sized using the following equations [13]:

$$\frac{Vo}{V_g} = \frac{1}{1-D} \tag{1}$$

$$L = \left(\frac{V_g{*}D}{f{*}CRF}\right) \tag{2}$$

$$R_o = \frac{V_o}{I_o} \tag{3}$$

$$C = \frac{D}{(f \times R_o \times VRF)} \tag{4}$$

where D is the duty cycle ratio, Vg is the input voltage to boost converter, Vo is the output voltage from boost converter, f is the switching frequency, VRF is voltage ripple factor (according to IEC harmonics standard, VRF should be bounded within 5%), CRF is the current ripple factor (according to IEC harmonics standard, CRF should be bounded within 30%) [14] and R_o is the load resistance. We introduce the different MPPT techniques below in an arbitrary order.

4.1 Incremental conductance algorithm

Inc. Cond based MPPTC is derived from the fact that there are three operating regions around MPP. Each operating region has unique characteristics represented

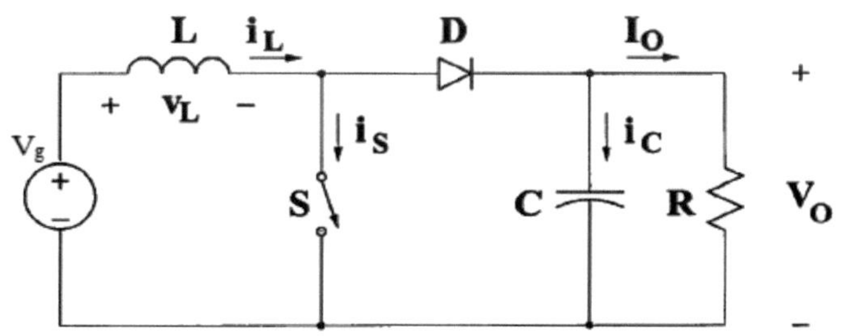

Figure 2.
Circuit diagram of boost converter.

in the ratio between the power change and the voltage change. Roughly speaking, it can be considered that Inc. Cond based MPPTC is based on the slope of the PV array power curve [15, 16].

$$\begin{cases} \frac{dP}{dV} = 0, \textit{at MPP} \\ \frac{dP}{dV} > 0, \textit{left of MPP} \\ \frac{dP}{dV} < 0, \textit{right of MPP} \end{cases} \tag{5}$$

Since

$$\frac{dP}{dV} = \frac{d(IV)}{dV} = 1 + V\frac{dI}{dV} = 1 + V\frac{\Delta I}{\Delta V} \tag{6}$$

$$\begin{cases} \frac{\Delta I}{\Delta V} = -\frac{I}{V}, \textit{at MPP} \\ \frac{\Delta I}{\Delta V} > -\frac{I}{V}, \textit{left of MPP} \\ \frac{\Delta I}{\Delta V} < -\frac{I}{V}, \textit{right of MPP} \end{cases} \tag{7}$$

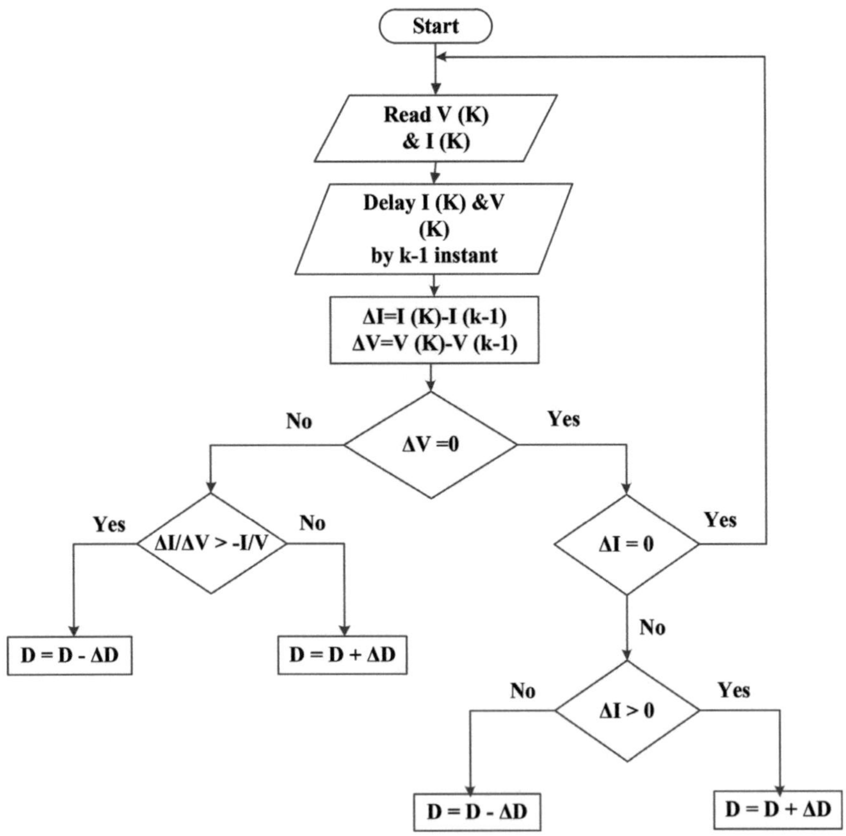

Figure 3.
Incremental conductance MPPT flowchart used for MATLAB simulation.

Thus, MPP can be tracked by comparing the instantaneous conductance (I/V) to the incremental conductance (ΔI/ΔV) as shown in the flowchart illustrated in **Figure 3**. The algorithm decrements or increments the duty cycle to track the new MPP. The increment size determines how fast the MPP is tracked.

4.2 Hill climbing/P&O algorithm

According to the sign of dP/dV where dP is the difference between power and dV is the difference between voltage of two succeeded point Hill climbing involves a perturbation in the duty ratio of the power converter [15–17]. The flow chart of the algorithm is shown in **Figure 4**. It is observed from P-V characteristic curve of the solar PV module that there are three main regions for operation. The first region is at the right hand side of MPP where the ratio between the power change over the voltage change (dP/dV) is negative. The second region is at the left hand side of MPP where the ratio dP/dV is positive. Moreover, the third region is at MPP exactly where the ration dP/dV is zero. P&O based MPPTC decides whether to increase or decrease the duty cycle depending on these three regions of operation.

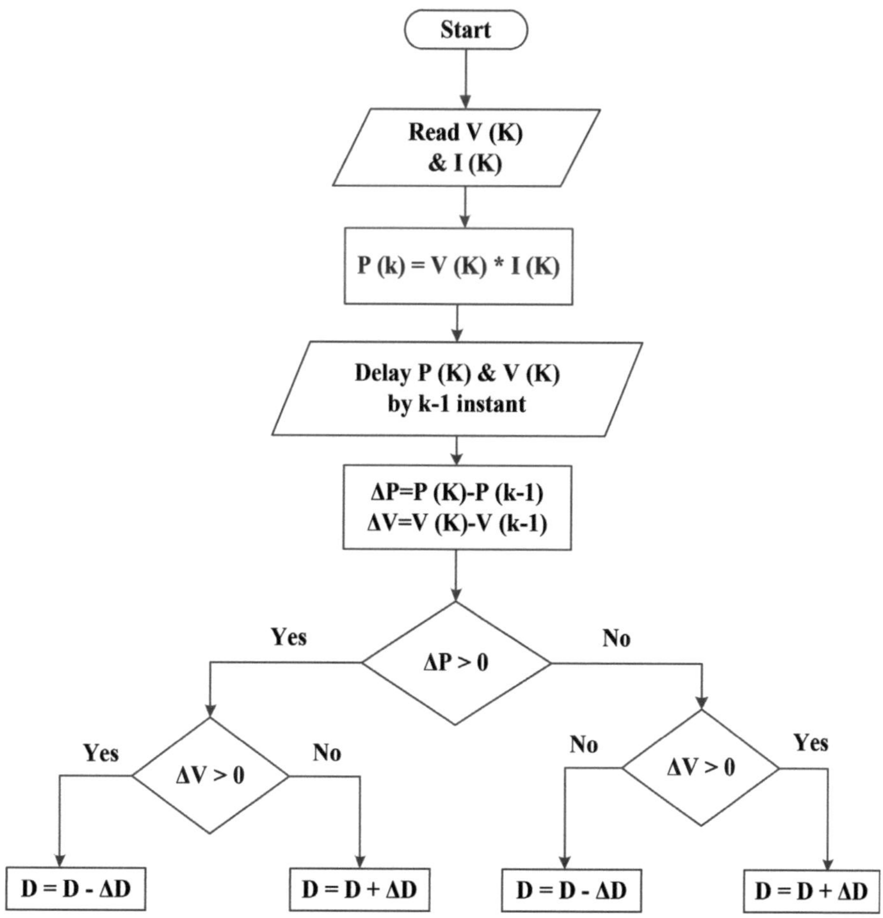

Figure 4.
Hill climbing/perturb and observe MPPT flowchart.

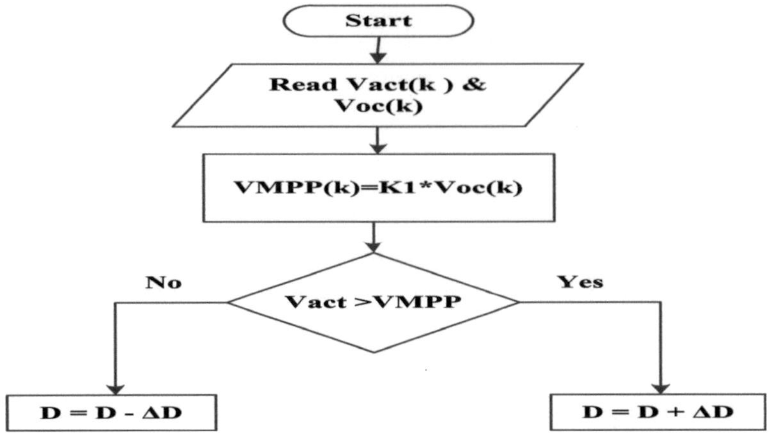

Figure 5.
Fractional open-circuit voltage algorithm.

4.3 Fractional open-circuit voltage algorithm

The linear characteristic of V_{OC} under various operating conditions paves the way for FOCV based MPPTC [15, 18].

$$V_{MPP} \cong K_1 \times V_{OC} \tag{8}$$

where K_1 is a constant of proportionality which depends on the characteristics of the PV panels. The algorithm of the fractional open circuit voltage is presented in **Figure 5**. The duty cycle is reduced or increase by comparing V_{MPP} computed from V_{OC} and the actual voltage V_{act}. The factor K_1 ranges between 0.71 and 0.78.

5. Fractional order PID control

FOPID control is proven to provide more flexibility and ability to enhance modeling and control of systems' dynamics [19]. The transfer function of FOPID is given by

$$G(S) = K_P \left\{ 1 + \frac{1}{T_i s^{\lambda}} + T_d\, S^{\mu} \right\} \tag{9}$$

where K_p, T_i and T_d are controller gains while λ and μ are the integral and differential power in real number. By changing the values of λ and μ, the controller can be configured to behave within the four possibilities presented in **Figure 6** [20].

Figure 6.
Control space of FOPID.

Figure 6 shows fractional PID control space. Recently, there are many optimization techniques are employed for solving engineering problems especially PI, PID, FOPI and FOPID based problems [19–39].

6. Grey wolf optimizer (GWO) technique

Grey wolves are considered as apex predators, meaning that they are at the top of the food chain [10]. **Figure 7** presents the social hierarchy of Grey wolves.

The mathematically model of the encircling behavior is represented by the following equations:

$$D = |CX_P - AX(t)| \tag{10}$$

$$X(T+1) = X_P(t) - AD \tag{11}$$

The vectors A and C are calculated as follows:

$$A = 2A\,r_1 - a \tag{12}$$

$$C = 2r_2 \tag{13}$$

Note that the random vectors (r_1 and r_2) allow wolves to reach any position between the points illustrated in **Figure 8**. So a grey wolf can update its position inside the space around the prey in any random location by using Eqs. (10) and (11).

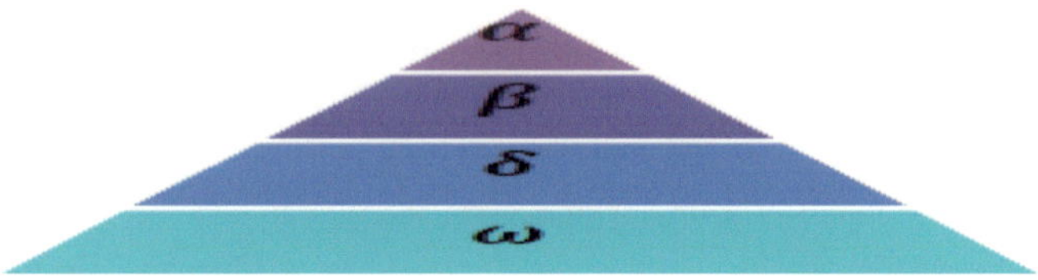

Figure 7.
Hierarchy of grey wolf (dominance decreases from top down).

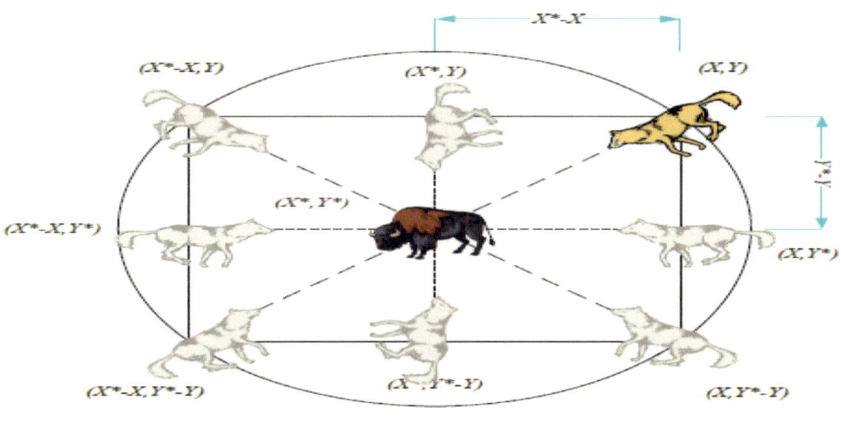

Figure 8.
Position vectors and their possible next locations.

7. Simulation results and comparison

To validate the effectiveness of the proposed MPPTCs, the system under study quipped with only one MPPTC (P&O, Inc. Cond, FOCV, and FSCC) at a time is simulated. Wide range of operating temperature and irradiance is considered in this chapter to prove the superiority of GWO based MPPTCs over the conventional ones. The simulation results spot the light on the output voltage as well as power.

7.1 Perturb and observe method

The system equipped with P&O based MPPTC is simulated under small as well as large variations in temperature and irradiance. **Figure 9** demonstrates the dynamic response of the output voltage. The time response of the output voltage presents small voltage ripples during the rapid changes of temperature and irradiance. A proper filter can be employed to remove these ripples. In **Figure 10**, the dynamic time response for the output power is presented. The features of the time response for the system output power in case of P&O based MPPTC interprets that the P&O based MPPTC smoothly tracks MPP but with some oscillations especially at the transition intervals (high to low or low to high temperature and irradiance variations).

7.2 Incremental conductance method

The time response for the system voltage and output power is presented in **Figures 11** and **12** respectively. The dynamic response for the system output power in case of Inc. Cond based MPPTC is significantly improved compared to P&O case even in case of rapid variations in temperature and irradiance. **Figure 12** spotted the light on how Inc. Cond based MPPTC supersedes the P&O in smoothly tracking MPP.

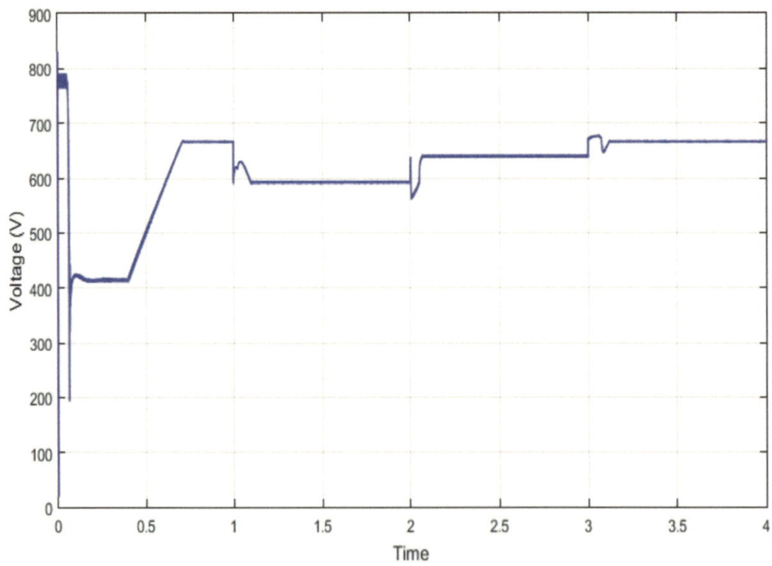

Figure 9.
P&O based MPPTC's output voltage waveform at different radiation and temperature.

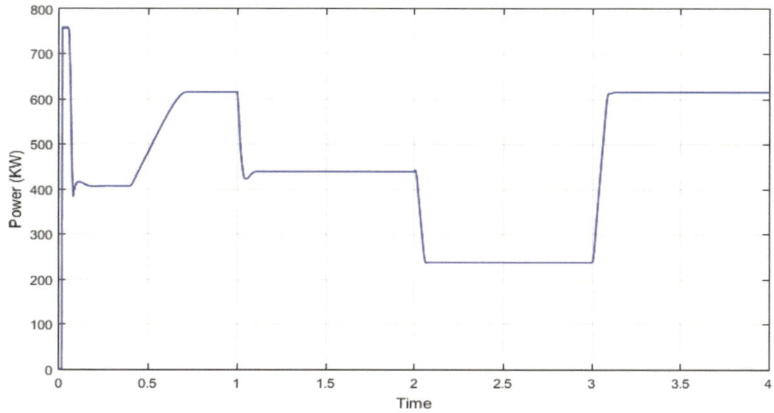

Figure 10.
P&O based MPPTC's output power waveform at different radiation and temperature.

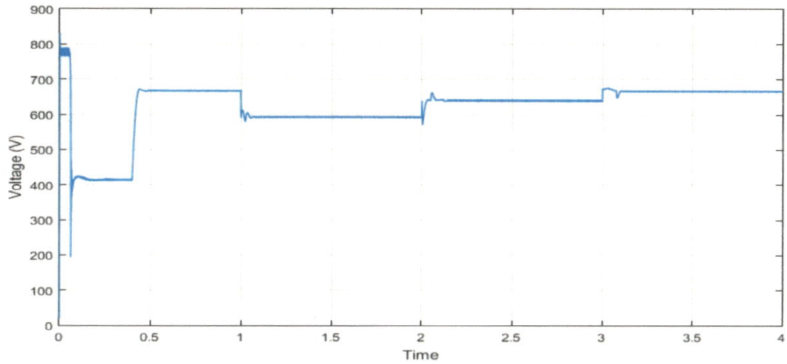

Figure 11.
Inc. Cond based MPPTC's output voltage waveform at different radiation and temperature.

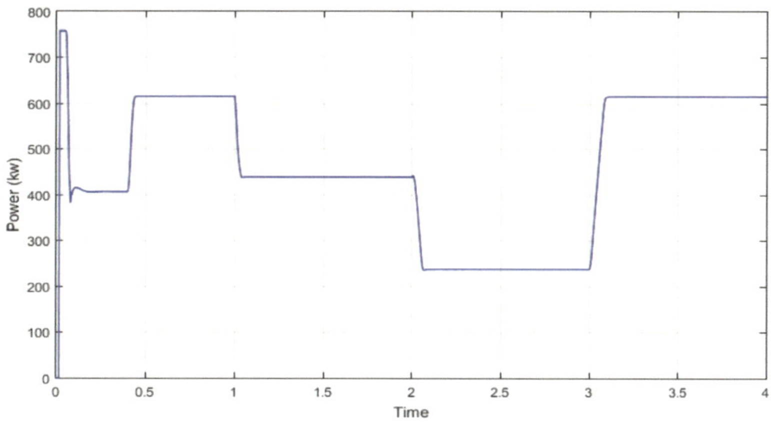

Figure 12.
Inc. Cond based MPPTC's output power waveform at different radiation and temperature.

7.3 Open-circuit voltage method

The time response of the voltage and output power for the system equipped with FOCV based MPPTC is shown in **Figures 13** and **14** respectively. It is evident from the simulation results that the system response is poor especially in case of rapid changes in the operating conditions.

Table 1 presents a comparative study between the various applied MPPT techniques. It is worth mentioning that although Inc. Cond MPPT technique has good tracking response but it requires voltage and current measurements. Moreover, its implementation complexity is higher than P&O and fractional open circuit voltage methods.

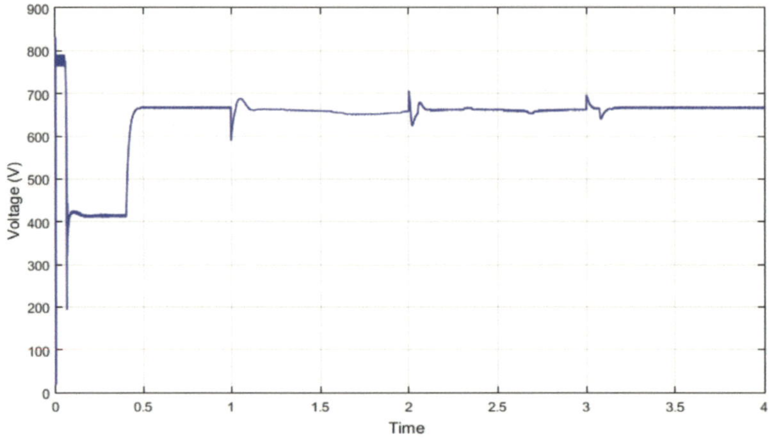

Figure 13.
FOCV based MPPTC's output voltage waveform at different radiation and temperature.

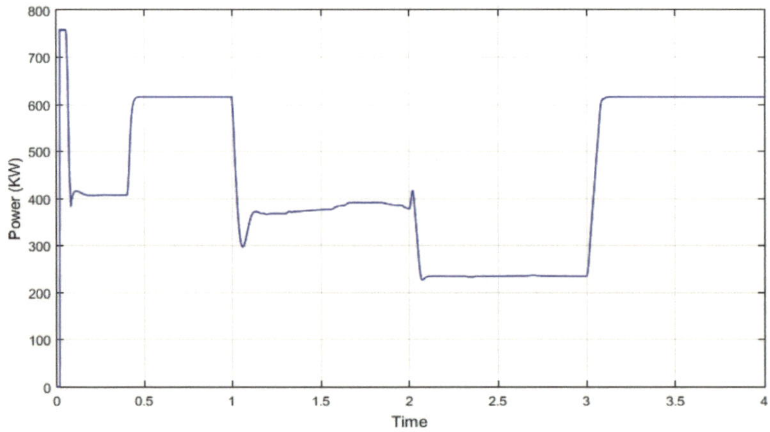

Figure 14.
FOCV based MPPTC's output power waveform at different radiation and temperature.

MPPT techniques	Parameters		
	Convergence speed	**Implementation complexity**	**Sensed parameters**
P & O method	Varies	Low	Voltage
Inc Cond method	Varies	Medium	Voltage, current
Fractional V_{oc} method	Medium	Low	Voltage

Table 1.
Comparative study for various MPPT techniques.

8. Conclusion

In this paper, four MPPT algorithms are implemented using the Boost converter. The models are simulated using MATLAB/SIMULINK. The simulation results show that P&O and Inc. Cond MPPTCs have better efficiency than FOCV and FSCC MPPTCs. Although, Inc. Cond provides good performance but its implementation has some challenges. Moreover, FOCV and FSCC based MPPTCs are very simple but both controllers lack to the accuracy due to their dependency on constant gains. Therefore, solar cell performance is significantly improved in the presence of MPPTCs. Hence, MPPTCs improvement has vital role in expanding the utilization of PV based systems.

Acknowledgements

The authors gratefully acknowledge the support of the Egyptian high education ministry, The Science and Technology Development Fund (STDF), and the French Institute in Egypt (IFE).

Conflict of interest

The authors of this chapter did not have 'conflict of interest' for publishing.

Author details

Mohamed Ahmed Ebrahim[1*] and R.G. Mohamed[2]

1 Faculty of Engineering at Shoubra, Benha University, Cairo, Egypt

2 Eastern Company, Giza, Egypt

*Address all correspondence to: mohamed.mohamed@feng.bu.edu.eg

References

[1] Wolfsegger C, Fraile D, Philbin P, Teske S. Solar generation-solar electricity for over one billion people and two million jobs by 2020. In: European Photovoltaic Industry Association, Renewable Energy House, Belgium and Greenpeace International, The Netherlands, Report No. 5. 2008

[2] Mohamed RG, Ibrahim DK, Youssef HK, Rakha HH. Optimal sizing and economic analysis of different configurations of photovoltaic systems. International Review of Electrical Engineering (I.R.E.E.). January–February 2014;**9**(1):146-156

[3] Mohamed RG, Ibrahim DK. Optimal sizing and economic analysis of stand-alone photovoltaic system. In: The 2012 World Congress on Power and Energy Engineering (WCPEE'12); Apr. Vol. 4. 2012. pp. 188-195

[4] Renewable Capacity Statistics 2018 (IRENA). https: //www.irena.org/publications/2018/Mar/Renewable-Capacity-Statistics-2018

[5] Annual Report 2018 (NREA). http: //www.nrea.gov.eg/Content/reports/English%20AnnualReport.pdf

[6] Rekioua D, Achour AY, Rekioua T. Tracking power photovoltaic system with sliding mode control strategy. Renewable and Sustainable Energy Reviews. 2013:219-230

[7] Mukhopadhyay S, Chen YQ, Singh A, Edwards F. Fractional order plasma position control of the STOR-1M Tokamak. In: Joint 48th IEEE Conference on Decision and Control and 28th Chinese Control Conference. 2009. pp. 422-427

[8] Li Y, Chen Y, Ahn H-S. A generalized fractional-order iterative learning control. In: 50th IEEE Conference on Decision and Control and European Control Conference (CDC-ECC). 2011. pp. 5356-5361

[9] Wang CY, Luo Y, Chen YQ. Fractional order proportional integral (FOPI) and [Proportional Integral] (FO [PI]) controller designs for first order plus time delay (FOPTD) systems. In: Chinese Control and Decision Conference. 2009. pp. 329-334

[10] Mirjalili S, Mirjalili SM, Lewis A. Advances in engineering software. Renewable and Sustainable Energy Reviews. 2014:46-61

[11] Khalid H, Mohamed TB, Ibrahim, Saad NB. Boost converter design with stable output voltage for wave energy. International Journal of Information Technology and Electrical Engineering. February 2013;**2**(1)

[12] Sathya P, Natarajan R. Design and implementation of 12V/24V closed loop boost converter for solar powered LED by lighting system. International Journal of Engineering and Technology (IJET). Feb-Mar 2013;**5**(1)

[13] Debashis DS, Pradhan SK. Modeling and Simulation of PV Array with Boost Converter. Rourkela: National Institute of Technology; 2011

[14] Hieu Nguyen X, Nguyen MP. Mathematical modeling of photovoltaic cell/module/arrays with tags in Matlab/Simulink. Environmental System Research. 2015;**1**(1):1-13

[15] Esram T, Chapman PL. Comparison of photovoltaic array maximum power point tracking techniques. IEEE transactions on Energy Conversion. June 2007

[16] Stamatescua I, Făgărășana I, Stamatescua G. Design and

implementation of a solar-tracking algorithm. 24th DAAAM International Symposium on Intelligent Manufacturing and Automation, 2013. Procedia Engineering. 2014;**1**(1): 500-507

[17] Femia N, Giovanni Petrone M, Spagnuolo G, Vitelli M. Optimization of perturb and observe maximum power point tracking method. IEEE Transactions on Power Electronics. July 2005;**20**(4):963-973

[18] Sai Babu C. Design and analysis of open circuit voltage based maximum power point tracking for photovoltaic system. International Journal of Advances in Science and Technology. 2011;**2**(2):51-86

[19] Tajjudin M, Arshad NM, Adnan R. A design of fractional-order PI controller with error compensation. International Journal of Computer, Electrical, Automation, Control and Information Engineering. 2013;7(6):727-735

[20] Lachhab N, Svaricek F, Wobbe F, Rabba H. Fractional order PID controller (FOPID)-toolbox. In: European Control Conference (ECC) July 17–19, Zürich. 2013. pp. 3694-3699

[21] Aouchiche N, Aitcheikh MS, Becherif M, Ebrahim MA. AI-based global MPPT for partial shaded grid connected PV plant via MFO approach. Solar Energy. 2018;**171**:593-603

[22] Ebrahim MA, Becherif M, Abdelaziz AY. Dynamic performance enhancement for wind energy conversion system using moth-flame optimization based blade pitch controller. Sustainable Energy Technologies and Assessments. 2018;**27**: 206-212

[23] Benmouna A, Becherif M, Depernet D, Ebrahim MA. Novel energy management technique for hybrid electric vehicle via interconnection and damping assignment passivity based control. Renewable Energy. 2018;**119**: 116-128

[24] Maher M, Ebrahim MA, Mohamed EA, Mohamed A. Ant-lion inspired algorithm based optimal design of electric distribution networks. In: 2017 Nineteenth International Middle East Power Systems Conference (MEPCON); IEEE. 2017, December. pp. 613-618

[25] Aouchiche N, Cheikh MA, Becherif M, Ebrahim MA, Hadjarab A. Fuzzy logic approach based MPPT for the dynamic performance improvement for PV systems. In: 2017 5th International Conference on Electrical Engineering-Boumerdes (ICEE-B). October; IEEE. 2017. pp. 1-7

[26] Mohamed RG, Ebrahim MA, Bendary FM, Osman SAA. Transient stability enhancement for 20 MW PV power plant via incremental conductance controller. International Journal of System Dynamics Applications (IJSDA). 2017;**6**(4): 102-123

[27] Maher M, Ebrahim MA, Mohamed EA, Mohamed A. Ant-lion optimizer based optimal allocation of distributed generators in radial distribution networks. International Journal of Engineering and Information Systems. 2017;**1**(7):225-238

[28] Mousa ME, Ebrahim MA, Hassan MM. Optimal fractional order proportional—integral—differential controller for inverted pendulum with reduced order linear quadratic regulator. In: Fractional Order Control and Synchronization of Chaotic Systems. Cham: Springer; 2017. pp. 225-252

[29] Ebrahim MA, AbdelHadi HA, Mahmoud HM, Saied EM, Salama MM. Optimal design of MPPT controllers for

grid connected photovoltaic array system. International Journal of Emerging Electric Power Systems. 2016; **17**(5):511-517

[30] Mousa ME, Ebrahim MA, Hassan MM. Stabilizing and swinging-up the inverted pendulum using PI and PID controllers based on reduced linear quadratic regulator tuned by PSO. International Journal of System Dynamics Applications. 2015;**4**(4):52-69

[31] Jagatheesan K, Anand B, Ebrahim MA. Stochastic particle swarm optimization for tuning of PID controller in load frequency control of single area reheat thermal power system. International Journal of Energy and Power Engineering. 2014;**8**(2): 33-40

[32] Ebrahim MA, El-Metwally KA, Bendary FM, Mansour WM, Ramadan HS, Ortega R, et al. Optimization of proportional-integral-differential controller for wind power plant using particle swarm optimization technique. International Journal of Emerging Technologies in Science and Engineering. 2011

[33] Ahmed M, Ebrahim MA, Ramadan HS, Becherif M. Optimal genetic-sliding mode control of VSC-HVDC transmission systems. Energy Procedia. 2015;**74**:1048-1060

[34] Jagatheesan K, Anand B, Dey N, Ebrahim MA. Design of proportional-integral-derivative controller using stochastic particle swarm optimization technique for single-area AGC including SMES and RFB units. In: Proceedings of the Second International Conference on Computer and Communication Technologies. New Delhi: Springer; 2016. pp. 299-309

[35] Ali AM, Ebrahim MA, Hassan MM. Automatic voltage generation control for two area power system based on particle swarm optimization. Indonesian Journal of Electrical Engineering and Computer Science. 2016;**2**(1):132-144

[36] Ebrahim MA, Elyan T, Wadie F, Abd-Allah MA. Optimal design of RC snubber circuit for mitigating transient overvoltage on VCB via hybrid FFT/ wavelet genetic approach. Electric Power Systems Research. 2017;**143**: 451-461

[37] Soued S, Ebrahim MA, Ramadan HS, Becherif M. Optimal blade pitch control for enhancing the dynamic performance of wind power plants via metaheuristic optimizers. IET Electric Power Applications. 2017;**11**(8): 1432-1440

[38] Ebrahim MA. Towards robust non-fragile control in wind energy engineering. Indonesian Journal of Electrical Engineering and Computer Science. 2017;**7**(1):29-42

[39] Ebrahim MA, Ramadan HS. Interarea power system oscillations damping via AI-based referential integrity variable-structure control. International Journal of Emerging Electric Power Systems. 2016;**17**(5): 497-509

Chapter 9

PV Plant Influence on Distribution Grid in Terms of Power Quality Considering Hosting Capacity of the Grid

Ivan Ramljak and Drago Bago

Abstract

Photovoltaic plants penetrate rapidly in distribution grid. Problems with their integration in distribution grid can exist in terms of load flow, protection settings, power quality, etc. This chapter analyzes influence of photovoltaic plants connection in distribution grid (0.4 and 10 kV voltage level) on power quality. The main focus will be on influence of photovoltaic plant connection point on distribution grid (hosting capacity—strength of the grid) in terms of power quality. Norms and regulations about influence of photovoltaic plants on distribution grid in terms of power quality will be analyzed. Influence of photovoltaic plants on distribution grid in theoretical aspects will be presented. Several case studies then will be described. Those case studies present different connection points of photovoltaic plants on distribution grid. Comparison of theoretical assumptions and real case studies will be compared. Some observations of real case studies and their impact on theoretical aspects, norms, and regulations about photovoltaic plant influence on distribution grid will be introduced.

Keywords: photovoltaic plant, distribution grid, power quality, hosting capacity, connection point, legal regulations

1. Introduction

Renewable sources (RES) (or distributed generation (DG) or embedded generation (EG)) are today one of the most actual fields in power engineering area. RES are very important factor in areas of environment protection, development of new technologies and smart grids. In the last several years, there is high penetration of RES on power grid regardless of voltage level. Those RES are with different technologies, but with same assignment: generation of renewable electrical energy. Considering source for electrical energy generation, RES can be based on sun radiation, wind power, waste utilization, exploitation of hydropower, etc. Rated power span of power plants based on RES can vary from 5 kW (microgeneration) to 150 MW (large generation) [1]. Nonrenewable sources (thermal and gas power plants and nuclear power plants) generally are connected on high voltage (HV) grid. It is due to their great rated power. In terms of that, middle- and low-voltage (MV and LV) distribution grid is generally without nonrenewable sources presence.

Without RES, distribution grid is a passive grid which transfers electrical energy from HV grid to consumers. But, penetration of RES makes distribution grid active. It is due to RES presence on all voltage levels (from HV to LV). That is due to great span of RES rated power. It means that RES can have connection in LV grid (e.g., PV plant with 10 kW) and in HV grid too (e.g., windfarm with more than 100 MW of rated power). Simply, conclusion can be derived. It is that conception of power system with RES penetration is changed. In area of electrical energy generation, RES today have great share in total electrical energy generation. But, electrical energy from, for example, PV plants and wind farms is uncontrollable and still difficult to foresee. That fact increased effort of researchers in area of storage systems development. RES created the most changes in distribution grid. Distribution grid, as already stated, became active grid. That fact made changes in distribution grid in terms of load flow conception, system protection, power quality, etc. Those changes are challenge for distribution system operators (DSOs). Distribution system operation and management becomes very difficult for those DSO. A lot of research about RES penetration in distribution grid is carried out. The results are presented in various publications. Some observations can be found in [2–7]. References [3, 4] give fundamental knowledge about RES penetration in distribution grid. Basics, in terms of grid connection, power quality, system protection, economy, and reliability concepts, are described. In [4], influence of RES on power quality and system stability was analyzed. The conditions for RES connecting were described. The authors showed that carefully selected and placed DG can have positive impact on distribution grid in terms of improving voltage profile and system losses. In [5], technical requirements for the connection of the new DG resources on the distribution grid have been analyzed. Some experiences from Greek grid were presented. In [6], technical requirements and assessment criteria for RES connection on distribution grid for power quality and system protection-related issues are presented. Reference [7] introduces statistics modeling for planning purposes of RES integration in distribution grid. One of the greatest challenges in RES connection on distribution grid is in terms of power quality. Simply, before RES penetration in distribution grid, voltage drop was the highlighted problem in distribution grids. With RES, opposite, voltage increasing is the most prominent problem in distribution grids than voltage drop. This is only one of the evidences of conception change in today's distribution grid. RES impact on power quality in distribution grid is analyzed in [2, 8–12]. Reference [8] analyzes short circuit power level with respect to power quality. Paper proposes a methodology to correlate rapid voltage changes to the load variations and to the short circuit power in point of common coupling (PCC). Presented analysis is valid for MV and high voltage (HV) grid. In [9], network impedance influence on power quality results in PCC is analyzed. Assessment procedures for connection of RES on distribution grid are presented. Registration and processing of power quality data in power grid with integrated RES is presented in [10]. In [11, 12], power quality monitoring and classification methods in distribution grids with penetrated RES are presented. There are various power quality parameters. RES penetration in distribution grid influences the most on voltage quality parameters. The impact of RES on distribution grid among voltage quality parameters such as harmonics, flicker, and voltage magnitude is presented in [4–6]. In [9], analyzed voltage parameters are voltage changes, flickers, harmonics, and commutation notches. Reference [11] analyzes voltage fluctuations and harmonics. In [12], frequency, magnitude of voltage, flicker, unbalance, and harmonics are analyzed. In [13], harmonic, unbalance, and flicker are presented. The most analyzed voltage quality parameters, among literature, on which RES can have influence on distribution grid are voltage variations, flicker, and harmonics. As the share of RES increases, the development and use of legal regulations becomes very important factor for proper function of distribution grid. Those legal regulations depend usually on country, due to each country specifics. But, there are some common international regulations. The best known is EN 50160, made by CENELEC in the middle 1990s. The other one is IEC 61000 series regulation on power quality. The IEEE regulations are present too in engineering practice, but in Europe, IEEE regulations usually cannot have legal status. In [3], EN 50160, IEC 61000 series, and some IEEE Standards are mentioned. PV plants are power sources that rapidly penetrate in power system. Generally, they use energy of the Sun and convert it to electrical energy [2]. PV plants are present on all voltage levels, from LV (10 kW rated power) to HV (more than 100 MW rated power). This book chapter deals with influence of PV plant connection on distribution grid in terms of power quality. This chapter puts focus on dependence between strength of the grid (connection point of PV plant in the distribution grid) and PV plant rated power. In this chapter, theoretical assumptions will be analyzed. Further on, in order to reconsider theoretical assumptions, case studies will be presented too. The chapter structure is as follows:

a. Influence of PV plant on power quality in connection point (PCC). Literature review (theoretical assumptions) and validation of it on case study.

b. Legal regulations of power quality considering RES influence on distribution grid. Literature review and current state.

c. Hosting capacity of the grid (strength of the grid) considering rated power of PV plant. Literature review and multiple case studies.

d. Comparison of theoretical and practical results of PV plant influence on distribution grid. Proposal of theoretical calculation on PV plant influence on distribution grid considering practical (measured) results. Verification on case study.

First, review of already known facts in this area will be given. Further, new conclusions will be introduced. The idea of further work will be presented based on new findings. At the end of the chapter, summary conclusions will be given.

In this chapter, power quality presents voltage quality parameters.

2. Influence of photovoltaic plant on power quality

PV plants can have connection in distribution grid in various PCC. That PCC potentially can be every bus in distribution grid. Generally, influence of connected PV plant in PCC (in terms of power quality) will be presented. This influence is analyzed among various literature. In [14, 15], influence of PV plants on waveform distortion in PCC was analyzed. It is concluded that if size of PV system is relatively low with respect to the short circuit power of the grid, there is no significant influence on the grid voltage quality. But, if short circuit power is relatively low, regarding PV system rated power, some problems considering power quality can be observed. According to [16], PV plants should not seriously degrade quality of supply with regard to harmonics. In [17], measurements in LV grid in PCC of PV plant connection are performed. Conclusion is that influence of PV plant in PCC is marginal in case of flicker and harmonics. This statement is valid for each measured point in grid (not only PCC). Same results for harmonics are obtained in [18]. Further on, in [19], PV plant connection on MV grid is analyzed. The result is that connected PV plant has positive contribution to the reduction of harmonic voltages distortion (with respect to the harmonic distortion occurring without connected PV plant). Power quality problems with connected PV plants are more often in rural grids (then in urban grid) due to usually longer feeders and small cross section of conductors [20]. In [21], this is presented as a potential problem too. Limitations of PV plants penetration in distribution grids apply only in weak grids [22]. But, PV plant penetration in weak grids can cause even better performance of those grids (improving voltage profile). Each case should consider individually, as stated in [22]. Simulations in [23] with PV plant connection in distribution grid showed no violation in the harmonic values. According to [24], in period of low irradiance, harmonic distortion is greater than in period with high irradiance.

2.1 Case study

Considering literature review, case study results will be presented. The goal is to compare theoretical assumptions and real case study results. A measurement of power quality for one PV plant was performed. Measurement lasted a week without PV plant and a week with connected PV plant, at PCC. Measuring (periods) was performed according to EN 50160:2011 issued by CENELEC (later analyzed). The most interesting voltage quality parameters will be presented:

- voltage unbalance—u (%);
- total harmonic distortion of voltage—THDU (%);
- slow voltage variations—U (V); and
- long-term flicker—P_{lt}.

The question about influence of PV plant on voltage quality in PCC is imposed. Rated power of PV plant is 23 kVA. Short circuit power of grid at PCC is 0.95 MVA. The PV plant is located at rural area connected to LV grid, on feeder with insulated overhead lines. The grid is in good condition (nearly reconstructed). There are two three-phase inverters in this PV plant. The goal is to find influence of PV plant connection on voltage quality at PCC. Analyzed results were in scope of 100% of measuring time periods. It is mandatory condition, so that total influence of PV plant on PCC in distribution grid can be seen.

2.1.1 Voltage unbalance

Figure 1 presents voltage unbalance (%) before and after PV plant connection. Due to clearness, 1-day results (24 hours, 10 min periods) of voltage unbalance before/after PV plant connection are presented in **Figure 1**. Similar is for all week too.

It is quite clear that any correlation of unbalance before and after PV plant connection cannot be found. Maximum (max.) and minimum (min.) value of unbalance is quite similar before (b.c) and after PV plant connection (a.c). It is shown in **Table 1** (for all week period).

From **Table 1**, it is visible that unbalance has lower value after PV plant connection!

2.1.2 Total harmonic distortion of voltage (THDU, %)

Figure 2 presents THDU values for L1 phase regarding P/P_n (%) values of PV plant. P/P_n presents ratio of generating power (P) and rated power (P_n) of PV plant. Due to clearness, 1-day results (24 hours, 10 min periods) of THDU (%) before/

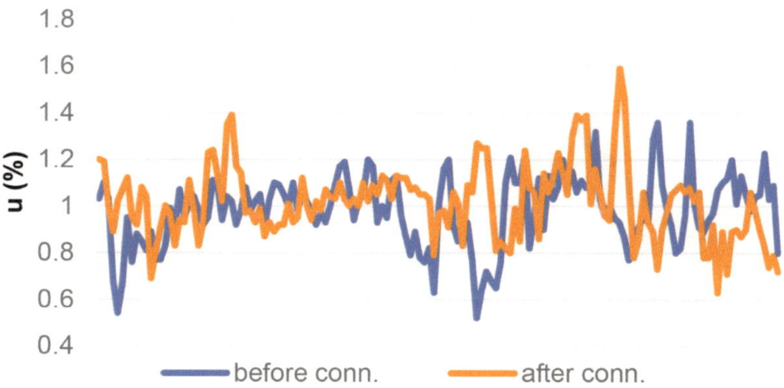

Figure 1.
Unbalance (%) before and after PV plant connection.

		b.c	a.c
u (%)	Maximum	1.98	1.60
	Minimum	0.52	0.50

Table 1.
Unbalance (%) values before and after PV plant connection.

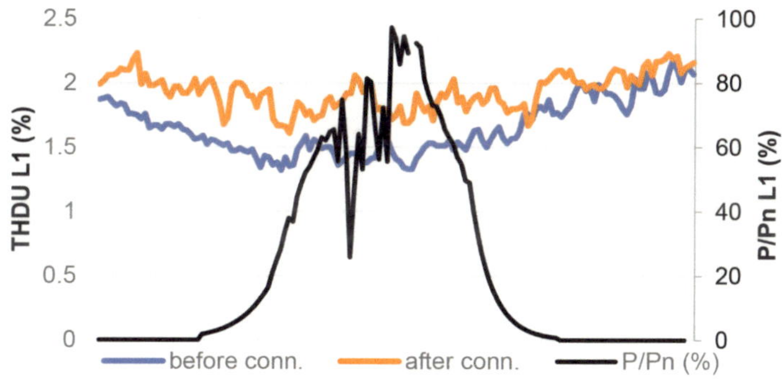

Figure 2.
THDU (%) for L1 phase before and after PV plant connection.

after PV plant connection are presented in **Figure 1**. Generally, results are similar for each day. Results for other two phases are similar.

It is quite obvious that after PV plant connection, THDU values are greater. Those increasing in THDU are the most presented mainly while PV plant produces electrical energy. But, analyzing 10 min values for all week period (all three phases), it is visible that max. and min. values of THDU are quite similar before (b.c) and after PV plant connection (a.c). It is shown in **Table 2**.

Both (min. and max.) values are slightly increased after PV plant connection in this case. It is mainly in period when PV plant produces electrical energy. In percent

		b.c	a.c
THDU (%)	Maximum	2.67	2.70
	Minimum	1.15	1.25

Table 2.
THDU (%) values before and after PV plant connection.

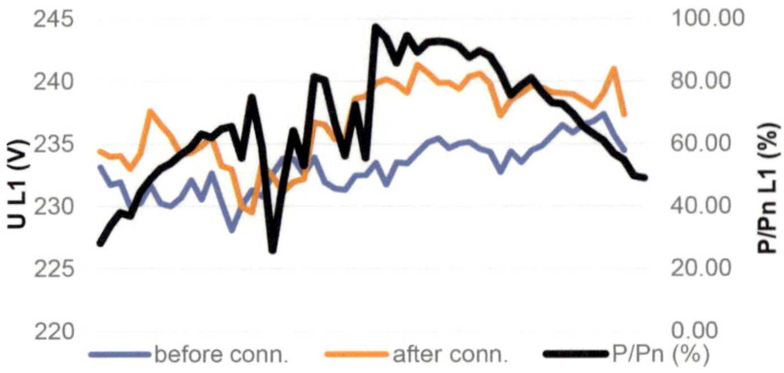

Figure 3.
Voltage values (V) for L1 phase before and after PV plant connection.

value, for a week period, it is negligible. Generally, THDU values before and after PV plant connection are in scope of permitted values regarding any regulation.

2.1.3 Slow voltage variations

Voltage values in V for L1 phase, compared to P/P_n value, are presented in **Figure 3**. Those voltage values are 10 min interval for one PV plant generation day, between 8.00 and 17.00 hours.

It is clear that PV plant generation increases voltage. As P/P_n is closer to 100%, voltage difference in V increases considering period before and after PV plant connection.

But, analyzing 10 min values for all week period (in all three phases), it is visible that max. and min. values of voltage values in V are quite similar before (b.c) and after PV plant connection (a.c). It is shown in **Table 3**.

In this case, both (min. and max.) values are slightly increased after PV plant connection. In percent value, for a week period, that increase looks negligible. But, 246.36 V can be potentially high voltage, depending on regulations!

Figure 4 presents voltage values of phase L2 in V after PV plant connection, considering P/P_n (%) values. Those values present 10 min interval for 1 day of PV plant generation, between 8.00 and 17.00 hours.

It is obvious that voltage value in V is related to P/P_n (%) ratio. Generally, greater P/P_n (%) ratio means greater voltage in V (voltage increase).

2.1.4 Long-term flicker

Figure 5 presents P_{lt} values before and after PV plant connection. On **Figure 5** is also represented current I (A) of PV plant phase L3 (I_{L3}). Measurement period is 10 min, from 20:00 to 20:00 (24 hours).

In night period, flicker values are practically the same with/without PV plant. At day period, those values are greater with/without PV plant. There was not found any correlation between PV plant energy generation and flicker values. **Table 4** presents P_{lt} values in PCC before/after PV plant connection for period of 2 weeks.

		b.c	a.c
U (V)	Maximum	243.17	246.36
	Minimum	225.56	227.19

Table 3.
Voltage values in V before and after PV plant connection.

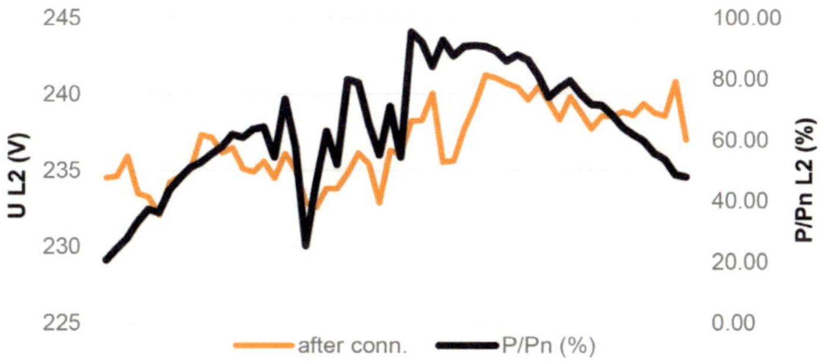

Figure 4.
Voltage values (V) for L2 phase considering P/P_n (%) values.

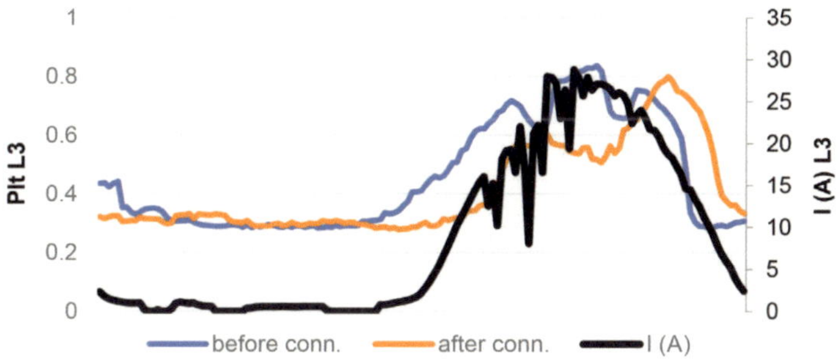

Figure 5.
P_{lt} L3 values before and after PV plant period regarding I_{L3} in A.

		b.c	a.c
P_{lt}	Maximum	1.391	1.244
	Minimum	/	/

Table 4.
P_{lt} values before and after PV plant connection.

It is obvious that PV plant does not contribute to increasing of P_{lt} at PCC. P_{lt} values have max. value before PV plant connection (not after connection) for this case study.

2.2 General conclusions of PV plant connection for case study

Some general conclusions for PV plant influence on voltage quality at PCC can be stated:

- Unbalance is quite similar before and after PV plant connection. From **Table 1** is visible that max. 10 min value of unbalance has lower value after PV plant connection. There is no correlation between unbalance values and PV plant generation.

- THD values are slightly increased after PV plant connection in this case. It is mainly in period when PV plant produces electrical energy. In percent value, for a week period, it is negligibly increasing. But, PV plant did have influence on THDU increasing (even slightly).

- It is obvious that voltage value in V is related to P/P_n (%) ratio. Generally, greater P/P_n (%) ratio means greater voltage in V (voltage increasing). PV plant has influence on voltage increasing due to its electrical energy generation.

- It is obvious that PV plant does not contribute to increasing of P_{lt} at PCC. P_{lt} values in this case have their max. before PV plant connection.

3. Legal regulations of power quality

Power quality is regulated with norms/standards. But, this area is not unique considering regulations. For this chapter, some of the most exploited norms/standards will be analyzed. Most used international norms/standards are as follows:

- EN 50160 by CENELEC,

- IEC standards, and

- IEEE standards.

These standards usually apply to normal weather conditions and exclude situations like storms, etc. Some of the most used norms/standards in terms of power quality in distribution grids will be shortly presented below. But, most of the countries exist with some legal regulations, like grid codes, in which voltage quality is exploited too.

3.1 EN 50160

In Europe, the most mentioned standard is EN 50160, published by CENELEC (European Committee for Electrotechnical Standardization). First published version of this norm is from the middle 1990s. It is probably the first norm dealing with voltage quality problems. This norm gives quantitative characteristics of voltage. The norm describes characteristics of voltage parameters regarding permissible deviations. EN 50160 analyzes voltage characteristics considering frequency, waveform, symmetry, and magnitude. Measurement of voltage must be performed for 7 days. Instruments for power quality measurement are described by IEC 61000-4-30 standard. Procedure for power quality parameters measuring, according to EN 50160, is defined by IEC 61000-4-x standards. This norm gets a new issue every few years. Voltage quality limits, interesting for this analysis, are presented in **Table 5**. In **Table 5**, compliance limits for statistical evaluation of 95% of time are shown.

Supply voltage parameter	Compliance limit
Voltage variations	±10%
Voltage unbalance	<2%
Long-term flicker	≤1
Harmonic voltage (THDU)	<8%

Table 5.
EN 50160—voltage quality parameters.

Supply voltage parameter	Compliance limit
Long-term flicker	≤0.8
Harmonic voltage (THDU)	<5%

Table 6.
IEEE—voltage quality parameters.

3.2 IEEE standards

There is set of standards that consider voltage quality area, issued by IEEE (Institute of Electrical and Electronics Engineers). List of those standards can partly be found in [25]. Some voltage quality limits (available to authors) in scope for this analysis are presented in **Table 6**. In **Table 6**, compliance limits for statistical evaluation of 95% of time are shown. Standard that analyzes flicker is IEEE Std. 1453 and standard that analyzes harmonics is IEEE Std. 519.

3.3 IEC standards

IEC standards are issued by International Electrotechnical Commission. In scope of voltage quality are IEC 61000 Series Standards. Some voltage quality limits prescribed by IEC (available to authors) are shown in **Table 7**.

Now, short overview considering **Tables 5–7** can be done. It is interesting that all three published standards have some unity elements. Mainly, 95% of time, voltage parameters must be in limited area. Slow voltage variations are usually limited 10% above/below nominal value. Unbalance is limited to 2%. P_{lt} is limited to 0.8 or 1. The greatest difference is, however, in THDU limits. The limited value varies from 5 to 8%.

Regarding the above-analyzed voltage quality parameters, voltage variations are the most exploited parameter. This is due to increased sensitivity of consumers on voltage values. It is a parameter of voltage quality on which consumers usually have the most complaints. Unfortunately, RES can potentially create even greater problems. Beside norms, in most of the countries exist local regulations (grid codes or similar) about permissible voltage values considering voltage level. Some of them are similar to the above-described norms, but some not.

Those regulations, based on internal country documents, can be divided into two classes. First regulation class presents permissible voltage values, which should be considered during distribution grid planning. Second regulation class considers limited voltage values during distribution grid exploitation. According to [3], permissible voltage values in LV grid are −10 to +6% of nominal value in France. In MV grid, this value is −5 to +5%. In [3] is stated that in the United States, for MV grid, permissible voltage values are from −8 to +8% of nominal value. For Canada, those values are from −6 to +6%. It is interesting according to [3], that in Brazil, in

Supply voltage parameter	Compliance limit
Voltage variations	±10%
Long-term flicker	≤0.8
Voltage unbalance	<2%
Harmonic voltage (THDU)	THD < 8%

Table 7.
IEC 61000 series—voltage quality parameters.

LV grid, permissible voltage variations are from −14 to +6%. The examples below are for the case of grid planning. Below, some regulations for distribution grid in exploitation will be presented. For LV customers in Hungary, 95% of the 10 min voltages should be within 7.5% of the nominal voltage value, 100% of the 10 min voltages should be within 10% nominal voltage value. Moreover, 100% of the 1 min voltages should be within 15% nominal voltage value [3]. In Norway, 100% of the 1 min voltage values should be within 10% of the nominal voltage values [3]. For Spain, 95% of the 10 min voltage values should be within 7% of the nominal voltage value [3]. Medium voltage customers in France can sign a contract that gives them a voltage within 5% of a nominal voltage [3].

It is visible great difference between individual country regulations, regarding existence of norms issued by CENELEC, IEEE, and IEC.

3.4 Legal regulations of power quality and RES

If one closely looks norms/standards mentioned above, the logical question can be asked. Where are RES in those regulations? Indeed, slow voltage variations are usually limited 10% above/below nominal value, no matter is or is not RES presented in PCC. Considering Section 2.1, this can be a potential problem. If voltage value is, for example, +9% above nominal value, PV plant connection can potentially contribute to voltage increasing of, for example, +2%. It means that now voltage increasing is 11% in that PCC—out of limit. On the other side, if voltage value is, for example, +5% above nominal value, PV plant connection can potentially contribute to voltage increasing of, for example, +2%. It is now voltage increasing of 7%—permitted value. But, what if on the same feeder two more PV plants should be connected. This can lead to unallowed voltage increasing. This is valid for other voltage quality parameters. Norms/standards that describe influence of RES connection in PCC, in terms of voltage quality, are not present as such.

This area is usually controlled with regulations (documents), like grid codes, where influence of RES on voltage quality in electrical grid is defined. Those regulations vary between countries, regarding RES type, grid voltage level, current state of grid, etc. Literature review of permissible voltage quality deviation in PCC by RES connection is relatively modest.

Voltage increase after RES connection in PCC can be up to max. 3% [1, 4, 5]. P_{lt} value can be up to 0.65 according to [1] and 0.46 according to [4]. According to [13], compatibility level of P_{lt} in LV grid must be up to 0.8, and planning level of max. 0.4. P_{lt} must be up to 0.7 for MV grid [11]. Installed rated power of RES in LV grid can be up to 10% of rated power of transformer MV/LV [4].

It is clear that there is a wide literature approach about RES influence on voltage quality in distribution grid.

4. Hosting capacity of distribution grid and PV plant connection

Hosting capacity as a term can have different definitions. Considering voltage increasing, hosting capacity is max. permissible voltage increasing with connected DG, which causes max. voltage value strictly at the overvoltage limit. This leads to a conclusion that, considering voltage increasing, the hosting capacity is the amount of generation that gives a voltage increasing equal to the overvoltage margin [3].

The hosting capacity is the max. amount of generation that can be connected in some PCC without resulting in an unacceptable voltage quality or reliability for other customers. Hosting capacity can be defined as capacity of grid resistance on voltage quality changes due to RES connection in PCC. Distance between the distributed generator and the transformer is inversely proportional to the hosting capacity. The hosting capacity is smaller as distributed generator is further away from the transformer. The feeder size and length determine strongly hosting capacity of the grid [3].

Simply, hosting capacity of distribution grid presents strength of the grid. In other words, it presents capacity potential for RES connection in one PCC. Practically, it is not the same to connect one amount of RES rated power in urban distribution grid and on the end of long rural feeder made by small cross section of conductors.

On the other side, to be able to estimate the hosting capacity, the max. permissible voltage in grid should be known. Max. permissible voltage values are analyzed in previous subtitle.

It is very probably that greater amount of RES can be connected in urban LV grid than in rural LV grid, in PCC quite distant from transformer MV/LV. But, those are assumptions that should be verified by live case studies.

4.1 PV plant connection and hosting capacity of distribution grid: case studies

This section presents summarized results of voltage quality measurements in LV PCC for four PV plants. Considering hosting capacity of LV grid, PV plants are located in different places (in LV grid). Those grids are in different areas, with overhead lines and underground cables. It is all presented in **Figure 6**. PV plants are named with letters: A, B, C, and D.

PV plants A and C are placed in rural areas. PV plant A is placed not far from transformer MV/LV. PV plant C is placed at the end of the LV feeder. PV plants B and D are placed in urban areas. PV plant B is connected at LV buses of transformer station. PV plant D is set in city center near to transformer MV/LV.

Hosting capacity of LV grid in this case will be presented by three-phase short circuit power (S_{sc}) and PV plant rated power (S_r) ratio. Greater S_{SC}/S_r ratio should mean stronger grid—greater hosting capacity. In other words, as PCC is closer to transformer, greater distributed generator can be connected. **Table 8** presents main characteristics of PV plants and related LV grids in PCC.

Table 9 presents summary of voltage quality results for analyzed PV plants. Measurements were performed a week before and after PV plant connection. Measurements are in accordance with EN 50160:2011. Analyzed voltage quality parameters include slow voltage variations, voltage unbalance, long-term flicker, and total harmonic distortion of voltage. Symbol Δ presents difference of max. values before/after PV plant connection.

Before and after PV plant connection, max. values of voltage quality parameters are given. It is due to examination of worst practical cases.

Figure 7 presents voltage quality parameter results for PV plants, sorted by PCC weakness (S_{SC}/S_r). Lower S_{SC}/S_r value in PCC means weaker grid (smaller hosting capacity).

Voltage increase is proportional to weakness of LV grid. As shown in **Figure 7a**, as grid becomes stronger, voltage increase is lower (there can be decreasing too). For PV plants B and D, voltage decrease has occurred. It means that voltage has lower values after PV plant connection and PV plant does not have any influence on voltage variations (values). The restriction is measurement period of only 1 week before and after PV plant connection. But, the probability of crucial result changing with longer measurement period is minor. For all other parameters such as Δu (%), ΔP_{lt}, and ΔTHDU, there is no correlation between results with/without PV plant and hosting capacity of the grid.

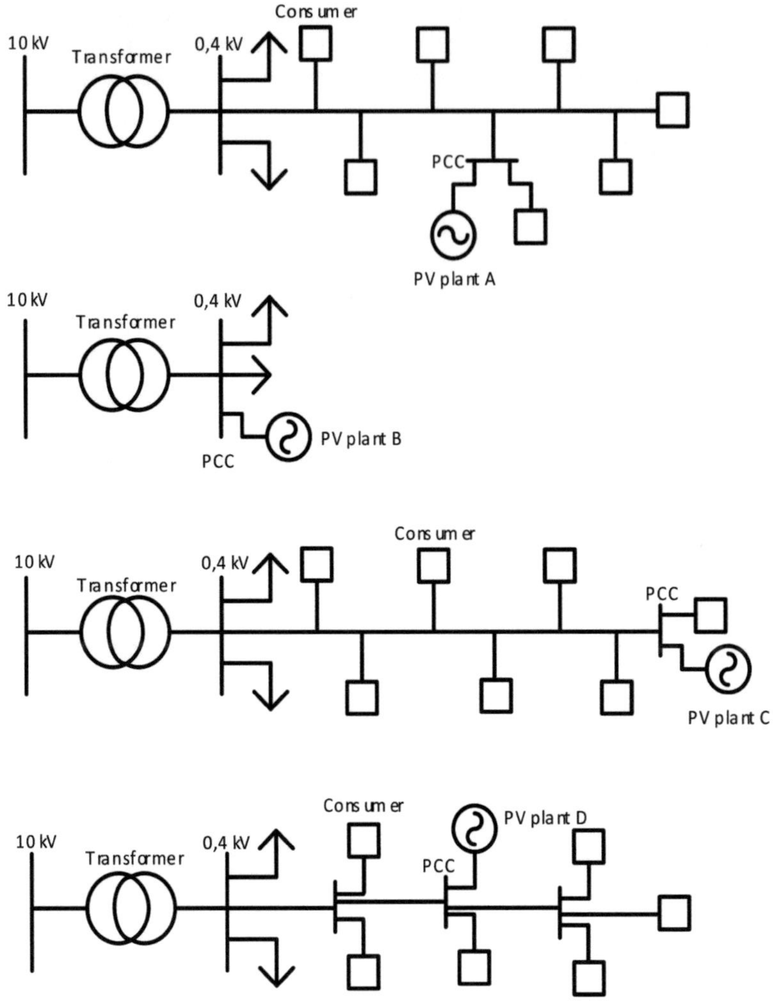

Figure 6.
Connection point of analyzed PV plants.

	A	B	C	D
Sr (MVA)	0.023	0.085	0.023	0.023
Ssc (MVA)	0.95	8.85	0.21	10
Ssc/Sr	41.30	104.11	9.13	434.78

Table 8.
Main characteristics of PV plants and related LV grids in PCC.

Conclusion is that strength of the grid has influence on voltage variations in distribution grid. It means that weak grid, regarding S_{SC}/S_r ratio, can lead to voltage quality problems in subjected distribution grid.

This cannot be concluded for other analyzed voltage quality parameters. **Figure 8** presents max. voltage values (voltage increase in % of rated voltage value) for PV plants before and after connection. Green and red lines present max. permissible voltage increase in PCC (in % of nominal value), assumed 5 and 7%. Those are realistic values in practice. If permitted max. voltage increase is 5%, there could be some technical problems. For PV plant B, even before PV plant connection,

		A		B		C		D	
		b.c	**a.c**	**b.c**	**a.c**	**b.c**	**a.c**	**b.c**	**a.c**
U (%)	Maximum	4.99	5.11	5.52	5.32	4.03	6.45	1.4	1.04
	Δ	0.12		−0.2		2.42		−0.36	
u (%)	Maximum	1.33	1.34	0.45	0.48	1.23	1.26	0.82	0.87
	Δ	0.01		0.03		0.03		0.05	
P_{lt}	Maximum	0.94	0.98	0.92	0.61	3.45	4.15	1.16	1.17
	Δ	0.04		−0.31		0.7		0.01	
THDU (%)	Maximum	2.67	2.7	2.65	2.7	1.81	1.85	2.39	2.41
	Δ	0.03		0.05		0.04		0.02	

Table 9.
Summary of voltage quality results for analyzed PV plants.

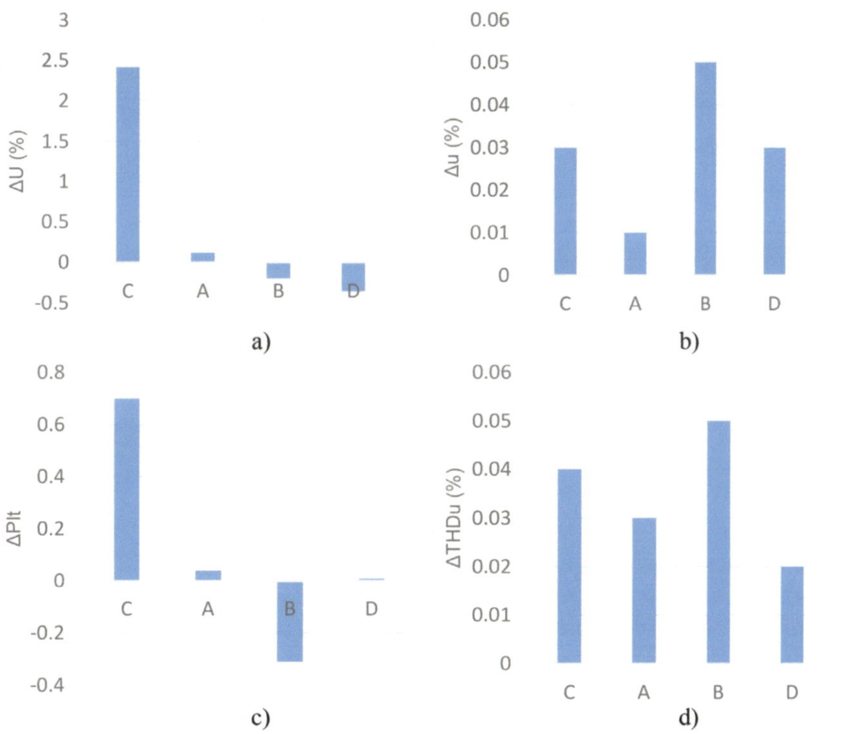

Figure 7.
Voltage quality parameter results for PV plants: (a) ΔU (%), (b) Δu (%), (c) ΔP_{lt}, and (d) ΔTHDU.

voltage values are increased. For PV plant A, voltage values are on margin. For PV plant C, after PV plant connection voltage values are increased. For permitted voltage increase of 7%, voltage values are in permitted area.

Regarding voltage variations, **Figure 9** shows max. voltage value increase (ΔU (%)) in analyzed LV grids after PV plant connection. Permitted voltage value increase of 2 and 3% is assumed. It means that voltage increase after PV plant connection can be 2 or 3%. Those values are the most exploited in literature.

It means that if max. voltage increase in LV grid is 3%, PV plant C can operate. But, if max. voltage increase is 2%, PV plant cannot stay connected on the grid (some changes must be performed). Those changes should be in way of increasing

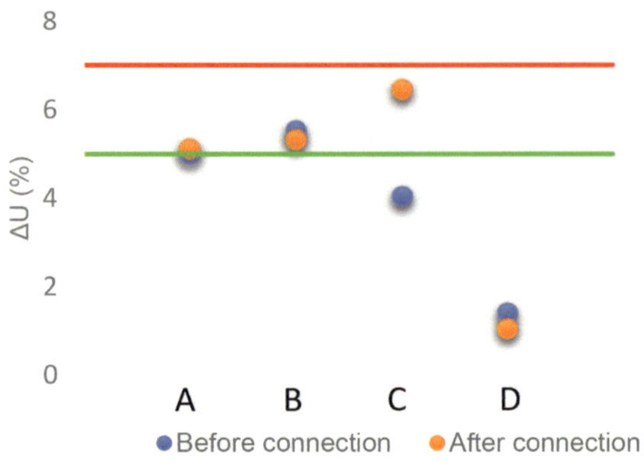

Figure 8.
Max. voltage values for PV plants before and after connection.

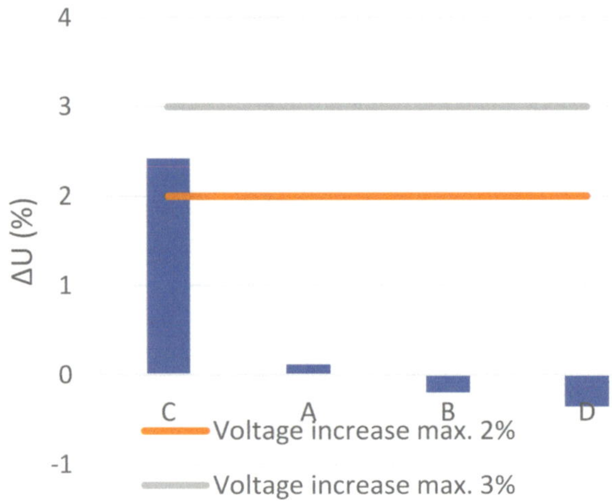

Figure 9.
Voltage increase in analyzed LV grids with connected PV plants.

short circuit current of the grid (greater hosting capacity of the grid) or in voltage regulations. Increasing of short circuit current in the grid means additional investment in grid (expending cross section of conductors or building new feeder for PV plant connection). Voltage regulation can be performed with inverters. This means regulation of power factor of PV plant (cos φ(P) or Q(U), for example). It is possible, but it usually means additional power losses in grid.

For example, let it be max. permitted voltage increase of 3% (considering nominal value) after PV plant connection. On the other side, let it be max. permitted voltage value increase of 5% (considering nominal value) in LV grid. It is obvious that PV plant B is out of permitted area for normal operation (**Figures 8** and **9**). But, even before PV plant connection, voltage values over permit value were increased. Moreover, after PV plant connection, voltage values were decreased. Another example is opposite. Let it be max. permitted voltage value increase of 2% for PV plant connection. On the other side, let it be max. permitted voltage value increase of 7% (considering nominal value) in LV grid. It is obvious that PV plant C is out of permitted area for normal operation (**Figures 8** and **9**). But, after PV plant connection, voltage values in PCC have permit value (regarding increase of voltage value more than 2% in PCC). This is area of sharing hosting capacity. It will be a challenge for further research.

4.2 General conclusions of PV plant connection regarding hosting capacity of LV distribution grid

Some conclusions for influence of PV plant on voltage quality at PCC, considering hosting capacity of that LV distribution grid, are as follows:

- PV plant has influence on voltage increase in PCC, considering hosting capacity of distribution grid. As the distribution grid is weaker, voltage increase is higher.
- THDU values, unbalance, and long-term flicker are apparently independent on hosting capacity of the grid.
- Contradiction is possible, in the way that voltage value increase after PV plant connection has permitted value, but in PCC, voltage value is unacceptable (increased); and in opposite, voltage value increase after PV plant connection has unacceptable value (increased voltage value), but voltage value in PCC is in limit values.

5. Comparison of theoretical and practical influence of PV plant connection in PCC considering slow voltage variations: case study

In this section, comparison of theoretical and practical results of PV plant influence in PCC considering slow voltage variations will be presented. This is important comparison, since in previous section, slow voltage variations (voltage increase) in PCC are turned out as potential problem for PV plant connection in distribution grid.

Practical results of measurements before and after PV plant connection are presented. A MV feeder (nominal voltage of 10 kV) with connected PV plant is observed. This PV plant will be named by letter E. Some basic parameters of PV plant and MV distribution grid are shown in **Table 10**.

Measurement site at 10 kV voltage level is shown in **Figure 10**.

Theoretically, steady-state slow voltage variation (voltage increase) in PCC can be formulated as follows:

$$\Delta u(\%) = 100 \cdot \frac{S_r}{S_{SC}} \cdot \cos\left(\psi_k + \phi\right) \tag{1}$$

where ψ_k is phase angle of the network impedance and ϕ is the phase angle of RES output current. Eq. 1 is simplified, but accurate enough for practical purposes [5]. Worst case considering voltage increase in PCC is for $\cos\left(\psi_k + \phi\right) = 1$. In this case study, result for voltage increase is **3.60%** (after PV plant connection) for $\cos\left(\psi_k + \phi\right) = 1$.

	E
Sr (MVA)	0.920
Ssc (MVA)	25.46
Ssc/Sr	27.67

Table 10.
Basic parameters of PV plant and MV distribution grid.

Figure 10.
Measurement site look for PV plant E.

In **Figures 11** and **12**, practical measurement of line voltages (in kV) for a week before and after PV plant connection is shown.

Table 11 presents max. voltage values measured according to EN 50160:2011 (10 min value) before and after PV plant connection.

It is visible that voltage increase due to PV plant connection has max. value of **2.18%**.

Max. voltage value increase after PV plant connection is close to 5%, that is, max. value for acceptable voltage value increase in MV grids according to [3].

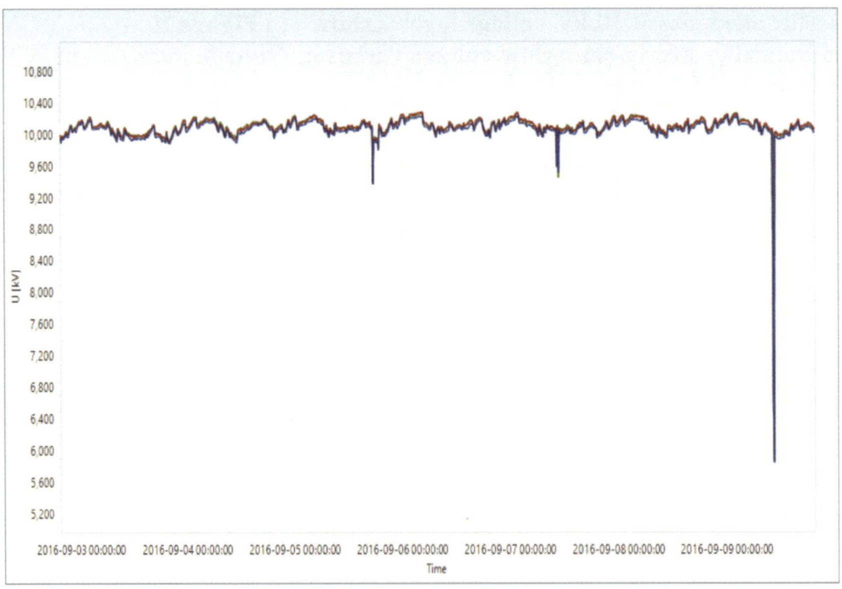

Figure 11.
Voltage measurement in PCC before PV plant connection.

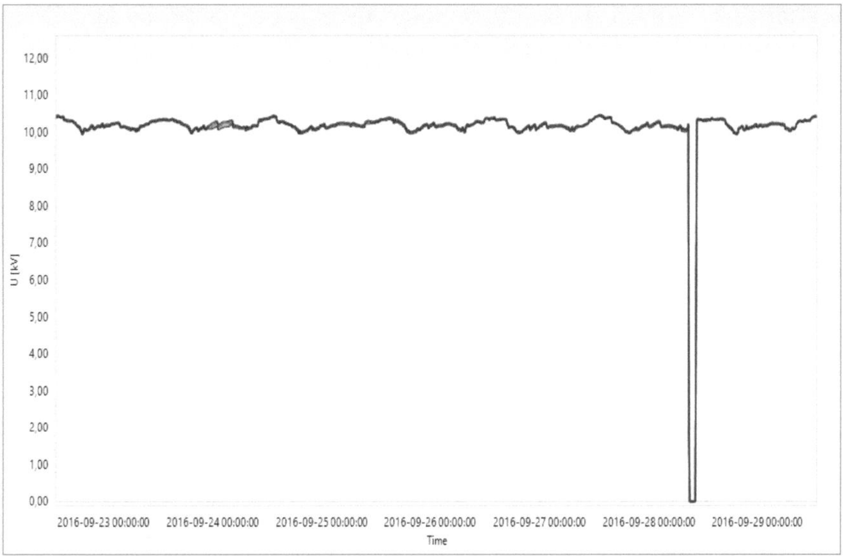

Figure 12.
Voltage measurement in PCC after PV plant connection.

		E	
		b.c	**a.c**
U (%)	Maximum	2.73	4.91
	Δ	2.18	

Table 11.
Slow voltage variation max. values before and after PV plant connection.

Comparing of theoretical and practical voltage value increase in PCC, after PV plant connection, can be performed. It is theoretical voltage increase of 3.60% comparing to 2.18% measured in practice. It is obvious that theoretical voltage value increase is greater due to assumption $\cos(\psi_k + \phi) = 1$. This assumption obviously leads to max. voltage value increase, which is hard to achieve in practice.

So, if theoretical result is acceptable, there should not be any problem in practice!

Table 12 shows other voltage quality results (u, P_{lt}, THDU) before and after PV plant E connection.

		E	
		b.c	**a.c**
u (%)	Maximum	0.40	0.40
	Δ	0	
P_{lt}	Maximum	1.57	1.70
	Δ	0.13	
THDU (%)	Maximum	2.29	2.51
	Δ	0.22	

Table 12.
Voltage quality parameters for PV plant E, before and after connection.

It is clear that for each showed parameter, there is no practically any difference in results before and after PV plant connection.

6. Conclusion

This chapter deals with PV plant influence on distribution grid. Literature review about PV plant influence on distribution grid is summarized. After that, case studies based on appropriate measurements are presented. The connection point of PV plant considering hosting capacity of distribution grid in terms of power quality was analyzed. Considering content of this chapter, following conclusions can be implemented:

- Power quality is regulated with some norms/standards. However, this area is not unique considering present regulations, which is presented in this chapter. There is lack of restriction condition documents for RES connection on distribution grid. In most of the countries exist local legal regulations, like grid codes, in which voltage quality is exploited in terms of RES influence on distribution grid.

- A case study of PV plant connection in LV grid is presented. The influence of PV plant on distribution grid in PCC, considering basic voltage quality parameters, was analyzed. Unbalance is quite similar before and after PV plant connection. THDU values are slightly increased after PV plant connection in this case. It is mainly in period when PV plant produces electrical energy. In percent value, for a week period of measurement, it is negligibly increasing. It is obvious that voltage value in V is related to P/P_n (%) ratio. Generally, greater P/P_n (%) ratio (greater generation of PV plant) means greater voltage value in volts (voltage increasing). PV plant has influence on voltage increasing due to its electrical energy generation. It was found that main potential problem for PV plant connection in distribution grid can be voltage increase. Finally, it is obvious that PV plant does not contribute to increasing of P_{lt} at PCC.

- A comparison of PV plants influence on distribution grid for different connection points of PV plants, considering strength of the grid, is presented. Strength of the grid is in fact hosting capacity of the grid. Four PV plants with different connection points (different S_{SC}/S_r ratio) were compared. PV plant connection point has influence on voltage increasing in PCC, considering hosting capacity of distribution grid. As the distribution grid is weaker, voltage increase is higher. THDU values, unbalance, and long-term flicker are apparently independent on hosting capacity of the grid. Contradiction is possible, in the way that voltage increase by PV plant connection has permitted value, but in PCC, voltage is increased. And in opposite, voltage increase by PV plant connection is increased, but voltage value in PCC is in limit values. This statement is valid for LV and MV grid. This area of contradiction falls in area of hosting capacity sharing and it is challenge for future work.

- Comparing of theoretical and practical voltage increase values in PCC was presented. Theoretical voltage value increase is greater due to the assumption $\cos(\psi_k + \phi) = 1$. This assumption obviously leads to max. voltage increase, which is hard to achieve in practice. So, if theoretical result is acceptable, there should not be any problem in practice. But, this analysis should not be only conducted analysis. It should be framework for more detailed analysis of PV plant influence on distribution grid. For each new PV plant connection, this analysis should be implemented.

Main contributions of this chapter can be summarized as follows:

- Proving of correlation of voltage value increase considering P/P_n (%) ratio increase. As the distribution grid is weaker (low S_{SC}/S_r ratio), voltage increase is higher.

- Voltage value increase in practice should not be greater than increase obtained by calculation considering assumption that $\cos(\psi_k + \phi) = 1$.

- PV plant does not contribute in increasing of P_{lt} and unbalance at PCC. Increasing of THDU in PCC is negligible after PV plant connection. These parameters are apparently independent of hosting capacity of the grid.

Acknowledgements

The authors would like to thank to JP Elektroprivreda HZ HB d.d Mostar for helping in issuing this chapter.

Author details

Ivan Ramljak* and Drago Bago
J. P. Elektroprivreda HZ HB d.d Mostar, Bosnia and Herzegovina

*Address all correspondence to: ivan.ramljak@ephzhb.ba

References

[1] Sikorski T, Rezmer J. Distributed generation and its impact on power quality in low-voltage distribution networks. In: Luszcz J, editor. Power Quality Issues in Distributed Generation. Rijeka, Croatia: IntechOpen; 2015. pp. 1-39. DOI: 10.5772/59895

[2] Jenkins N, Allan P, Crossley D, Kirschen D, Strbac G. Embedded Generation. 1st ed. London: The Institution of Engineering and Technology; 2000. 292 p. DOI: 10.1049/PBPO031E

[3] Bollen M, Hassan F. Integration of Distributed Generation in the Power System. 1st ed. Chichester: Wiley; 2011. 528 p. ISBN: 978-1-118-02902-2

[4] Mgaya EV, Muller Z. The impact of connecting distributed generation to the distribution system. Acta Polytechnica. 2007;**47**:96-101. ISSN: 1805-2363

[5] Papathanassiou SA, Hatziargyriou ND. Technical requirements for the connection of dispersed generation to the grid. In: Proceedings of IEEE PES Summer Meeting; 15-19 July 2001; Canada. New York: IEEE; 2002. pp. 749-755

[6] Papathanassiou SA. A technical evaluation framework for the connection of DG to the distribution network. Electric Power Systems Research. 2007;**77**:24-34. DOI: 10.1016/j.epsr.2006.01.009

[7] Toubeau JF, Klonari V, Lobry J, De Greve Z, Vallee F. Planning tools for the integration of renewable energy sources. In: Cao W, Hu Y, editors. Renewable Energy. Rijeka, Croatia: IntechOpen; 2016. pp. 227-260. DOI: 10.5772/61758

[8] Allegranza V, Ardito A, De Berardinid E, Delfanti M, Lo Schiavo L. Assessment of Short Circuit Power Levels in HV and MV Networks with Respect to Power Quality [Internet]. 2007. Available from: http://www.cired.net/publications/cired2007/pdfs/CIRED2007_0161_paper.pdf [Accessed: October 10, 2018]

[9] Meyer J, Ammeter U, Hanzlik J, Zierlinger J. Methods for the assessment of network disturbances in distribution networks. In: 3rd International Conference on Harmonics and Quality of Power; 28 September-1 October 2008; Australia. New York: IEEE; 2008. pp. 1-6

[10] Leonowicz Z, Rezmer J, Sikorski T, Szymanda J, Kotyla P. Wide-area system of registration and processing of power quality data in power grid with distributed generation: Part I. System description, functional tests and synchronous recordings. In: 14th International Conference on Environment and Electrical Engineering; 10-12 May 2014; Poland. New York: IEEE; 2005. pp. 175-181

[11] Golovanov N, Lazatoiu GC, Roscia M, Zaninelli D. Monitoring power quality in small scale renewable energy sources supplying distribution systems. In: Zobaa AF, editor. Power Quality Issues. Rijeka, Croatia: IntechOpen; 2013. pp. 227-260. DOI: 10.5772/53464

[12] Cobben S. Power quality monitoring and classification. In: Eberhard A, editor. Power Quality. Rijeka, Croatia: IntechOpen; 2011. pp. 103-128. DOI: 10.5772/595

[13] Hernandez JC, Ortega MJ, De la Cruz J, Vera D. Guidelines for the technical assessment of harmonic, flicker and unbalance emission limits for PV-distributed generation. Electric Power Systems Research. 2011;**81**: 1247-1257. DOI: 10.1016/j.epsr.2011.03.012

[14] Chicco G, Schlabbach J, Spertino F. Experimental assessment of the waveform distortion in grid-connected photovoltaic installations. Solar Energy. 2009;**83**:1026-1039. DOI: 10.1016/j.solener.2009.01.005

[15] Ramljak I, Bago D. Influence of PV plant connection on voltage quality parameters considering connection point in distribution grid. In: Proceedings of the First International Colloquium on Smart Grid Metrology (SmaGriMet); 24-27 July 2018; Croatia. New York: IEEE; 2018. pp. 1-5

[16] Eltawil MA, Zhao Z. Grid-connected photovoltaic power systems: Technical and potential problems—A review. Renewable and Sustainable Energy Reviews. 2009;**14**:112-129. DOI: 10.1016/j.rser.2009.07.015

[17] Kopicka M, Ptacek M, Toman P. Analysis of the power quality and the impact of photovoltaic power plant operation on low-voltage distribution network. In: Proceedings of Electric Power Quality and Supply Reliability Conference; 11-13 June 2014; Estonia. New York: IEEE; 2014. pp. 99-102

[18] Kontogiannis KP, Vokas GA, Nanou S, Papathanassiou S. Power quality field measurements on PV inverters. International Journal of Advanced Research in Electrical, Electronics and Instrumentation Engineering;**2013**(2): 5301-5314. ISSN: 2278-8875

[19] Barbu V, Chicco G, Corona F, Golovanov N, Spertino F. Impact of a photovoltaic plant connected to the MV network on harmonic distortion: An experimental assessment. Electrical Engineering. 2013;**75**:179-193. ISSN: 2286-3540

[20] Canova A, Giaccone L, Spertino F, Tartaglia M. Electrical impact of photovoltaic plant in distributed network. In: Proceedings of IEEE Industry Application Annual Meeting; 23-27 September 2007; USA. New York: IEEE; 2007. pp. 1450-1455

[21] Ramljak I, Ramljak I. PV plant connection in urban and rural LV grid: Comparison of voltage quality results. In: 2nd ISPQ Symposium; 21-24 June 2018; Bosnia and Herzegovina. 2018. pp. 1-8

[22] Ivanovici TD, Ionel M, Dogaru-Ulieru V, Mihaescu S. The influence of photovoltaic systems on low-voltage grids. WSEAS Transactions on Environment and Development. 2011;**3**: 65-74. ISSN: 1790-5079

[23] Papaioannou IT, Bouhouras AS, Marinopoulus AG, Alexiadis MC, Demoulias CS, Labridis DP. Harmonic impact of small photovoltaic systems connected to the LV distribution network. In: Proceedings of 5th International Conference on the European Electricity Market; 28-30 May 2009; Portugal. New York: IEEE; 2008. pp. 1-6

[24] Patsalides M, Evagorou D, Makrides G, Achillides Z, Georghiou GE, Stavrou A, et al. The Effect of Solar Irradiance on the Power Quality Behaviour of Grid Connected Photovoltaic System [Internet]. 2007. Available from: http://icrepq.com/icrepq07/284-patsalides.pdf [Accessed: September 15, 2018]

[25] Tokić A, Milardić V. Power Quality. 1st ed. Tuzla: PrintCom; 2015. 319 p. (in Croatian)

Chapter 10

Electrical Field Distribution along HVDC GIL Spacer in SF_6/N_2 Gaseous Mixture

Boxue Du, Jin Li and Hucheng Liang

Abstract

Many researchers have proposed a variety of mathematical models to simulate the surface charge accumulation process of DC-GIS/GIL spacers. However, few of them took the gas collision ionization and charge trapping-detrapping process into consideration. This chapter combined the plasma hydrodynamics and charge transport equations and built a modified model. Some conclusions are shown as follows: for the basin-type spacer, the surface charge has the same polarity as the applied voltage on the lower surface but the opposite polarity on the upper surface. For the disc-type spacer, the surface charge has the same polarity as the applied voltage near the shell but the opposite polarity near the conductor under negative voltage. But under positive voltage, negative charge exists almost on the whole surface. The most serious distortion of the electric field occurs at the triple junction of epoxy spacer. Under load condition, there is an obvious temperature rise on the conductor due to joule heating, which has a great influence on the electric field distribution. The application of shielding electrodes has the function of field grading at the triple junction, which can be referred in the DC GIS/GIL design.

Keywords: DC-GIS/GIL, epoxy spacer, temperature, surface charge accumulation, electric field distribution, shielding electrode

1. Introduction

Over the last few decades, the high voltage direct-current (HVDC) transmission technology have been developed to carry out long distance and high capacity power transmission [1, 2]. Thanks to their high reliability and low footprint, Gas Insulated Switchgears and Gas Insulated Lines (GIS/GIL) have been widely used in AC power system but scarcely in DC system [3, 4]. One important reason is that surface charge is easy to accumulate on the spacer surface under DC voltage, which leads to a substantial electric field distortion and even flashover faults [5]. Therefore, the surface charge accumulation and electric field distribution along the GIS/GIL spacers need to be understood.

In order to improve the reliability of GIS/GIL, researches aiming on the surface charge measurement and simulation have been conducted worldwide [6–8]. Some researchers built a scaled-down GIS/GIL and measured the surface charge distribution on the spacers [9–12]. Their measurements were conducted offline because the applied voltage on the conductor can easily cause damage to the electrostatic probe. The poor accuracy of offline measurement makes their measurements differ from each other. Besides, the accumulated surface charge at the triple junction can hardly be detected by the probe, which actually has a very significant effect on the electric field distribution.

Simulation is another way to study the surface charge distribution on spacers. Many researchers have proposed a variety of mathematical models to simulate the surface charge accumulation process [13–17]. However, few of them took the gas collision ionization and charge trapping-detrapping process into consideration. This paper detailed the corona discharge and charge trapping-detrapping process by combining the plasma hydrodynamics and charge trapping-detrapping equations together. Under load conditions, GIS/GIL has an obvious temperature rise on the conductor. Considering there are many temperature-dependent factors in the surface charge accumulation process, the heat transfer equations are also coupled in this model to obtain the temperature distribution. Besides, the effects of shielding electrodes on the distorted electric field distribution are also studied.

2. Mathematical model

2.1 Geometric model

Figure 1 shows the geometric models of GIS with different spacer structures. There are two commonly used types of spacers as shown in **Figure 1a** which is shaped like a basin and **Figure 1b** which is shaped like a disc. In the simulation, the external diameter of the central conductor is set to 60 mm and the internal diameter of the shell is set to 330 mm. DC ±320 kV is applied on the central conductor and the shell is grounded.

2.2 Corona discharge in SF_6/N_2 mixture

In this paper, the SF_6/N_2 mixture is set as the insulating gas in GIS. The percentage of SF_6 is set to 50% and the pressure is set to 0.5 MPa.

The drift-diffusion process of the electron density n_e and electron energy density n_ε can be described as [18],

$$\frac{\partial}{\partial t}(n_e) + \nabla \cdot \left[-n_e\left(\mu_e \cdot \vec{E}\right) - D_e \cdot \nabla n_e\right] = R_e \tag{1}$$

$$\frac{\partial}{\partial t}(n_\varepsilon) + \nabla \cdot \left[-n_\varepsilon\left(\mu_\varepsilon \cdot \vec{E}\right) - D_\varepsilon \cdot \nabla n_\varepsilon\right] + E \cdot \Gamma_e = R_\varepsilon \tag{2}$$

where the electron mobility μ_e, electron diffusivity D_e, energy mobility μ_ε and energy diffusivity D_ε are computed by solving the Two-term Boltzmann

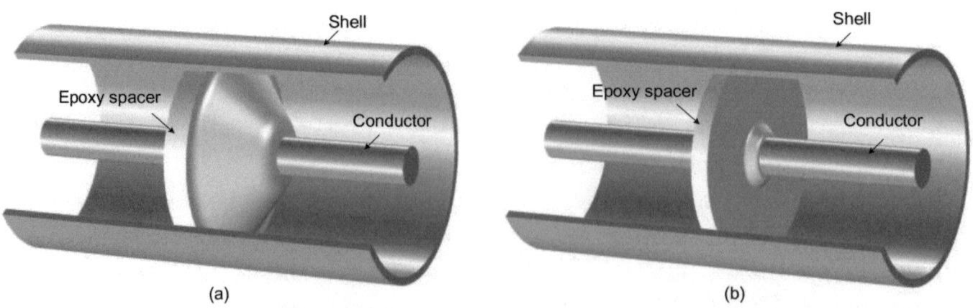

Figure 1.
The geometric models of GIS with different spacer structures: (a) basin-type and (b) disc-type.

equation [19]. The electron source R_e and energy source R_ε can be obtained from Eqs. (3) and (4). This paper takes into consideration M electron-participating reactions (N two-body reactions and M-N three body reactions) and P energy-losing reactions,

$$R_e = \sum_{j=1}^{N} x_j k_j N_n n_e \Delta n_{ej} + \sum_{j=N+1}^{M} x_{j1} x_{j2} k_j N_n{}^2 n_e \Delta n_{ej} \tag{3}$$

$$R_\varepsilon = \sum_{j=1}^{P} x_j k_j N_n n_e \Delta \varepsilon_j \tag{4}$$

$$N_n = p/k_B T \tag{5}$$

where x_j, x_{j1} and x_{j2} are the mole fractions of the target species for reaction j; N_n is the total neutral density (1/m^3); p is the gas pressure (Pa); T is the temperature (K); Δn_{ej} is the electron increment of reaction j; $\Delta \varepsilon_j$ is the energy loss of reaction j (V) and k_j is the rate coefficient for reaction j (m^3/s), which is also obtained from the two-term Boltzmann equation [19].

For heavy species, the space–time dependent mass fraction of species k is controlled by the following equations,

$$\rho \frac{\partial}{\partial t}(w_k) - \nabla \cdot \vec{j_k} = R_k \tag{6}$$

$$\vec{j_k} = \rho w_k \vec{V_k} \tag{7}$$

$$\vec{V_k} = D_k \left(\frac{\nabla w_k}{w_k} + \frac{\nabla M_n}{M_n} \right) - z_k \mu_k \vec{E} \tag{8}$$

where ρ is the gas density (kg/m^3); w_k is the mass fraction of species k; j_k is the flux of species k; R_k is the rate expression for species k (kg/(m^3·s)); V_k is the multicomponent diffusion velocity of species k (m/s); D_k is the mixture averaged diffusion coefficient (m^2/s); M_n is the mean molar mass of the gas mixture (kg/mol); z_k is the charge number of species k; μ_k is the mixture averaged mobility of species k (m^2/(V s)); E is the electric field strength (V/m). This paper takes into consideration M reactions (N two-body reactions and M-N three-body reactions) that change the mass fraction of species k,

$$R_k = \frac{M_k}{N_A} \left(\sum_{j=1}^{N} k_j x_{j1} x_{j2} N_n{}^2 \Delta n_{kj} + \sum_{j=N+1}^{M} k_j x_{j1} x_{j2} x_{j3} N_n{}^3 \Delta n_{kj} \right) \tag{9}$$

$$x_k = \frac{M_n w_k}{M_k} \tag{10}$$

where x_{j1}, x_{j2} and x_{j3} are the mole fractions of the species involved in reaction j; N_A is the Avogadro constant; Δn_{kj} is the species k increment of reaction j; M_k is the molar mass of species k (kg/mol). Eq. (9) shows the relationship between the mole fraction and mass fraction of species k.

The mixture averaged diffusion coefficient D_k is defined as Eq. (11) and the mixture averaged mobility μ_k can be calculated by Eq. (12) according to the Einstein's relation,

$$D_k = \frac{1 - w_k}{\sum_{j \neq k}^{Q} x_j / D_{k,j}} \tag{11}$$

$$\mu_k = \frac{eD_k}{k_B T} \tag{12}$$

where e is the unit charge (C); k_B is the Boltzmann's constant (J/K); T is the gas temperature (K) and $D_{k,j}$ is the binary diffusion coefficient between species k and j.

When electrons reach the conductor, the shell and the spacer surface, the boundary conditions for the electrons and electron energy flux can be defined as,

$$\vec{n} \cdot \vec{\Gamma}_e = \frac{1}{4} v_{e,th} n_e - \sum_p \gamma_p \left(\vec{\Gamma}_p \cdot \vec{n} \right) \tag{13}$$

$$\vec{n} \cdot \vec{\Gamma}_\varepsilon = \frac{5}{12} v_{e,th} n_\varepsilon - \sum_p \varepsilon_p \gamma_p \left(\vec{\Gamma}_p \cdot \vec{n} \right) \tag{14}$$

$$v_{e,th} = \left(\frac{8 k_B T_e}{\pi m_e} \right)^{1/2} \tag{15}$$

$$T_e = \frac{2 n_\varepsilon}{3 n_e} \tag{16}$$

The second term on the right-hand side of Eqs. (13) and (14) are the gain of electrons and electron energy due to secondary emission effects; γ_p is the secondary emission coefficient, which is set to 0.1 for the cathode and 0 for the anode; ε_p is the mean energy of the secondary electrons, which is set to 5 eV; $v_{e,th}$ is the electron thermal velocity (m/s); T_e is the electron temperature (eV).

For ions, the boundary condition can be described as,

$$\vec{n} \cdot \vec{j}_k = \frac{1}{4} v_{k,th} \rho w_k + \alpha \rho w_k z_k \mu_k \left(\vec{E} \cdot \vec{n} \right) \tag{17}$$

$$v_{k,th} = \left(\frac{8 k_B T}{\pi m_k} \right)^{1/2} \tag{18}$$

where $v_{k,th}$ is the thermal velocity of species k (m/s); T can be considered to be the gas temperature in GIS (K); $\alpha \bullet z_k / |z_k| = 0$ if the electric field is directed away from the boundary and $\alpha \bullet z_k / |z_k| = 1$ if the electric field is directed toward the boundary.

The discharge processes of SF_6/N_2 mixture are pretty complex and some of them are still unclear. In this simulation, only those primary reactions are taken into consideration. **Table 1** lists some typical physicochemical reactions considered in this paper after some reduction. The cross sections and energy losses of those collision reactions are extracted from papers [20–23].

2.3 Charge transport in epoxy spacer

When ions and electrons reach the spacer surface, charges are considered to be injected into the skin layer of epoxy owing to the surface reactions ($N_2^+ \rightarrow N_2$, $SF_6^+ \rightarrow SF_6$, $SF_6^- \rightarrow SF_6$) [14]. The charge transport process in epoxy spacer volume is mainly controlled by the following equations (19)-(21) [24–26]. Note that the epoxy spacer in this numerical model is considered to be clean enough, the surface current due to surface transmission can be ignored since there is no special surface treatment to the spacer [27]. Therefore, the surface transmission process is not taken into consideration in this paper.

No.	Formula	Type	$\Delta\varepsilon$ (eV)	Δn_e
1	$N_2 + e \rightarrow 2e + N_2^+$	Ionization	−16	1
2	$SF_6 + e \rightarrow 2e + SF_6^+$	Ionization	−15.8	1
3	$N_2 + e \rightarrow e + N_2$	Excitation	−8	0
4	$SF_6 + e \rightarrow e + SF_6$	Excitation	−10	0
5	$N_2 + e \rightarrow e + N_2$	Elastic	0	0
6	$SF_6 + e \rightarrow e + SF_6$	Elastic	0	0
7	$SF_6 + e \rightarrow SF_6^-$	Attachment	—	−1
8	$SF_6^+ + e \rightarrow SF_6$	Reaction	—	−1
9	$SF_6^+ + SF_6^- \rightarrow 2SF_6$	Reaction	—	—
10	$e + N_2^+ + N_2 \rightarrow 2N_2$	Reaction	—	−1
11	$2e + N_2^+ \rightarrow N_2 + e$	Reaction	—	−1

Table 1.
Table of some physicochemical reactions in the corona discharge.

$$\frac{\partial n_{mb}}{\partial t} + \nabla \cdot \vec{J}_c^{e} = n_{tr}P_{de} - n_{mb}P_{tr} - R_1 n_{mb} h_{tr} - \frac{1}{\tau}\Delta n \tag{19}$$

$$\frac{\partial n_{tr}}{\partial t} = n_{mb}P_{tr} - R_2 n_{tr} h_{mb} \tag{20}$$

$$\vec{J}_c^{e} = \mu_{e,2} n_{mb} \cdot \nabla V - D_{e,2} \cdot \nabla n_{mb} \tag{21}$$

$$\frac{\partial h_{mb}}{\partial t} + \nabla \cdot \vec{J}_c^{h} = h_{tr}P_{de} - h_{mb}P_{tr} - R_2 n_{tr} h_{mb} - \frac{1}{\tau}\Delta h \tag{22}$$

$$\frac{\partial h_{tr}}{\partial t} = h_{mb}P_{tr} - R_1 n_{mb} h_{tr} \tag{23}$$

$$\vec{J}_c^{h} = -\mu_h h_{mb} \cdot \nabla V - D_h \cdot \nabla h_{mb} \tag{24}$$

where n_{mb} is the mobile electron density (1/m^3); n_{tr} is the trapped electron density (1/m^3); h_{mb} is the mobile hole density (1/m^3); h_{tr} is the trapped hole density (1/m^3); $\vec{J}_c^{e}$ and $\vec{J}_c^{h}$ are the flux of mobile electrons and holes (1/(m^2•s)); $\mu_{e,2}$ and μ_h are the mobility of electrons and holes in epoxy spacer (m^2/(V•s)); $D_{e,2}$ and D_h are the diffusion coefficient of electrons and holes (m^2/s); V is the potential (V); P_{tr} and P_{de} are the trapping and de-trapping coefficient of electrons and holes (1/s); Δn and Δh are the non-equilibrium carrier density in epoxy spacer (1/m^3); R_1 and R_2 are the recombination coefficients (m^3/s); τ is the lifetime of non-equilibrium carriers (s).

In this paper, the volume conductivity, carrier mobility and diffusion coefficient of spacer are assumed as following:

$$\sigma = 40000 \cdot e^{-12860/T} \tag{25}$$

$$\mu_{e,2} = \mu_h = \left(1.5 \times 10^{-4}\right) \cdot e^{-6470/T} \tag{26}$$

$$D_{e,2} = D_h = \mu_{e,2} \cdot \frac{k_B T}{e} \tag{27}$$

where T is the temperature of spacer (K). Supposing the mobile electron density n_0 and hole density h_0 have the same value, the intrinsic carrier density can be described as,

$$n_0 = h_0 = \frac{\sigma}{2e\mu_{e,2}} \tag{28}$$

Inspired from the theories on the non-equilibrium carriers in the semiconductor physics, this paper considers the product of n_O and h_O as a constant in the epoxy spacer under thermal equilibrium condition. Supposing the electron and hole density n_{mb} and h_{mb} are in an unbalanced state, Δn and Δh can be obtained by solving the Eq. (29). Positive Δn and Δh represent the electron–hole recombination process, negative Δn and Δh represent the generation of electron–hole pairs.

$$n_0 h_0 = (n_{mb} - \Delta n)(h_{mb} - \Delta h) \tag{29}$$

$$\Delta n = \Delta h = \frac{1}{2}\left[(n_{mb} + h_{mb}) - \sqrt{(n_{mb} - h_{mb})^2 + 4n_0 h_o}\right] \tag{30}$$

At the gas–solid interface, the boundary condition can be described as,

$$-\vec{n} \cdot \vec{\Gamma}_{n_{mb}} = \vec{n} \cdot \vec{j_e} \tag{31}$$

$$-\vec{n} \cdot \vec{\Gamma}_{h_{mb}} = \vec{n} \cdot \vec{j_i} \tag{32}$$

where $\vec{J_e}$ and $\vec{J_i}$ is the electron and ion flux through the gas–solid interface in the corona discharge model (1/m^2·s).

At the metal-insulation interface, the boundary condition can be described as,

$$-\vec{n} \cdot \vec{j_e} = \alpha \cdot J_{sch-e} - (1-\alpha) \cdot \vec{n} \cdot \vec{E} \cdot \mu_{e,2} \cdot n_{mb} \tag{33}$$

$$-\vec{n} \cdot \vec{j_h} = (1-\alpha) \cdot J_{sch-h} - \alpha \cdot \vec{n} \cdot \vec{E} \cdot \mu_h \cdot h_{mb} \tag{34}$$

$$J_{sch-h} = J_{sch-h} = \frac{AT^2}{e} \cdot \exp\left(\frac{-\phi + \sqrt{E \cdot e^3/4\pi\varepsilon_0\varepsilon_r}}{k_B T}\right) \tag{35}$$

where $J_{sch\text{-}e}$ and $J_{sch\text{-}h}$ are the Schottky injection electron and hole flux (1/(m^2•s)); A is the Richardson constant (C/m^2·s·K^2), ϕ is the contact potential barrier between the spacer and metal (eV); α = 0 if the electric field is directed away from the boundary and α = 1 if the electric field is directed toward the boundary. **Table 2** lists some parameters of epoxy spacer used in the charge transport model.

Parameters	Value
Trapping and de-trapping coefficients, 1/s	
P_{tr}	7.0×10^{-3}
P_{de}	7.7×10^{-5}
Recombination coefficients, m^3/s	
R_1	8.0×10^{-19}
R_2	8.0×10^{-19}
Potential barrier, eV	
ϕ	1.25

Table 2.
Table of some parameters in the charge transport model.

2.4 Poisson equation

The corona discharge model comprises p species (electrons, positive ions and negative ions) and the charged number of specie k is z_k. Thus, in the SF_6/N_2 gas, Poisson equation can be described as,

$$-\nabla^2 V = e\left(\sum_{1}^{p} n_k z_k\right)/\varepsilon_0 \tag{36}$$

where e is the elementary charge (C); n_k is the density of species k (1/m^3); ε_0 is the vacuum dielectric constant (F/m).

In the epoxy spacer, Poisson equation is described as,

$$-\nabla^2 V = e(h_{mb} + h_{tr} - n_{mb} - n_{tr})/(\varepsilon_0 \varepsilon_r) \tag{37}$$

where ε_r is the relative dielectric constant of epoxy spacer, which is set to 4.

2.5 Heat transfer in GIS

In GIS, the heat transfer is controlled by the following equation [28],

$$\rho C_p \frac{\partial T}{\partial t} + \rho C_p \vec{u} \cdot \nabla T = \nabla \cdot (k \nabla T) + Q_\rho \tag{38}$$

$$\rho \frac{\partial \vec{u}}{\partial t} + \rho\left(\vec{u} \cdot \nabla\right) \vec{u} = \nabla \cdot \left[-p \vec{I} + \mu\left(\nabla \vec{u} + \left(\nabla \vec{u}\right)^T\right) - \frac{2}{3}\mu\left(\nabla \cdot \vec{u}\right) \vec{I}\right)] + \vec{G} \tag{39}$$

$$\frac{\partial \rho}{\partial t} + \nabla \cdot \left(\rho \vec{u}\right) = 0 \tag{40}$$

where ρ is the density of SF_6/N_2 gas, epoxy spacer or shell (kg/m^3); C_p is the heat capacity (J/(kg·K)); k is the thermal conductivity (W/(m·K)); T is the thermodynamic temperature (K); $\boldsymbol{u}$ is the natural convection velocity field of SF_6/N_2 gas (m/s); μ is the coefficient of kinetic viscosity, which is set to 1.7×10^{-5} Pa s; $\boldsymbol{G}$ is the gravity vector (N/m^3); Q_ρ is the heat source in the spacer (W/m^3), which is defined as,

$$Q_\rho = eE^2\mu(n_{mb} + h_{mb}) \tag{41}$$

The density of SF_6/N_2 gas in GIS is defined as,

$$\rho = \frac{M_n p}{RT} \tag{42}$$

where M_n is the mean molar mass (kg/mol).

The heat radiation of conductor, spacer and shell is described as,

$$-\vec{n} \cdot (-k\nabla T) = \varepsilon\sigma\left(T_{amb}^4 - T^4\right) \tag{43}$$

where σ is the Avogadro Boltzmann constant; T_{amb} is the environment temperature; ε is the radiation coefficient, which is set to 0.8 for the spacer, 0.05 for the conductor and 0.15 for the shell.

In addition to the heat radiation at the shell-air interface, the natural convection of air can be described as,

Parameters	Spacer	Conductor/shell	SF_6/N_2
Thermal conductivity k, W/(m K)	0.71	165	0.019
Heat capacity C_p, J/(kg K)	700	900	817
Density ρ, kg/m^3	2200	2700	—

Table 3.
Table of some parameters in the heat transfer model.

$$-\vec{n}\cdot(-k\nabla T) = h(T_0 - T) \quad (44)$$

where T_0 is the environment temperature (K) and h is the heat transfer coefficient, which is set to 10 W/(m^2·K). **Table 3** lists some parameters of GIS used in the heat transfer model.

Note that many simulation parameters such as $\mu_{e,2}$, μ_h, $D_{e,2}$, D_h, P_{de}, n_0, $v_{k,th}$, $J_{sch\text{-}e}$, $J_{sch\text{-}h}$, ρ and so on are all temperature-dependent. As a result, the temperature distribution has a great influence on the surface charge accumulation process. In this paper, the environment temperature is set to 293 K and the temperature of central conductor is respectively set to 330, 350 and 370 K.

3. Results and discussion

This paper built a temporal numerical model to simulate the surface charge accumulation process. But what should be cared about is the steady-state surface charge and electric field distributions. After many calculations, it can be drawn that the temporal change in surface charge density slows down with time under both negative and positive voltages. However, it takes a much longer time for the surface charge density to reach the steady state under negative voltage than that under positive voltage. For convenience, the simulation time is set to be 1 hour. After 1 hour, the increment of surface charge density has been slow enough. The difference between the surface charge distribution and the steady state is less than 10%. As a result, the surface charge density after 1 hour can be approximately considered as the steady-state distribution. It is necessary to note that it takes more than 20 hours for the measured surface charge distribution to reach the steady state [11, 12], which is much longer compared with the simulation. This phenomenon is mainly caused by the non-uniformity of material properties in the surface charge measurement.

3.1 Effects of surface charge

When GIS is under load condition, the central conductor has an obvious temperature rise due to joule heating. In this paper, the room temperature is set to 293 K. **Figure 2** shows the simulated temperature distribution and gas convection in DC-GIS when the conductor temperature is set to 350 K. The temperature goes down from the conductor to the shell and the shell temperature is about 17 K higher than the room temperature. Because of the gas convection, the spacer temperature is higher on the lower surface than that on the upper surface. There is a large temperature range along the epoxy spacers, thus it is important to study the temperature effects on the surface charge and electric field distribution.

Figure 3 shows the simulated surface charge distribution on the basin-type spacer, which is similar to the measured results in paper [9, 11]. The conductor temperature is set to 350 K. The color variation from red to blue represents the

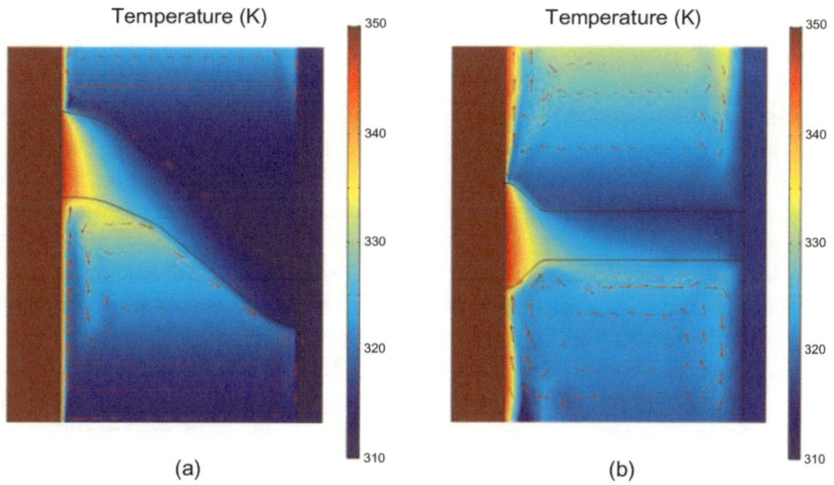

Figure 2.
The simulated temperature distribution and gas convection in DC-GIS.

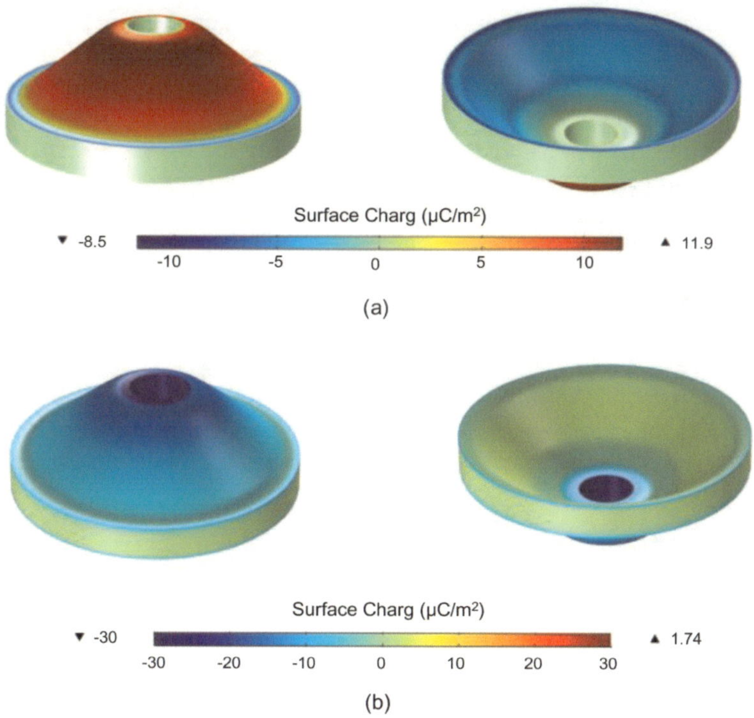

Figure 3.
The simulated surface charge distribution on the basin-type spacer at 350 K: (a) negative voltage and (b) positive voltage.

conversion of surface charge polarity from positive to negative. In this paper, the convexity and concave of spacer are defined as the upper and lower surface. Generally speaking, the surface charge has the same polarity as the applied voltage on the lower surface but the opposite polarity on the upper surface. However, a density drop or even a polarity reversal of surface charge shows up near the central conductor.

Figure 4 shows the mechanism of the surface charge accumulation process, which can be used to explain the surface charge distribution in **Figure 3**. Usually, three processes are considered to influence the surface charge accumulation on GIS spacers. (1) Gas transmission: Ions and electrons produced by corona discharge in the insulating gas transport to the spacer surface by electric force; (2) Volume transmission: Carriers in the spacer volume transport to the spacer interface by electric force; (3) Surface transmission: Carriers transport along the spacer surface by electric force. In this paper, the spacer is considered to have a pretty clean surface, the surface transmission is thus not as influential as the gas and volume transmission.

As shown in **Figure 4** under negative voltage, electrons and negative ions in the SF_6/N_2 mixture transport from the conductor to the shell and accumulate on the lower surface of spacer. Accordingly, the positive ions transport in the opposite direction and accumulate on the upper surface. In this case, the gas transmission takes the main part in the surface charge accumulation process. As a result, the surface charge has the same polarity as the applied voltage on the lower surface but the opposite polarity on the upper surface. When the conductor temperature is set to 350 K, there is a large temperature range along the spacer. According to the Eq. (25), the volume conductivity of spacer has been greatly increased near the central conductor. In this case, the volume transmission takes a more important role in the surface charge accumulation process. Electrons transport away from the conductor in the spacer volume and accumulate at the upper interface. Holes transport in the opposite direction and accumulate at the lower interface. This could explain why the density drop or the polarity reversal of surface charge shows up near the central conductor.

Figure 5 shows the electric field distribution along the basin-type spacer under both negative and positive voltages. The temperature of conductor is set to 350 K. The coordinate r from 30 to 167 mm represents the spacer surface from the central conductor to the shell. In this paper, the electric field distribution without corona discharge is defined as the normal field. And the field distribution with corona discharge is defined as the distorted field. It can be seen that the normal field near the conductor is much weaker than that in the surrounding zone. This is induced by the greatly enhanced volume conductivity of spacer due to temperature rise on the

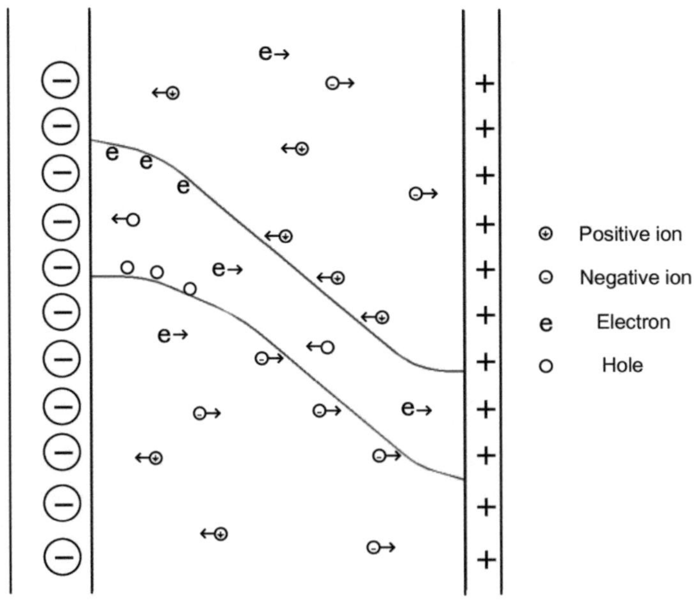

Figure 4.
Mechanism of the surface charge accumulation process.

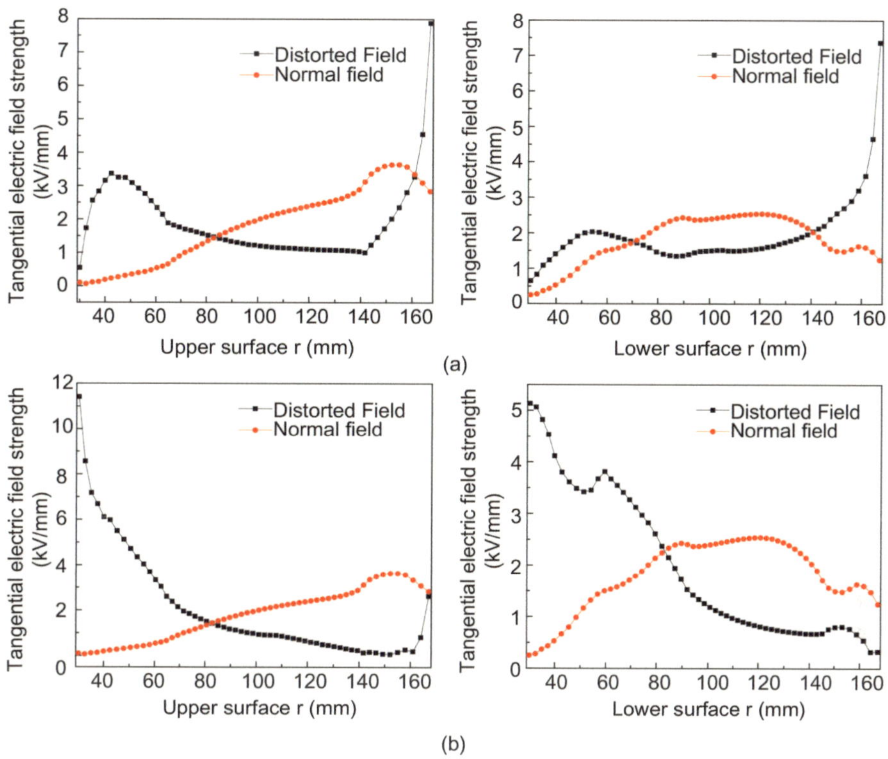

Figure 5.
The electric field distribution along the basin-type spacer at 350 K: (a) negative voltage and (b) positive voltage.

conductor. When the corona discharge is taken into consideration, ions and electrons in the SF_6/N_2 mixture will transport under electric force and accumulate on the spacer surface. Surface charge due to gas transmission can evidently distort the electric field. **Figure 5a** presents the distorted field distribution under negative voltage. The most serious distortion of electric field shows up in the zone close to the shell. From **Figure 3a** it can be learned that there is an annular zone with negative charge close to the shell. When negative voltage is applied on the conductor, the inner wall of the shell carries positive charges. Thus the existence of the negative charge zone greatly strengthens the electric field. **Figure 5b** presents the distorted field distribution under positive voltage. The most serious distortion of electric field shows up in the zone close to the conductor. Similarly, the negative charge zone surrounding the conductor should be responsible.

Figure 6 shows the surface charge and the corresponding electric field distribution along the disc-type spacer under both negative and positive voltages. The temperature of conductor is set to 350 K. As shown in **Figure 6a** under negative voltage, the surface charge has the same polarity as the applied voltage near the shell but the opposite polarity near the conductor. But under positive voltage, negative surface charge exists almost on the whole surface. **Figure 6b** presents the normal and distorted field distribution under both negative and positive voltages. Without considering the corona discharge, the normal field near the conductor is weaker than that in the surrounding zone, which is similar as the field distribution along the basin-type spacer. The serious field distortion under negative voltage is induced by the negative charge zone close to the shell. And the serious field distortion under positive voltage is induced by the negative charge zone surrounding the conductor.

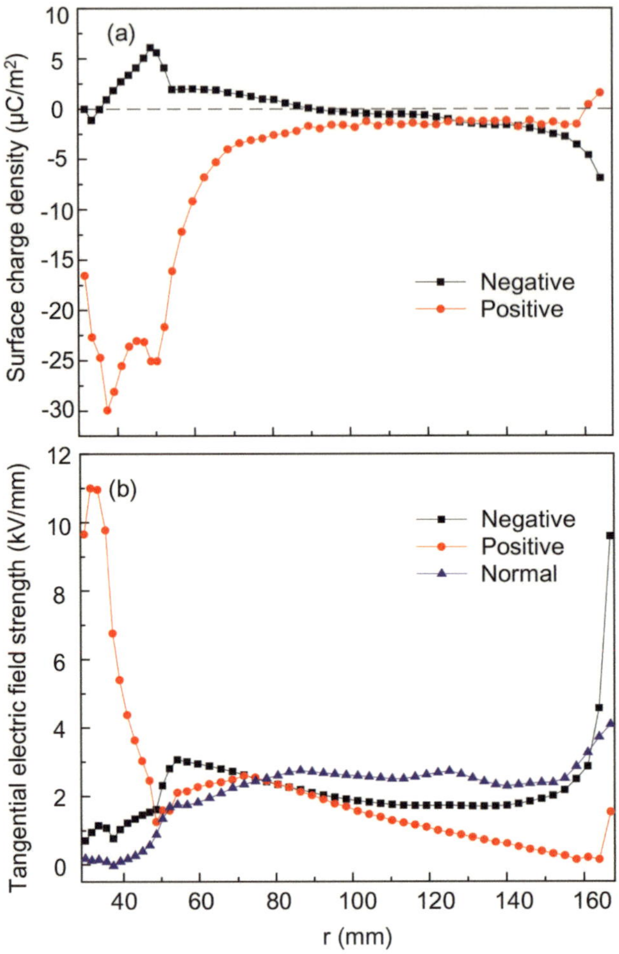

Figure 6.
Surface charge and the corresponding electric field distribution along the disc-type spacer at 350 K: (a) surface charge and (b) electric field.

3.2 Effects of temperature

Under different load conditions, the conductor has different temperatures, leading to different temperature distributions in GIS. In this paper, many factors taken into consideration are temperature-dependent, such as carrier transport coefficient, carrier concentration, ion transport coefficient, Schottky injection current, gas density, thermal conductivity and so on. Therefore, temperature has a great influence on the surface charge accumulation process and the electric field distribution.

Figure 7 shows the surface charge distribution along the basin-type spacer at different temperatures. Under negative voltage as shown in **Figure 7a**, less positive charges accumulate on the upper surface as the temperature increases. On the lower surface, the density of negative surface charge decreases with temperature and more positive charges are observed near the conductor. As the temperature increases, the conductivity of epoxy spacer grows up sharply. The volume transmission takes a more important role in the surface charge accumulation process. Similarly, under positive voltage as shown in **Figure 7b**, the volume transmission causes more positive charges to transport to the upper surface. Thus the negative surface charge due to gas transmission decreases with temperature. On the lower

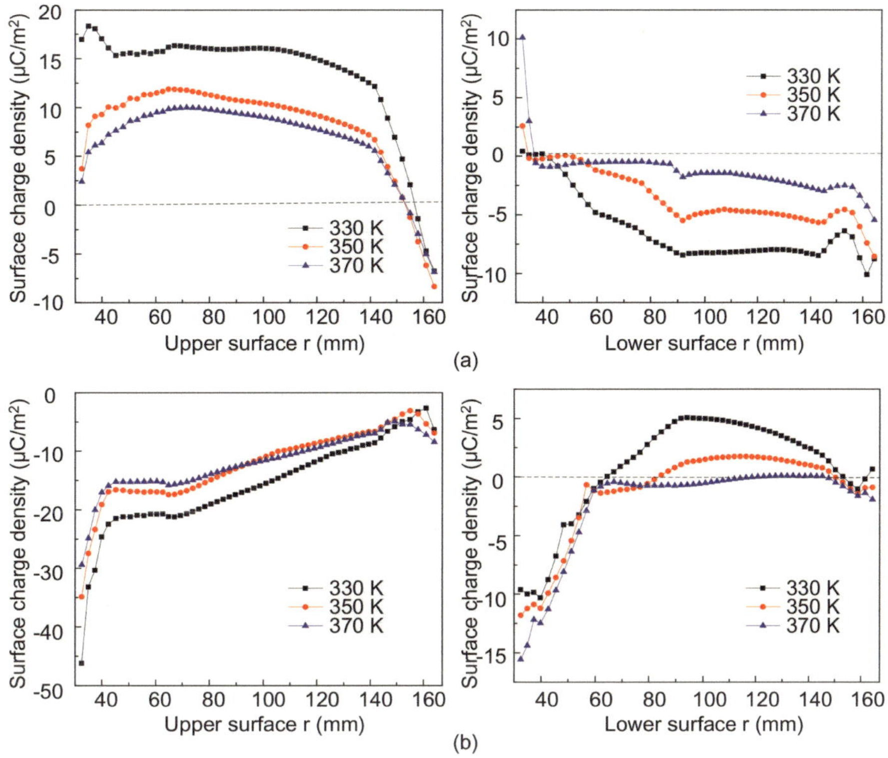

Figure 7.
The surface charge distribution along the basin-type spacer at different temperatures: (a) negative voltage and (b) positive voltage.

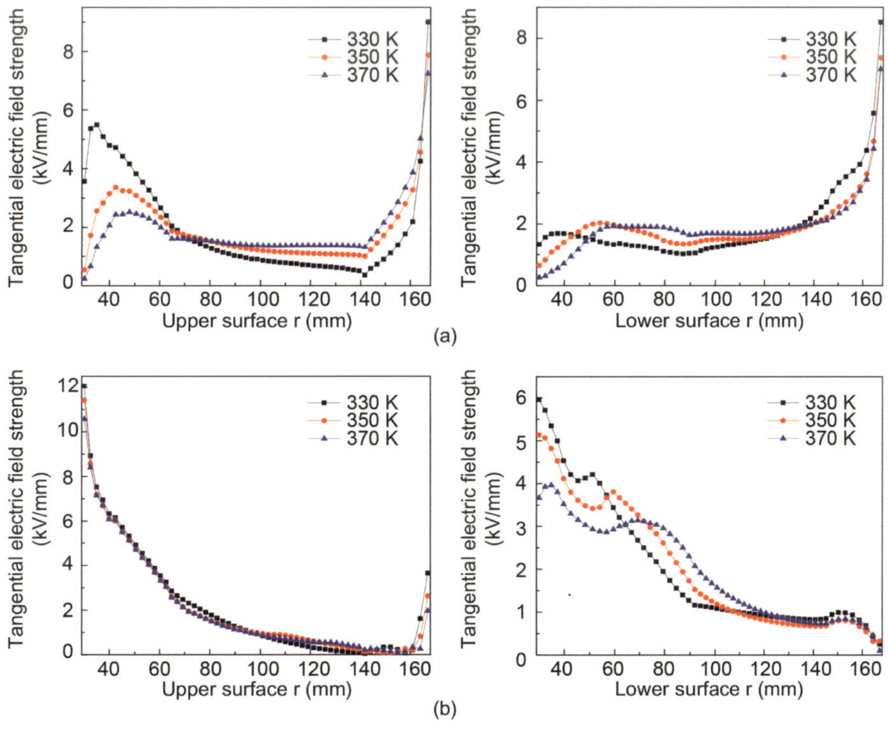

Figure 8.
The distorted electric field distribution along the basin-type spacer at different temperatures: (a) negative voltage and (b) positive voltage.

surface, more negative charges and less positive charges are observed due to the enhanced volume transmissions as the temperature increases.

Figure 8 shows the distorted electric field distribution along the basin-type spacer at different conductor temperatures. Under negative voltage as shown in **Figure 8a**, the field strength near the conductor is obviously decreased as the temperature increases. As the temperature goes up, the spacer has a huge temperature rise near the conductor, leading to the sharp increase in the volume conductivity. As a result, the grown volume conductivity greatly weakens the electric field near the conductor. Under positive voltage as shown in **Figure 8b**, the electric field near the conductor is also weakened with the increasing temperature, especially on the lower spacer surface. It seems that the heavy load on GIS is helpful to uniform the electric field distribution under positive voltage.

Figure 9 shows the distorted electric field distribution along the disc-type spacer at different conductor temperatures. Similarly, the electric field is weakened near the conductor with the increasing temperature under both negative and positive voltages. This is induced by the temperature-dependent volume conductivity of epoxy spacer. When positive voltage is applied on the disc-type spacer, the huge temperature rise on the conductor is also considered to have the effect of field grading.

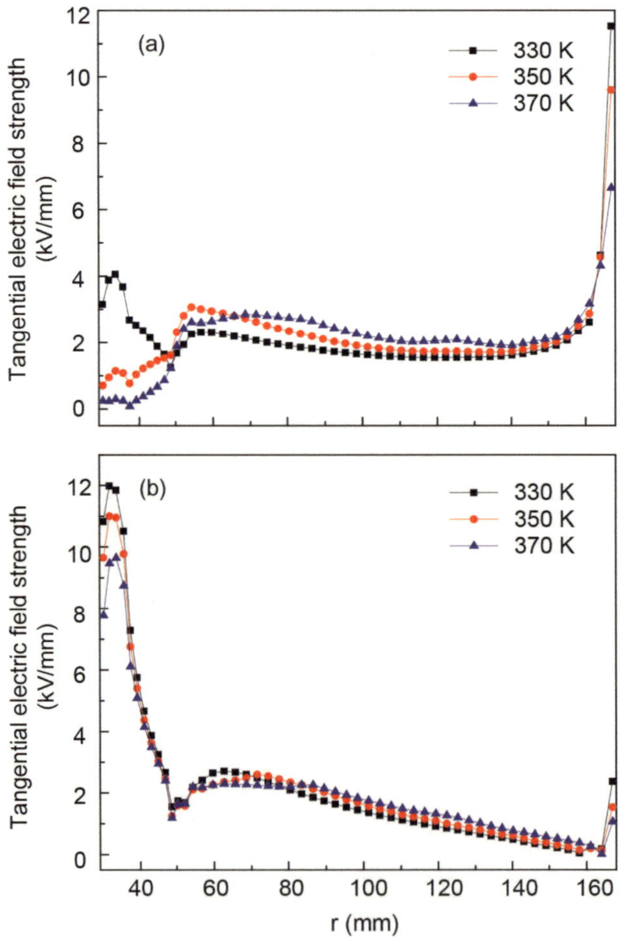

Figure 9.
The distorted electric field distribution along the disc-type spacer at different temperatures. (a) negative voltage, (b) positive voltage.

3.3 Effects of shielding electrode

From the results in Section 3.1 it can be inferred that the most serious distortion of electric field occurs at the triple junction in GIS. On various occasions, the triple junction is always the cause of field distortion and partial discharge. In order to solve this problem, both the conductor and shell are designed with shielding structures as shown in **Figure 10**. Note that the grounded shielding electrode is designed

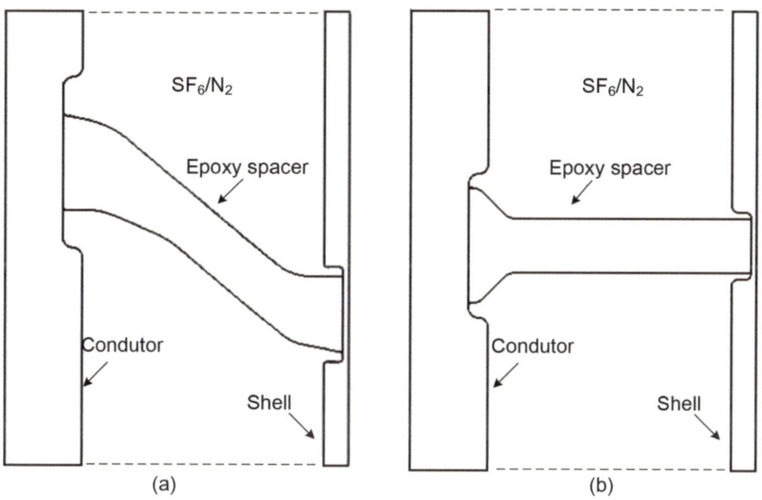

Figure 10.
GIS with shielding structures on the conductor and shell. (a) basin-type spacer, (b) disc-type spacer.

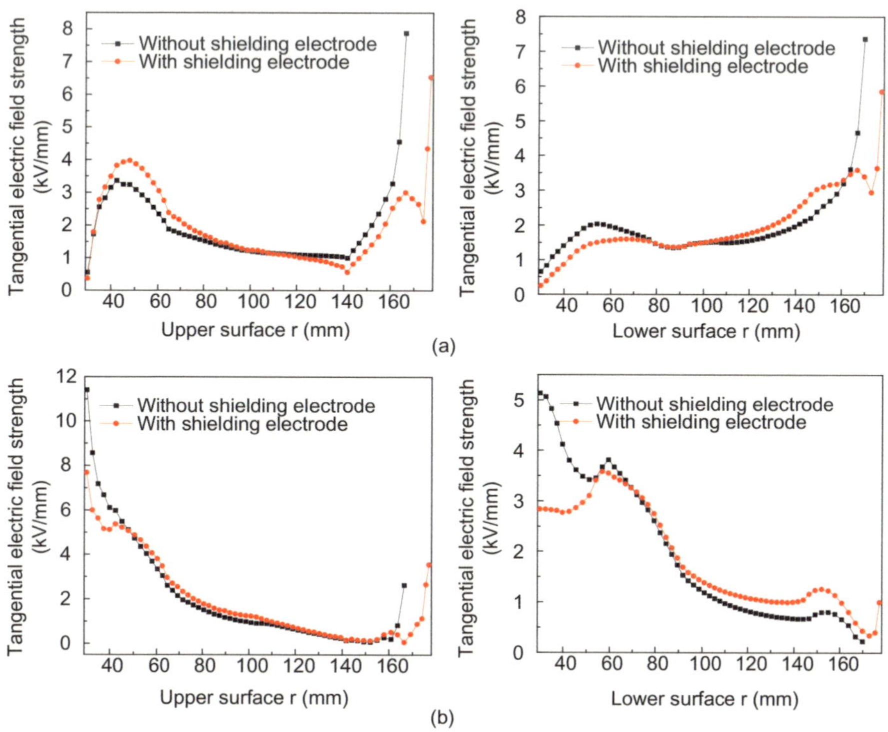

Figure 11.
The electric field distribution along the basin-type spacer at 350 K with and without shielding electrodes: (a) negative voltage and (b) positive voltage.

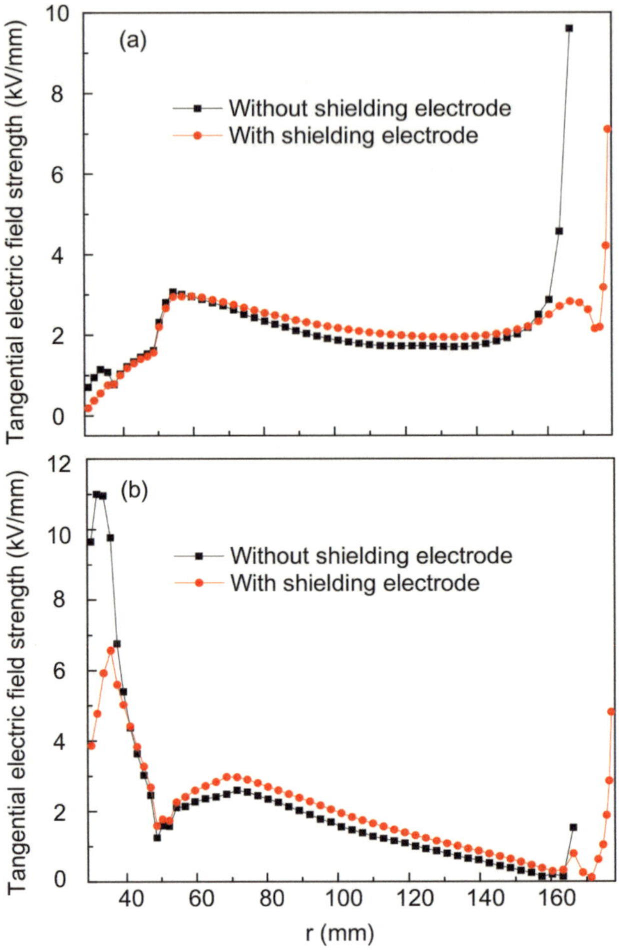

Figure 12.
The electric field distribution along the disc-type spacer at 350 K with and without shielding electrodes: (a) negative voltage and (b) positive voltage.

to be concave on the shell. As a result, the distance along the spacer surface from the conductor to the shell becomes a little longer.

Figure 11 presents the comparison of electric field between the GIS with and without shielding electrodes. It is easy to find that the electric field distribution is greatly improved by the shielding electrodes. As mentioned above, the most serious distortion of electric field is induced by the negative surface charges at the triple junction. The existence of shielding electrodes prevents the charges from reaching the spacer surface at the triple junction. Besides, the uniformed electric field at the triple junction is considered to suppress the partial discharge, less surface charges are produced. **Figure 12** presents the electric field distribution along the disc-type spacer with and without shielding electrodes. Also, the electric field distribution along the disc-type spacer is uniformed by the shielding electrodes.

4. Conclusions

This paper built a numerical model to simulate the surface charge accumulation process and the electric field distribution, the main conclusions can be drawn as follows:

1. Under load condition, there is a huge temperature rise on the GIS conductor due to joule heating. The wide variation of temperature along the epoxy spacers has a great influence on the surface charge accumulation process.

2. For the basin-type spacer, the surface charge has the same polarity as the applied voltage on the lower surface but the opposite polarity on the upper surface due to gas transmission. Because of the increasing spacer conductivity with temperature, the volume transmission takes a more important role in the surface charge accumulation process near the conductor. A density drop or even a polarity reversal of surface charge is observed near the conductor. For the disc-type spacer, the surface charge has the same polarity as the applied voltage near the shell but the opposite polarity near the conductor under negative voltage. But under positive voltage, negative charge exists almost on the whole surface.

3. The most serious distortion of the electric field distribution along the epoxy spacers occurs at the triple junction. As the conductor temperature increases, the field strength near the conductor is weakened owing to the temperature-dependent volume conductivity.

4. The application of shielding electrodes has the function of field grading at the triple junction, which can be referred in the DC-GIS/GIL design.

Author details

Boxue Du, Jin Li* and Hucheng Liang
Key Laboratory of Smart Grid of Education Ministry, School of Electrical and Information Engineering, Tianjin University, Tianjin, China

*Address all correspondence to: lijin@tju.edu.cn

References

[1] Delucchi MA, Jacobson MZ. Providing all global energy with wind, water, and solar power, Part II: Reliability, system and transmission costs, and policies. Energy Policy. 2011; **39**(3):1170-1190. DOI: 10.1016/j.enpol.2010.11.045

[2] Liu Y, Li C, Mu Q, et al. Side-by-side connection of LCC-HVDC links to form a DC grid. In: European Conference on Power Electronics and Applications. IEEE. 2015. pp. 1-10. DOI: 10.1109/EPE.2015.7311787

[3] Du BX, Li A. Effects of DC and pulse voltage combination on surface charge dynamic behaviors of epoxy resin. IEEE Transactions on Dielectrics & Electrical Insulation. 2017;**24**(4):2025-2033. DOI: 10.1109/TDEI.2017.006164

[4] Du BX, Li A, Li J. Effects of AC and pulse voltage combination on surface charge accumulation and decay of epoxy resin. IEEE Transactions on Dielectrics & Electrical Insulation. 2016;**23**(4): 2368-2376. DOI: 10.1109/TDEI.2016.7556515

[5] Lin C, Li C, He J, et al. Surface charge inversion algorithm based on bilateral surface potential measurements of cone-type spacer. IEEE Transactions on Dielectrics & Electrical Insulation. 2017; **24**(3):1905-1912. DOI: 10.1109/TDEI.2017.006496

[6] Du BX, Liang HC, Li J, et al. Temperature-dependent surface potential decay and flashover characteristics of epoxy/SiC composites. IEEE Transactions on Dielectrics & Electrical Insulation. 2018;**25**(2): 631-638. DOI: 10.1109/TDEI.2017.006872

[7] Liang HC, Du BX, Li J, et al. Effects of nonlinear conductivity on charge trapping and de-trapping behaviors in epoxy/SiC composites under dc stress. IET Science Measurement & Technology. 2018;**12**(1):83-89. DOI: 10.1049/iet-smt.2016.0528

[8] Liang HC, Du BX, Li J, et al. Numerical simulation on the surface charge accumulation process of epoxy insulator under needle-plane corona discharge in air. IET Science Measurement & Technology. 2018; **12**(1):9-16. DOI: 10.1049/iet-smt.2017.0164

[9] Vu-Cong T, Zavattoni L, Vinson P, et al. Surface charge measurements on epoxy spacer in HVDC GIS/GIL in SF6. IEEE Electrical Insulation and Dielectric Phenomena. 2016. DOI: 10.1109/CEIDP.2016.7785506

[10] Nakanishi K, Yoshioka A, Arahata Y, et al. Surface charging on epoxy spacer at DC stress in compressed SF_6. Gas. IEEE Power Engineering Review. 1983;**PER-3**(12):46-46. DOI: 10.1109/MPER.1983.5520154

[11] Zhang B, Qi Z, Zhang G. Charge accumulation patterns on spacer surface in HVDC gas-insulated system: Dominant uniform charging and random charge speckles. IEEE Transactions on Dielectrics & Electrical Insulation. 2017;**24**(2):1229-1238. DOI: 10.1109/TDEI.2017.006067

[12] Zhang B, Qi Z, Zhang G. Thermal gradient effects on surface charge of HVDC spacer in gas insulated system. IEEE Electrical Insulation and Dielectric Phenomena. 2016. DOI: 10.1109/CEIDP.2016.7785563

[13] Zhou HY, Ma GM, Li CR, et al. Impact of temperature on surface charges accumulation on insulators in SF6-filled DC-GIL. IEEE Transactions on Dielectrics & Electrical Insulation. 2017;**24**(1):601-610. DOI: 10.1109/TDEI.2016.005838

[14] Sato S, Zaengl WS, Knecht A. A numerical analysis of accumulated

surface charge on DC epoxy resin spaces. IEEE Transactions on Electrical Insulation. 2007;**22**(3):333-340. DOI: 10.1109/TEI.1987.298999

[15] Lutz B, Kindersberger J. Surface charge accumulation on cylindrical polymeric model insulators in air: Simulation and measurement. IEEE Transactions on Dielectrics & Electrical Insulation. 2012;**18**(6):2040-2048. DOI: 10.1109/TDEI.2011.6118642

[16] Winter A, Kindersberger J. Stationary resistive field distribution along epoxy resin insulators in air under DC voltage. IEEE Transactions on Dielectrics & Electrical Insulation. 2012; **19**(5):1732-1739. DOI: 10.1109/ TDEI.2012.6311522

[17] Winter A, Kindersberger J. Transient field distribution in gas-solid insulation systems under DC voltages. IEEE Transactions on Dielectrics & Electrical Insulation. 2014;**21**(1): 116-128. DOI: 10.1109/ TDEI.2013.004110

[18] He W, Liu XH, Yang F, et al. Numerical simulation of direct current glow discharge in air with experimental validation. Japanese Journal of Applied Physics. 2012;**51**(51):6001. DOI: 10.1143/JJAP.51.026001

[19] Hagelaar GJM. Solving the Boltzmann equation to obtain electron transport coefficients and rate coefficients for fluid models. Plasma Sources Science & Technology. 2005; **14**(4):722-733. DOI: 10.1088/0963-0252/ 14/4/011

[20] Christophorou LG, Brunt RJV. SF_6// N_2 mixtures: Basic and HV insulation properties. IEEE Transactions on Dielectrics & Electrical Insulation. 2002;**2**(5):952-1003. DOI: 10.1109/ 94.469988

[21] Itikawa Y. Cross sections for electron collisions with nitrogen molecules. Journal of Physical & Chemical Reference Data. 2006;**35**(1): 31-53. DOI: 10.1063/1.1937426

[22] Gulley RJ, Cho H, Buckman SJ. The total elastic cross section for electron scattering from SF_6. Journal of Physics B: Atomic, Molecular and Optical Physics. 2000;**33**(8):L309-L315. DOI: 10.1088/0953-4075/33/8/105

[23] Phelps AV, Van Brunt RJ. Electron-transport, ionization, attachment, and dissociation coefficients in SF6 and its mixtures. Journal of Applied Physics. 1988;**64**(9):4269-4277. DOI: 10.1063/ 1.341300

[24] Zhou TC, Chen G, Liao RJ, et al. Charge trapping and detrapping in polymeric materials. Journal of Applied Physics. 2009;**106**(12):644-637. DOI: 10.1063/1.3273491

[25] Chen G, Zhao J, Zhuang Y. Numerical modeling of surface potential decay of corona charged polymeric material. IEEE International Conference on Solid Dielectrics. 2010:1-4. DOI: 10.1109/ICSD.2010.5567997

[26] Min DM, Li ST, Li GC. The effect of charge recombination on surface potential decay crossover characteristics of LDPE. In: International Symposium on Electrical Insulating Materials. 2014. pp. 104-107. DOI: 10.1109/ISEIM.2014. 6870731

[27] Straumann U, Schuller M, Franck CM. Theoretical investigation of HVDC disc spacer charging in SF6 gas insulated systems. IEEE Transactions on Dielectrics & Electrical Insulation. 2013; **19**(6):2196-2205. DOI: 10.1109/ TDEI.2012.6396980

[28] Wang Q, Wang H, Peng Z, et al. 3-D coupled electromagnetic-fluid-thermal analysis of epoxy impregnated paper converter transformer bushings. IEEE Transactions on Dielectrics & Electrical Insulation. 2017;**24**(1): 630-638. DOI: 10.1109/TDEI.2016. 005641

Chapter 11

Electrodeposition Approaches to Deposit the Single-Phase Solid Solution of Ag-Ni Alloy

Brij Mohan Mundotiya, Wahdat Ullah and Krishna Kumar

Abstract

The Ag-Ni films have attracted the attention of material scientists and researchers due to its applications as magnetic materials, catalyst, optical materials, etc. In the bulk, the formation of a single-phase solid solution of the Ag-Ni alloy film is difficult due to the large differences in the atomic size and the positive enthalpy of mixing. However, the immiscibility in the Ag and Ni constituents is diminished in the nano-sized level. The electrodeposition method has established itself as a most suitable method for synthesis of single-phase Ag-Ni alloy films due to its time efficiency, cost-effectiveness, and ability to mass production of single-phase solid solutions. In this method, the miscibility of alloying elements (i.e., Ag and Ni) and the quality of the Ag-Ni film can also be easily controlled by tuning the electrodeposition process parameters such as the magnitude of the applied current density, temperature of the electrolyte, additives in electrolytes, etc. This chapter presents a detailed overview of the process parameters affecting the miscibility, morphology, and the quality of the single solid solution of Ag-Ni film. Furthermore, the nucleation and growth mechanism of Ag-Ni film and the effect of the curvature of the deposited film's particles in the miscibility of the Ag and Ni elements have also been discussed in detailed.

Keywords: deposition parameters, electrodeposition, immiscibility, morphology, single-phase Ag-Ni alloy

1. Introduction

In recent years, the researchers are looking with the hope that the metastable (non-equilibrium) alloys might be materials with potentially unique and exploitable functional properties. The alloys in the non-equilibrium state are generally made from the immiscible element system such as Co-Ag, Ag-Ni, Au-Co, Ag-Pt, Fe-Cu, Co-Cu, and Ag-Fe alloy systems. In bulk state, constituent elements of these systems exhibit strong immiscibility behavior even up to very high temperatures. However, miscibility of their constituent elements can be enhanced by mixing them in nanoscale level to form a nano-size system. The enhanced miscibility of constituent elements in the nano-size system also induces its exploitable functional properties due to the evolution of the unique type of microstructure [1]. In recent years, numerous amorphous and nanocrystalline single solid solution alloy systems of immiscible elements Ni-Co [2], Ni-W [3], Ni-Si [4], and Ni-Fe [5, 6] have all been explored and reported in scientific literatures. Among the metastable alloys systems, Ag-Ni alloy system is one of the very interesting system due to the presence of one magnetic (Ag) and one nonmagnetic (Ni) alloying elements and their possible applications in the field of catalysts [7], magnetic materials [8, 9], electrical materials [10], and optical materials [11]. But due to the alloying characteristics of the constituent elements of the Ag-Ni system, these systems have been explored very little.

The alloying elements of Ag-Ni system (i.e., Ag and Ni elements) have the same lattice structure (FCC), and the difference between their atomic sizes is very large (~14%). They are immiscible in both the solid and liquid states or even up to very high temperatures. The phase diagram of the Ag-Ni system is very simple (see **Figure 1**). As it is obvious from the phase diagram that Ag and Ni are mutually insoluble, even at 400°C, the solubility of Ni in Ag is only ~0.0219 at% [12]. The reasons of the high immiscibility of Ag and Ni alloying elements are (a) the large difference in their atomic sizes (~14%) and (b) a positive enthalpy of mixing (ΔH^{mix} for solid solution equiatomic Ag-Ni alloy is 23 kJ/mol) [13, 14]. However, an enhancement of miscibility between Ag and Ni atoms in nano-size level to form a nano-size system has been achieved by researchers [8–11]. Enhancement of miscibility of atoms of nanoparticles of Ag and Ni can be attributed to (a) the effect of increase of particles curvatures with the reduction of their sizes that facilitate miscibility in accordance with the Gibbs-Thomson effect and (b) the decrease of the driving force for the nucleation and growth of the second phase within a nano-sized particle [13].

In recent years, the single-phase Ag-Ni alloys have successfully synthesized by various techniques such as ion beam mixing [16], laser ablation [10], gamma-ray irradiation [17], etc. Zhang et al. [17] synthesized the single solid solution phase of Ag-Ni alloy by simultaneous reduction of Ag and Ni ions by gamma-ray irradiation. The authors obtained a single solid solution phase of the Ag-Ni system with the homogeneous distribution of compositions of Ag and Ni atoms inside the particles. The average size of the particle of a single solid solution of Ag-Ni system

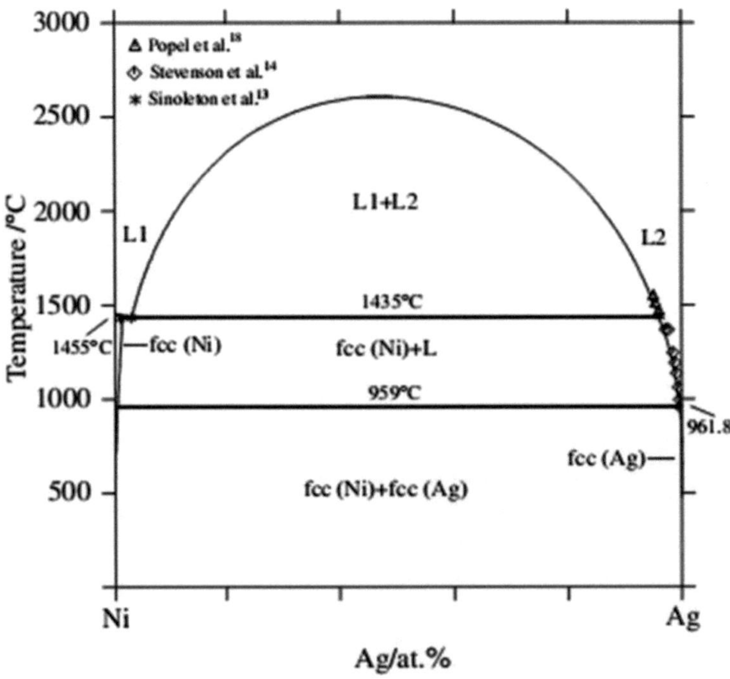

Figure 1.
Phase diagram of Ag-Ni system [15].

was 6 nm. In another study reported by Kumar et al. [18], an average particle size of 30 nm of a single solid solution of Ag-Ni system was synthesized by immersion of a film of thermally evaporated fatty acids sequentially in solutions containing Ag^+ ions and Ni^+ ions. Srivastava et al. [19] studied the size-dependent microstructural evolution for Ag-Ni alloy nanoparticles. The authors synthesized Ag-Ni nanoparticles by the co-reduction of Ag and Ni metal precursors. It was found that the Ag and Ni atoms were completely mixed in Ag-Ni alloy nanoparticles when the particle size was less than 7 nm. As the size of the particles was increased, the nanoparticles showed two separate regions: (a) a pure Ag region and (b) a solid solution of Ag-Ni alloy.

The electrodeposition method has established itself as a most suitable method for synthesis of single-phase Ag-Ni alloy films due to its time efficiency, cost-effectiveness, and ability to mass production of single-phase solid solutions [13, 20–26]. In the electrodeposition process, the apparatus used for the deposition of the film is less complicated than the vacuum-based methods such as molecular beam epitaxy or sputtering. In this method, miscibility of the alloying elements (Ag and Ni) to form a single-phase solution of Ag-Ni films and their quality can also be easily controlled by tuning the electrodeposition process parameters such as current density, electrolytes, electrolyte temperatures, additives in electrolytes, etc. In the following sections of this chapter, the overview of information related to the effects of these parameters on the miscibility of the alloying elements (Ag and Ni) and the quality of the deposited film are discussed.

2. Nucleation and growth mechanism of film

The deposition of the alloy films on the conductive substrate involves the nucleation and growth process. The nucleation process is very important because it decides the nature of the evolution of microstructure and crystallinity in the deposited alloy film. The alloy film deposition on the conductive substrate by electrodeposition method involves a heterogeneous process. The interface between the electrolyte solution and the substrate surface provides preferential sites for nucleation. As an example, consider the nucleation of a nucleus on the substrate surface (see **Figure 2**). On the substrate surface, nucleus forms in the shape of a half convex lens (spherical cap). The surface tension force acting on the surface-nucleus-electrolyte interfaces is determined by the formula given in Eq. (1). This equation is known as Young's equation [27]:

$$\gamma_{sl} = \gamma_{fs} + \gamma_{lf} \cos\theta \quad (1)$$

where γ_{lf}, γ_{fs}, and γ_{sl} are the surface or interface energy of electrolyte-nucleus, nucleus-substrate, and substrate-electrolyte interfaces, respectively, and θ is the contact angle that depends on nature of the surfaces involved to form interfaces.

In the heterogeneous nucleation process, the probability of the nucleation is determined by the critical nucleus radius r* and the activation energy barrier ΔG* that a nucleation process must be overcome to form a nucleus. The formulae for the r* and ΔG* are given by Eqs. (2) and (3), respectively [28]:

$$r^* = \left(\frac{2\pi\gamma_{lf}}{\Delta G_l}\right)\left(\frac{\sin^2\theta \cdot \cos\theta + 2\cos\theta - 2}{2 - 3\cos\theta + \cos^3\theta}\right) \quad (2)$$

$$\Delta G^* = \frac{16\pi\gamma_{lf}}{3(\Delta G_l)^2}\left(\frac{2 - 3\cos\theta + \cos^3\theta}{4}\right) \quad (3)$$

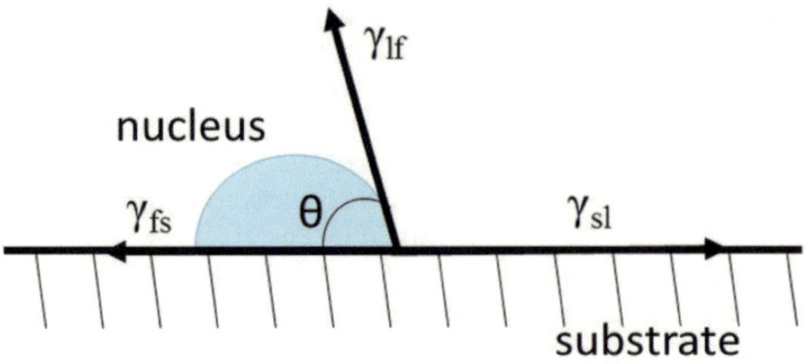

Figure 2.
Schematic diagram of a nucleus on a solid surface (substrate) with energy vectors and contact angle.

where ΔG_l is the Gibbs free change per unit volume. If the radius (r) of the nucleus is smaller than the critical nucleus radius (r*), the nucleus is not survived for further growth and redissolved.

From the atomic point of view, the electrodeposition of film occurs in two different stages. In the first stage, few metal atoms from the electrolyte solution adhere to the substrate. Normally, it is considered that the first stage may process via either of the two mechanisms: (a) step-edge site ion transfer or (b) terrace site ion transfer [29]. In the step-edge site ion-transfer mechanism, the metal adions transfer from the electrolyte solution to a kink site of a step edge of the substrate surface via two paths. In the first path, the metal adions directly transfer from the electrolyte solution to the kink site of a step edge of the substrate. While in the second path, the metal adions diffuse from the electrolyte solution to the step edge first and then finally transfer to the kink site. On the other hand, in the terrace ion-transfer mechanism, the metal adions transfer from the electrolyte solution to the flat surface of the terrace region. The bonds between the metal adions are stronger to each other than to the surface of terrace regions. Therefore, metal adions easily diffuse from the terrace and transfer to the step-edge site and then finally to a kink site. These clusters of metal adions on the different sites of the substrate form nuclei of ordered structure.

Following the first stage, the second stage of growth of the nucleus occurs. The experimental observations revealed that there are three basic kinds of growth modes of alloy film [30]: (a) island growth, (b) layer growth, and (c) island-layer growth. When the bound between the depositing species (i.e., metal adions) is stronger than to the substrate, the island growth occurs. As the deposition proceeds, the islands of deposit subsequently grow and merge to form a continuous film on the substrate. The layer growth of film occurs, when the depositing species (i.e., metal adions) are bound strongly to the substrate than to each other. In the layer growth mode, first of all, a complete monolayer of depositing species is formed on the substrate, then a second layer starts to deposit on the monolayer. The island-layer growth mode of alloy film is a combination of both island and layer growth modes. In the island-layer growth mode, first of all, a layer of deposit is grown on the substrate by the layer growth mechanism; this layer has some degree of residual stresses when the critical thickness of the layer is reached then due to an increase of the stresses on it, the island growth mechanism is activated.

3. Electrodeposition solution for Ag-Ni films

Metastable Ag-Ni alloy films are deposited by electrodeposition method by using their respective salts and some additives (such as thiourea, saccharin, trisodium

Coating/ nano-particles	Electrolyte bath	pH	Temperature (°C)	Current density (mA/ cm^2)	Substrate	Reference
Ag-Ni film	$C_6H_5Na_3O_7$: 0.3 M $NiSO_4$: 0.7 M $AgNO_3$: 2–5 mM	4.5–7.5	Room temperature	0.5–50	Brass plate	Eom et al. [20]
Ni-Ag coating	$NiSO_4{\cdot}6H_2O$: 49.93 g/L $AgNO_3$: 0–0.339 g/L $C_6H_5Na_3O_7{\cdot}2H_2O$: 0–41.160 g/L H_3BO_3: 12 g/L	5.6	30	5	Cu plate	Raghupathy et al. [21]
Ni-Ag layer	$C_6H_5Na_3O_7$: 0.26 mol/L $NiSO_4$: 0.7 mol/L $AgNO_3$: 0.002 mol/L	5.5	Room temperature	—	Cu plate	Schneider et al. [22]
Ag-Ni film	$AgClO_4$: 0.01 M $Ni(ClO_4)_2$: 0.15 M $NaClO_4$: 0.1 M H_3BO_3: 0.15 M Thiourea: 0.2 M	3	—	—	Cu sheet	Liang et al. [23]
Ag-Ni film	$AgNO_3$: 0.01 M $Ni(NO_3)_2{\cdot}6H_2O$: 0.08 M H_3BO_3: 0.003 M Thiourea: 0.045 M	3.2	Room temperature	25	Cu plate	Srivastava et al. [24]
Ag-Ni solid film	$AgNO_3$: 1.69 g $Ni(NO_3)_2{\cdot}6H_2O$: 0.25 g H_3BO_3: 0.02 g in 100 mL distilled water	12	Room temperature	100	Cu plate	Srivastava et al. [25]
Ag-Ni nano particles	$AgNO_3$: 1.69 g $Ni(NO_3)_2{\cdot}6H_2O$: 0.25 g H_3BO_3: 0.03 g Thiourea: 0.35 g in 100 mL distilled water	3.6	80	20, 70, and 100	Cu plate	Srivastava et al. [26]

Table 1.
Electrodeposition parameters used for the deposition of Ag-Ni alloy films/nanoparticles.

citrate, citric acid, sodium perchlorate, etc.). The electrolyte is the electrochemical solution of metal or alloy salts that is required to deposit at cathode (conductive substrate) surface during electrolysis. The deionized and distilled waters are used as a solvent for most of the electrolytes; therefore, these electrolytes are called aqueous electrolytes. The costs of the aqueous electrolytes are lesser than the nonaqueous electrolytes due to the easy availability of the water solvent. Deionized water is water that has been treated to remove all the dissolved mineral salts. While, the distilled water is prepared by boiling the normal water so that it evaporates and then recondensed, leaving most impurities behind [31].

Mostly sulfate and nitrate salts with some suitable additives (such as thiourea, saccharin, trisodium citrate, citric acid, sodium perchlorate, etc.) are used for the preparation of the electrolytes for deposition of the Ag-Ni systems. The sulfate and nitrate salts have high solubility in the water. The additives in the electrolyte act as grain refiners, complexing agents, brighteners, etc. The additive of sodium perchlorate is commonly used for reducing the resistance (i.e., viscosity) and increasing the ionic conductivity of the electrolytes [32]. And for improving quality of Ag in the deposited film, additives of boric acid and sodium gluconate are generally added in the electrolyte [33, 34].

List of some typical salts of metal ions with and without additives which are often used for deposition of Ag-Ni films/nanoparticles and the involved deposition process parameters are shown in **Table 1**.

4. Electrodeposition parameters for the Ag-Ni alloy films

The solid solubility, structure, and morphology of the deposited Ag-Ni alloy films depend on the several electrodeposition process parameters. Some of the crucial process parameters influencing the miscibility, morphology, and structure of the Ag-Ni alloy films are discussed in given below subsections.

4.1 Deposition current density

According to the electro-crystallization theory, whenever a high current density is applied, a high overpotential is generated at the cathode, as a consequence of this, the nucleation rate increases. The faster nucleation rate reduces the particle sizes of the deposited film [35–39], because the fine particle sizes increase the specific area or curvature that promotes the miscibility of the constituent elements [40, 41]. This phenomenon is known as the Gibbs-Thomson effect.

The composition of the Ag-Ni alloy film could be altered by changing the current density. Santhi et al. [32] have deposited metastable Ag-Ni film with the granular kind of morphology. The average particle size was ~50 nm. They observed that the Ni content in the deposited Ag-Ni film increases with increasing current densities. Bdikin et al. [42] reported that the particle size of the deposited Ag-Ni film decreases with increasing Ni contents in the film. For 70 wt% of Ni content, the Ag-Ni film particle size goes down to very small (~3.3 nm). Eom et al. [20] investigated the effect of the current density on the chemical composition and crystallography of the deposited Ag-Ni film. They deposited Ag-Ni films on a brass substrate using $NiSO_4$, $AgNO_3$, and $C_6H_5Na_3O_7$ electrolyte. The Ag content in the deposited Ag-Ni film was found to be higher at lower current density. The high content of Ag in the deposited film was due to its lower reduction potential. However, the content

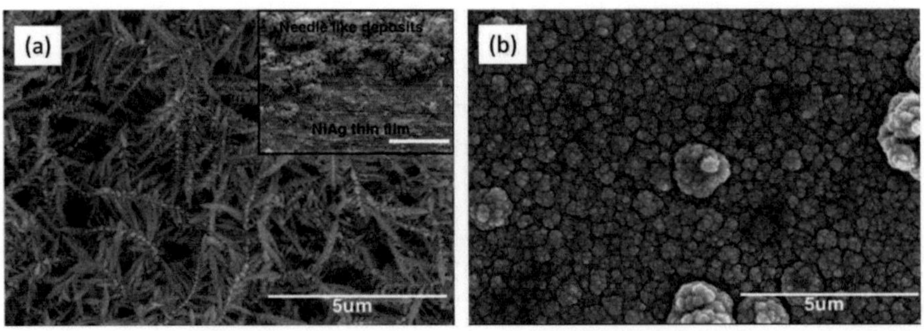

Figure 3.
SEM micrographs of Ag-Ni films deposited at current densities of (a) 0.5 mA/cm² and (b) 10 mA/cm² [20].

of Ni in the deposited films was found to be increased with increasing current density. At higher current density, Ni becomes the dominant component because of the more transfer limitation of Ag ions. Apart from the change in composition, the change in morphology as well as the grain size of the films with changing current densities was also reported by the authors. At a current density of 0.5 mA/cm^2, films had dendritic kind of morphology. But when the current densities were increased from 0.5 to 50 mA/cm^2, the morphologies were changed from dendritic to particle type (see **Figure 3**), and a significant reduction in grain sizes had also occurred.

The enhancement of miscibility of Ag and Ni elements to form a single-phase solid solutions of Ag-Ni alloy systems can be explained by Tafel equation (Eq. (4)) [43] and Eq. (5) [20] given below:

$$\eta = a + b \log i \tag{4}$$

$$J = K_1 exp\left(\frac{-gs\,\varepsilon^2}{zekT\eta}\right) \tag{5}$$

where η is the overpotential at cathode, a and b are the constants, i is the current density, J is the nucleation rate, K_1 is the rate constant, g is the factor relating the surface area S of the nucleus to the perimeter P ($g = P^2/4S$), s is the area occupied by one atom on the surface of the nucleus, T is the electrolytic temperature, k is the Boltzmann constant, and z, e, and ε are the number of electrons involved in the reaction, electron charge, and edge energy, respectively. According to Tafel equation (Eq. (4)), the reduction in the grain size is primarily due to the increase in the overpotential by increasing the current density. The relations between the nucleation rate, overpotential, and current density are given by Eqs. (4) and (5). From Eq. (5), it is clear that the nucleation rate has an exponential relationship with the overpotential. As the overpotential at the cathode increases, the nucleation rate also increases exponentially. In Eqs. (4) and (5), it can also be seen that the nucleation rate and current density are related to each other. Therefore, when current density increases, the overpotential at the cathode increases, as a result of this, the nucleation rate also increases. The increase in the nucleation rate produces finer grains in the deposited film. The finer grains increase the specific area or curvature which promotes an increase in the extent of miscibility of Ag and Ni atoms to form a single solid solution of Ag-Ni system.

Srivastava and Mundotiya [26] investigated the influence of the current density on the particles size of the deposited Ag-Ni alloy films. The Ag-Ni films were deposited on the Cu-substrate (30 mm X 10 mm X 0.3 mm) by using the electrolyte (0.17 g of $AgNO_3$, 0.25 g of $Ni(NO_3)_2{\bullet}6H_2O$, 0.03 g of H_3BO_3, and 0.35 g of thiourea in 100 ml of distilled water (pH ~ 3.6). During the electrodeposition, the temperature of the electrolyte was maintained at 80°C. The addition of an additive (complexing agent) of thiourea was chosen to deposit a granular kind of morphology in Ag-Ni film. They observed no significant change in morphologies of the deposited Ag-Ni films at current 20, 70, and 100 mA. At these current densities, the morphologies of the deposited films were similar granular morphology. However, the content of Ni in the Ag-Ni films increases, while the sizes of grains decrease with increasing current densities. These trends were in agreement with those reported in the scientific literature [20, 32, 42]. Thus, as a general statement, it can be concluded that the decrease in the grain size of the Ag-Ni film increases the miscibility of the constituent elements.

The miscibility behavior of the Ag-Ni system with different contents of Ni in Ag-Ni films can be explained by the X-ray diffraction (XRD) patterns shown in **Figure 4**. It can be seen in the patterns that the peaks of the Ag-Ni films shift to higher angles with increasing Ni content. The shifting of the peaks indicates substitution of the more Ag atoms by the relatively smaller Ni atoms with increasing content of Ni. That means the lattice constant of the Ag-Ni films decreases with increasing content of Ni. As the lattice constant decreases, the interplanar spacing of the Ag-Ni alloy films also decreases because the lattice constant is directly proportional to the interplane spacing. And the interplanar spacing is inversely proportional to the angle. Another important characteristic showed by the XRD profile is the broadening of the Ag-Ni films peaks with increasing Ni content. The broadening could be the result of decreasing grain size of the Ag-Ni films with increasing Ni content.

Therefore, the current density is the important parameter to control the miscibility of the Ag-Ni alloy system by controlling the size and composition.

4.2 Electrolyte temperature

The temperature of the electrolyte also influences the morphologies, the particles sizes, and the quality of the deposited single solid solution of the Ag-Ni films. This is due to the fact that the solubility of the metal salts in the electrolyte can be sufficiently increased by increasing the temperature of the electrolyte. As the temperature increases, the viscosity of the electrolyte decreases, and as a result of this, the mobility (diffusion and migration) of the metal ions toward the cathode increases [36]. Due to an increase in solubility and mobility of the metal ions in the electrolyte at a higher temperature, the conductivity of the electrolytes also improves, because the conductivity of the electrolyte depends on the degree of dissociation of dissolved metal ions in the electrolyte and the mobility of dissolved metal ions. The improvement in the conductivity of electrolyte at higher temperature decreases the overpotential at the cathode [44, 45]. A decrease in the cathodic overpotential with an increase in the electrolyte temperature above 55°C is reported by Rashidi and Amadeh [46]. The decrease in the cathodic overpotential slows down the nucleation rate; at low nucleation rate, the existing nuclei grow faster.

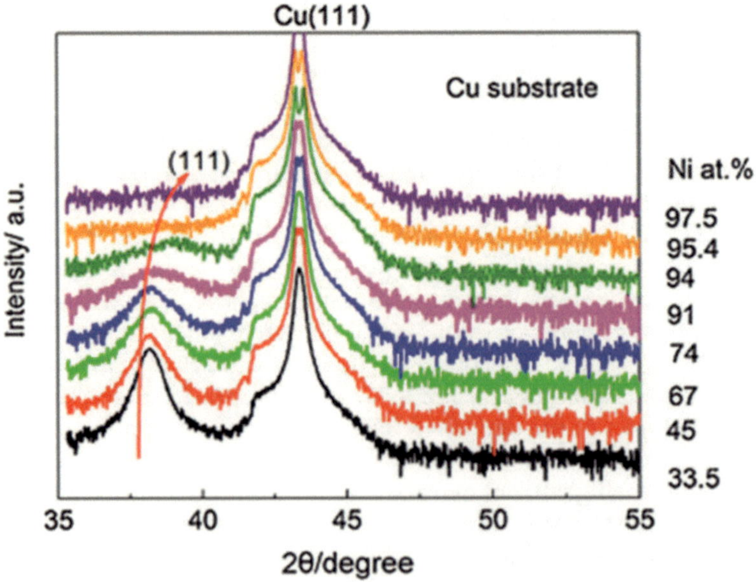

Figure 4.
XRD pattern of Ag-Ni alloy films with different Ni-content [23].

Thus it can be concluded that an increase in the temperature of the electrolyte may lead to increased particle/grain growth in the deposited Ag-Ni film [26, 46, 47].

The temperature of the electrolyte also influences the composition of the deposited film. The concentration of Ag in the deposited film increases with increasing electrolyte temperature [48]. To deposit the films of the same compositions at different electrolyte temperatures, very high overpotential at the cathode is required. But high overpotential at the cathode favors hydrogen evolution which is undesirable for deposition of high-quality alloy film because it creates pores and/or hydrogen embrittlement in the deposit films [48].

4.3 Additives

The addition of the suitable additives in the electrolyte affects the morphology and the quality of the deposited film. The additives in the electrolyte produce an additional overpotential at the cathode [36] which helps in the creation of a homogeneous film with a low level of roughness [49]. For improving the quality of the deposited film, many suggestions have been given by the researchers. Among those suggestions, the most common suggestion is related to the blocking of lattice sites by adsorption of additives on the surface of the substrate. This helps to prevent the growth of grains in the deposited film by inhibiting short-range surface diffusion [50].

The reduction of Ag and Ni takes place at a potential of +0.2 and −0.26 V, respectively. The difference between the reduction potentials of Ag and Ni is too large. Due to the existence of this large difference in the reduction potentials of Ag and Ni during electrodeposition of Ag-Ni film, morphological instabilities between Ag and Ni phases occur during the growth process of the Ag-Ni film. Therefore, synthesizing a perfect Ag-Ni film by co-deposition of Ag and Ni is difficult in this condition. The deposited Ag-Ni film generally has a porous and dendritic morphology. For minimizing the difference of the reduction potentials of Ag and Ni, some suitable additives such as sodium citrate ($C_6H_5Na_3O_7$), thiourea, etc. are commonly added with electrolyte [20, 34]. The morphology of the electrodeposited Ag-Ni film can be tuned by the addition of suitable additives and/or complexing agents in the electrolyte. These additives shift the reduction potentials of Ag and Ni more closely. The thiourea is one of the additives which has a strong complexing capacity with Ag [41] but no or weak complex with Ni. Due to its strong complexing capacity with Ag, it decreases the reduction potential of Ag into more negative potential without altering the reduction potential of Ni. The minimization of the difference of the reduction potentials of Ag and Ni by addition of additive in electrolyte also helps in suppressing the evolution of the hydrogen at the cathode. After the addition of the additive of thiourea in the electrolyte solution, a change of morphology of the feature of the deposited Ag-Ni films is observed [51]. The dendritic morphology changes to nanoparticles by adding additives in the electrolyte solution as shown in SEM micrographs in **Figure 5**. This transformation of dendritic morphology into nanoparticles after addition of thiourea additive in the electrolyte solution can be attributed to the minimization of the difference of the reduction potentials of the Ag and Ni elements.

The decrease of the reduction potential of Ag by adding additive of thiourea in the electrolytic solution is also reported by Liang et al. [23]. They deposited Ag-Ni films on the Cu substrate by using thiourea-based acidic solutions as an electrolyte. It was observed that the presence of thiourea in the electrolyte decreases the reduction potential of Ag via formation of the metal-organic complex. The average grain size in the deposited film was found to be lesser than 200 nm. It was also detected that the deposited films consist of a mixture of various phases such as a supersaturated solid solution of Ag-rich phase, pure Ni phase, and an amorphous phase at the boundaries. These phases were metastable. However, when the Ag-Ni alloy films were annealed at 600°C, these metastable phases were separated into Ag phase and Ni phase. Bdikin et al. [42] have also confirmed the separation of supersaturated solid solution (metastable phase) of Ag-Ni film into a single elemental phase after annealing the film at 600°C. They deposited the Ag-Ni films of different compositions on the copper substrates by using a single electrolyte containing Ag and Ni ions at room temperature and the current density 2–5.4 mA/cm^2. The authors detected that the deposited nanocrystalline Ag-Ni films were metastable. These deposited Ag-Ni films become unstable with increasing annealing temperatures and decompose into a single elemental phase when annealed at 600°C.

4.4 Curvature of the deposited film's particles

The curvature provided by the particles of the deposited films is also playing a very crucial role in increasing the miscibility of the constituent elements. Srivastava and Mundotiya [25] explored the possibility of changes in the extent of miscibility of Ag and Ni elements with changes in the system's morphology and tried to establish a correlation between size, morphology, and extent of miscibility in the Ag-Ni system. They deposited the Ag-Ni alloy films on the Cu substrate (30 mm × 10 mm × 0.3 mm) by using the electrolyte solution (1.69 g of $AgNO_3$, 0.25 g of $NiNO_3{\bullet}6H_2O$, and 0.02 g of H_3BO_3 in 100 mL of distilled water) at the current input of 100 mA and deposition time of 10 minutes. It was observed that the Ag-Ni films contain features of two distinct morphologies: (a) dendritic morphology (designated by type A feature) and (b) branched wire type morphology (designated by type B feature), as can be seen in **Figure 6(a)** and **(b)**, respectively. The dendritic morphology (i.e., type A feature) had high Ag content in single-phase Ag-Ni when compared to the branched wire type morphology (i.e., type B of the feature). The SAED pattern from the type A feature shows a regular array of diffraction spots, which indicates that the dendritic morphology is a single crystal with size in micrometer range (**Figure 6(c)**), while the SAED pattern from the region of the type B feature consists of series of concentric rings, which indicates that the wire type morphology is polycrystalline in nature, and contains a large number of nanoparticles, and the rings in SAED pattern result in falling of the sum of individual patterns of these particles on the concentric rings (**Figure 6(d)**). The SAED patterns (both from the regions of the type A and type B features) did not show any diffraction for Ag, Ni, or even their oxides. The measured d-spacing

Figure 5.
SEM micrograph of Ag-Ni alloy film deposited at 100 mA/cm² and an electrolyte temperature of 80°C for: (a) without thiourea additive and (b) with thiourea additive in the electrolyte [51].

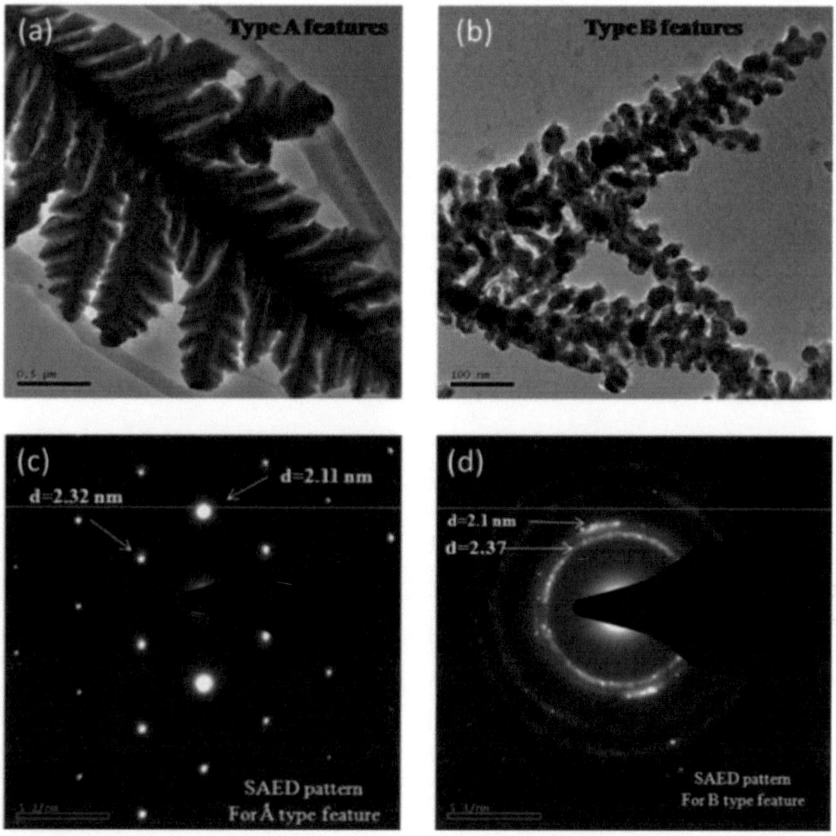

Figure 6.
(a and b) TEM bright field image of type A and type B features. (c and d) SAED pattern obtained from type A and type B features [25].

values from the diffraction spots or concentric rings did not match with any of the d-spacing values for standard FCC crystal structures (i.e., Ag and Ni crystal structures). This indicates that the Ag and Ni atoms arranged themselves in solid solution atomic configurations to form a single solid solution of Ag-Ni system. The average size of the nanoparticles in the branched wire type morphology was ~20 nm. It was observed that both features (i.e., type A and type B) contained Ag and Ni atoms in a single solid solution atomic arrangement, but the presence of Ni atoms in the feature made up of ~20-nm-sized particles (i.e., type B) was found to be larger than the type A feature. The greater extent of miscibility of immiscible Ag and Ni atoms in the type B feature can be attributed to the greater curvature provided by the nanoparticles.

Author details

Brij Mohan Mundotiya[1*], Wahdat Ullah[2] and Krishna Kumar[2]

1 National Institute of Technology (NIT), Hamirpur, Himachal Pradesh, India

2 Malaviya National Institute of Technology (MNIT), Jaipur, Rajasthan, India

*Address all correspondence to: brij2010iitrke@gmail.com

References

[1] Burton JJ, Hyman E, Fedak DG. Surface segregation in alloys. Journal of Catalysis. 1975;**37**:106-113. DOI: 10.1016/0021-9517(75)90138-4

[2] Zhu H, Yang S, Ni G, Yu D, Du Y. Fabrication and magnetic properties of $Co_{67}Ni_{33}$ alloy nanowire array. Scripta Materialia. 2001;**44**:2291-2295. DOI: 10.1016/S1359-6462(01)00761-8

[3] Yamasaki T. High-strength nanocrystalline Ni-W alloys produced by electrodeposition and their embrittlement behaviors during grain growth. Scripta Materialia. 2001;**44**:1497-1502. DOI: 10.1016/S1359-6462(01)00720-5

[4] Lee W, Lee J, Bae JD, Byun CS, Kim DK. Syntheses of Ni_2Si, Ni_5Si_2, and NiSi by mechanical alloying. Scripta Materialia. 2001;**44**:97-103. DOI: 10.1016/S1359-6462(00)00547-9

[5] Yichun L, Jiamin Z, Jikang Y, Jinghong D, Guoyou G, Jianhong Y. Direct electrodeposition of Fe-Ni alloy films on silicon substrate. Rare Metal Materials and Engineering. 2014;**43**:2966-2968. DOI: 10.1016/S1875-5372(15)60043-1

[6] Torabinejad V, Aliofkhazraei M, Assareh S, Allahyarzadeh MH, Rouhaghdam AS. Electrodeposition of Ni-Fe alloys, composites, and nano coatings—A review. Journal of Alloys and Compounds. 2017;**691**:841-859. DOI: 10.1016/j.jallcom.2016.08.329

[7] Guo H, Chen Y, Wen R, Yue GH, Peng DL. Facile synthesis of near-monodisperse Ag@Ni core-shell nanoparticles and their application for catalytic generation of hydrogen. Nanotechnology. 2011;**22**(19):195604. DOI: 10.1088/0957-4484/22/19/195604

[8] Rai RK, Srivastava C. Nonequilibrium microstructures for Ag-Ni nanowires. Microscopy and Microanalysis. 2015;**21**:491-497. DOI: 10.1017/S1431927615000069

[9] Kabir L, Mandal AR, Mandal SK. Polymer stabilized Ni-Ag and N-Fe alloy nanoclusters: Structural and magnetic properties. Journal of Magnetism and Magnetic Materials. 2010;**322**:934-939. DOI: 10.1016/j.jmmm.2009.11.027

[10] Van Ingen RP, Fastenau RHJ, Mittemeijer EJ. Formation of crystalline Ag_xNi_{1-x} solid solutions of unusually high supersaturation by laser ablation deposition. Physical Review Letters. 1994;**72**:3116-3119. DOI: 10.1103/PhysRevLett.72.3116

[11] Lee CC, Chen DH. Large-scale synthesis of Ni-Ag core-shell nanoparticles with magnetic, optical and anti-oxidation properties. Nanotechnology. 2006;**17**:3094-3099. DOI: 10.1088/0957-4484/17/13/002

[12] Tyler EH, Clinton JR, Luo HL. Electrical resistivity of metastable Ag-Ni alloys. Solid State Communications. 1973;**13**:1409-1411. DOI: 10.1016/0038-1098(73)90178-6

[13] Mundotiya BM, Srivastava C. Ag-Ni nanoparticles: Synthesis and phase stability. Electrochemical and Solid-State Letters. 2012;**15**(5):K41-K44. DOI: 10.1149/2.esl120008

[14] Singleton M, Nash P. The Ag-Ni (silver-nickel) system. Bulletin of Alloy Phase Diagrams. 1987;**8**(2):119-121. DOI: 10.1007/BF02873194

[15] Liu XJ, Gao F, Wang CP, Ishida K. Thermodynamic assessments of the Ag-Ni binary and Ag-Cu-Ni ternary systems. Journal of Electronic Materials. 2008;**37**:210-217. DOI: 10.1007/s11664-007-0315-1

[16] Li ZC, Liu JB, Li ZF. Interface assisted formation of a metastable hcp

phase by ion mixing in an immiscible Ag-Ni system. Journal of Physics: Condensed Matter. 2000;**12**:9231-9235. DOI: 10.1088/0953-8984/12/44/305

[17] Zhang Z, Nenoff TM, Huang JY, Berry DT, Provencio PP. Room temperature synthesis of thermally immiscible Ag-Ni nanoalloys. Journal of Physical Chemistry C. 2009;**113**:1155-1159. DOI: 10.1021/jp8098413

[18] Kumar A, Damle C, Sastry M. Low temperature crystalline Ag-Ni alloy formation from silver and nickel nanoparticles entrapped in a fatty acid composite film. Applied Physics Letters. 2001;**79**:3314-3316. DOI: 10.1063/1.1414298

[19] Srivastava C, Chithra S, Malviya KD, Sinha SK, Chattopadhyay. Size dependent microstructure for Ag-Ni nanoparticles. Acta Materialia. 2011;**59**:6501-6509. DOI: 10.1016/j.actamat.2011.07.022

[20] Eom H, Jeon B, Kim D, Yoo B. Electrodeposition of silver-nickel thin films with a galvanostatic method. Materials Transactions. 2010;**51**(10):1842-1846. DOI: 10.2320/matertrans.M2010126

[21] Raghupathy Y, Natarajan KA, Srivastava C. Anti-corrosive and anti-microbial properties of nanocrystalline Ni-Ag coatings. Materials Science and Engineering B. 2016;**206**:1-8. DOI: 10.1016/j.mseb.2016.01.005

[22] Schneider M, Krause A, Ruhnow M. Formation and structure of Ni-Ag layer by electrolytic deposition. Journal of Materials Science Letters. 2002;**21**:795-797. DOI: 10.1023/A:1015710111396

[23] Liang D, Liu Z, Hilty RD, Zangari G. Electrodeposition of Ag-Ni films from thiourea complexing solutions. Electrochimica Acta. 2012;**82**:82-89. DOI: 10.1016/j.electacta.2012.04.100

[24] Srivastava C, Mundotiya BM. Electron microscopy of microstructural transformation in electrodeposited Ni-rich Ag-Ni film. Thin Solid Films. 2013;**539**:102-107. DOI: 10.1016/j.tsf.2013.05.080

[25] Srivastava C, Mundotiya BM. Morphology dependence of Ag-Ni solid solubility. Electrochemical and Solid-State Letters. 2012;**15**:K10-K15. DOI: 10.1149/2.003202es

[26] Srivastava C, Mundotiya BM. Size and solid solubility in electrodeposited Ag-Ni nanoparticles. Materials Science Forum. 2013;**736**:21-26. DOI: 10.4028/www.scientific.net/MSF.736.21

[27] Bona AD. Characterizing ceramics and the interfacial adhesion to resin: II—The relationship of surface treatment, bond strength, interfacial toughness and fractography. Journal of Applied Oral Science. 2005;**13**:101-109. DOI: 10.1590/S1678-77572005000200002

[28] Cao G. Nanostructures and Nanomaterials: Synthesis, Properties, and Applications. London: World Scientific; 2004. 175 p. DOI: 10.1142/9781860945960_0005

[29] Paunovic M, Schlesinger M, Snyder DD. Fundamental considerations. In: Schlesinger M, Paunovic M, editors. Modern Electroplating. 5th ed. New Jersey: Wiley; 2011. pp. 1-32. DOI: 10.1002/9780470602638

[30] Markov IV. Crystal Growth for Beginners: Fundamentals of Nucleation, Crystal Growth and Epitaxy. 2nd ed. Singapore: World Scientific; 1995. 422 p. DOI: 10.1142/9789812386243_0003

[31] US Water System. Deionized Water vs Distilled Water. 2018. Available from: https://www.uswatersystems.com/deionized-water-vs-distilled-water. [Accessed: 2018-06-27]

[32] Santhi K, Karthick SN, Kim H, Nidhin M, Narayanan V, Stephen A. Microstructure analysis of the ferromagnetic Ag-Ni system synthesized by pulsed electrodeposition. Applied Surface Science. 2012;**258**:3126-3132. DOI: 10.1016/j.apsusc.2011.11.049

[33] Gomez E, Garcia-Torres J, Valles E. Study and preparation of silver electrodeposits at negative potentials. Journal of Electroanalytical Chemistry. 2006;**594**:89-95. DOI: 10.1016/j.jelechem.2006.05.030

[34] Kamel MM. Anomalous codeposition of Co-Ni: Alloys from gluconate baths. Journal of Applied Electrochemistry. 2007;**37**:483-489. DOI: 10.1007/s10800-006-9279-8

[35] Qu NS, Zhu D, Chan KC, Lei WN. Pulse electrodeposition of nanocrystalline nickel using ultra narrow pulse width and high peak current density. Surface and Coatings Technology. 2003;**168**:123-128. DOI: 10.1016/S0257-8972(03)00014-8

[36] Sharma A, Das S, Das K. Pulse electroplating of ultrafine grained tin coating. In: Aliofkhazraei M, editor. Electroplating of Nanostructures. Intech Open; 2015. pp. 105-129. DOI: 10.5772/61255

[37] Ibl N. Some theoretical aspects of pulse electrolysis. Surface Technology. 1980;**10**:81-104. DOI: 10.1016/0376-4583(80)90056-4

[38] Watanabe T. Nano-Plating Microstructure Control Theory of Plated Film and Data Base of Plated Film Microstructure. Oxford, UK: Elsevier Ltd; 2004

[39] Sharma A, Bhattacharya S, Das S, Das K. A study on the effect of pulse electrodeposition parameters on the morphology of pure tin coatings. Metallurgical and Materials Transactions A. 2014;**45**:4610-4622. DOI: 10.1007/s11661-014-2389-8

[40] Weissmueller J, Bunzel P, Wilde G. Two-phase equilibrium in small alloy particles. Scripta Materialia. 2004;**51**:813-818. DOI: 10.1016/j.scriptamat.2004.06.025

[41] Bellomo A, De Marco D, De Robertis A. Formation and thermodynamic properties of complexes of Ag(I) with thiourea as ligand. Talanta. 1973;**20**:1225-1228. DOI: 10.1016/0039-9140(73)80088-8

[42] Bdikin IK, Strukova GK, Strukov GV, Kedrov VV, Matveev DV, Zer'kov SA, et al. Growth, crystal structure and stability of Ag-Ni/Cu films. Materials Science Forum. 2006;**514-516**:1166-1170. DOI: 10.4028/www.scientific.net/MSF.514-516.1166

[43] Burstein GT. A hundred years of Tafel's equation: 1905-2005. Corrosion Science. 2005;**47**:2858-2870. DOI: 10.1016/j.corsci.2005.07.002

[44] Tan AC. Tin and Solder Plating in the Semiconductor Industry: A Technical Guide. 1st ed. London: Chapman & Hall; 1993. 326 p. ISBN: 0442317123

[45] Inamdar AI, Mujawar SH, Barman SR, Bhosale PN, Patil PS. The effect of bath temperature on the electrodeposition of zinc oxide thin films via an acetate medium. Semiconductor Science and Technology. 2008;**23**:085013. (6pp). DOI: 10.1088/0268-1242/23/8/085013

[46] Rashidi AM, Amadeh A. Effect of electroplating parameters on microstructure of nanocrystalline nickel coatings. Journal of Material Science and Technology. 2010;**26**:82-86. DOI: 10.1016/S1005-0302(10)60013-8

[47] Sahaym U, Miller SL, Norton MG. Effect of plating temperature on Sn surface morphology. Materials Letters. 2010;**64**:1547-1550. DOI: 10.1016/j.matlet.2010.04.036

[48] Garcia-Torres J, Valles E, Gomez E. Influence of bath temperature and bath composition on Co-Ag electrodeposition. Electrochimica Acta. 2010;**55**:5760-5767. DOI: 10.1016/j.electacta.2010.05.014

[49] Gomez H, Lizama H, Suarez C, Valenzuela A. Effect of thiourea concentration on the electrochemical behavior of gold and copper electrodes in presence and absence of Cu(II) ions. Journal of the Chilean Chemical Society. 2009;**54**:439-444. DOI: 10.4067/S0717-97072009000400026

[50] Schmidt WU, Alkire RC, Gewirth AA. Mechanic study of copper deposition onto gold surfaces by scaling and spectral analysis of in-situ atomic forces microscopic images. Journal of the Electrochemical Society. 1996;**143**:3122-3132. DOI: 10.1149/1.1837174

[51] Mundotiya BM. Phases and morphology in electrodeposited Ag-Ni film and nano-sized particles [thesis]. Indian Institute of Science Bangalore: Bangalore; 2012

Meet the editor

Professor Mohannad Al-Qudah obtained his medical degree from the University of Jordan in 1991. He started his otolaryngology residency in Jordan, then finished two years of clinical accredited fellowship programs in rhinology and endoscopic skull base surgery in the USA. He is a fellow of the American Academy of ORL-HNS, the American College of Surgeons, and the International College of Surgeons. Currently, he is a professor at JUST University and a senior consultant at KAUH Hospital, Irbid, Jordan. He has published more than 150 articles and abstracts in international journals. His research interests are sinusitis pathogenesis, new minimally invasive techniques, and indications in endoscopic sinus and skull base surgery. Professor Al-Qudah described many novel surgical techniques and pathological hypotheses in his specialty. In recognition of his medical achievements, he has been awarded the Royal Silver Jubilee medal.

Contents

Preface

Our fundamental understanding of rhinological disorders has improved substantially over the last three decades. Nowadays, the sinonasal region is considered as one physiological functionally separate unit that can be affected by disorders of different etiopathogenesis with variable structural histopathology manifestations.

This progression has been facilitated by advances in basic and clinical research, new methods of endoscopic visualization, high-resolution CT scans, the global exchange of experience, and the new guidelines agreed upon by various prestigious international scientific committees.

The umbrella of rhinology disorders is wide. It involves variations in anatomical structures, allergy, inflammation, cosmetic deformity, trauma, and tumors. These disorders have common indistinguishable and nonspecific presenting symptoms that require accurate knowledge and clinical skills to reach the correct diagnosis so the proper management plan can be set.

This book highlights certain topics concerning rhinological disorders that have been left unexplained by other literature. The first section of this book reviews the complex embryology and anatomy of the sinonasal region in a concise yet precise method covering all the essential clinical data that the reader needs to understand.

Nonallergic (intrinsic) rhinitis is a multifactorial common disorder that is usually misdiagnosed as allergic rhinitis because of the similarity between their symptoms. The peculiar aspect of this subject is thoroughly explained in the medical rhinology section. In another chapter in this section, the intricate relationship between rhinology disorders and neuropsychological health is unraveled, offering valuable insights and strategies to emphasize the importance of addressing both the physical and psychological aspects of rhinology disorders.

Effective management of refractory chronic rhinosinusitis and other complex sinonasal conditions by various extended sinus surgical approaches and techniques is explained in the surgical rhinology section. A separate chapter in this section discusses the application of medical lasers in rhinology and provides inspiration for exploring new ideas of medical lasers in this field.

Pathology in the sinonasal region is common, and its manifestations can present in different medical specialties. The present book covers certain rhinology disorders that are often neglected elsewhere.

Mohannad Al-Qudah, MD, FAAOHNS, FACS
Professor of ORL-HNS,
Jordan University of Science and Technology,
Irbid, Jordan

Section 1

Basic Rhinology

Chapter 1

Surgical Anatomy of the Nose and Paranasal Sinuses

Basma S. Almaadani

Abstract

This chapter delves into the intricate surgical anatomy of the nasal cavity and paranasal sinuses, highlighting their relevance in otolaryngological procedures. Understanding these anatomical details is crucial for performing safe and effective surgical procedures, minimizing complications, and enhancing patient outcomes. By integrating anatomical knowledge with clinical insights, this chapter aims to equip surgeons with a comprehensive understanding necessary for successful surgical intervention of nasal and sinus pathologies.

Keywords: anatomy, nasal cavity, paranasal sinuses, skull base, surgery

1. Introduction

The anatomical complexity of nasal cavity and paranasal sinuses is heightened by the proximity of significant surrounding structures, including the orbits and skull base which demand meticulous surgical navigation.

This chapter elucidates the basic anatomy of the nasal cavity and paranasal sinuses, emphasizing their clinical significance and the intricate relationships with adjacent critical structures. Mastery of this anatomical knowledge is imperative for the otolaryngologist to perform precise and safe surgical interventions, thereby optimizing patient outcomes and reducing intraoperative risks.

2. External nose anatomy

The external nose is composed of both bony and cartilaginous structures (osteo-cartiliginous framework), providing it with support and shape.

The bony part of the nose consists primarily of two nasal bones united in the midline by internasal suture. These nasal bones articulate with the frontal bone superiorly through the nasal process of the frontal bone and to the maxillary bones laterally through the frontonasal process of the maxilla (**Figure 1**).

The cartilaginous part of the nose consists of the nasal septum in the midline; in addition to the upper lateral cartilages and lower lateral cartilages (also known as allar cartilages).

The nasal septum divides the nasal cavity into two sides. It consists of the columellar septum which is the most anterior part of the septum, it is formed by the medial

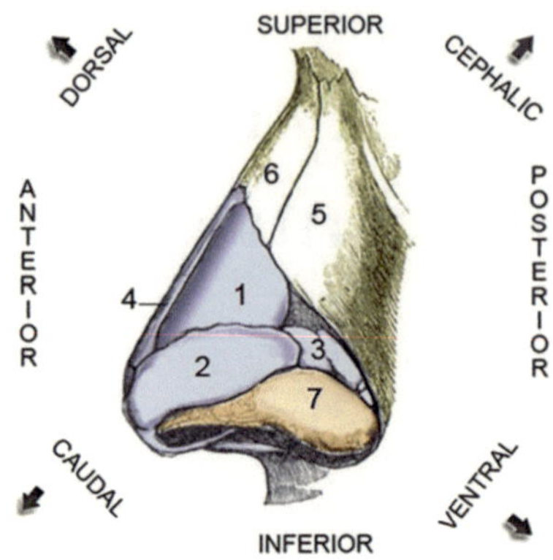

Figure 1.
Showing the terminology for nasal surgery topography and the external nasal osteocartilaginous framework. 1: upper lateral cartilage; 2: lower lateral cartilage; 3: accessory cartilage; 4: cartilaginous part of nasal septum; 5: frontonasal process of maxilla; 6: nasal bone; 7: fibroconnective tissue. Adapted from Ref. [1].

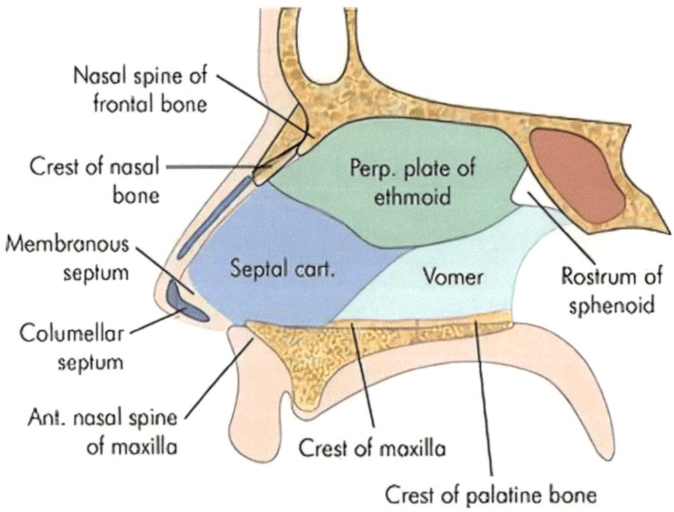

Figure 2.
Illustration of the nasal septum anatomy parts. Adapted from Ref. [2].

crura of alar cartilages which are connected by fibrous tissue and covered by skin. The membranous part of the septum, which lacks any bony or cartilaginous support, is situated between the columella and the lower edge of the septal cartilage. These two parts of the septum can be movable from side to side. The septum proper consists of an osteocartilaginous framework covered by nasal mucosa. The cartilaginous part is formed by the quadrangular septal cartilage anteriorly while the bony part is composed of the perpendicular plate of the ethmoid postero-superiorly and the vomer postero-inferiorly (**Figure 2**).

3. Blood supply and lymphatics

The nose is richly supplied with blood, with arterial contributions from both the internal carotid (including the anterior and posterior ethmoid arteries from the ophthalmic artery) and the external carotid (including the sphenopalatine, greater palatine, superior labial, and angular arteries).

The dorsal nasal artery, a branch of the ophthalmic artery, traverses the medial palpebral ligament to reach the dorsum of the nose, primarily supplying the bony portion of the external nose. Meanwhile, the external carotid artery gives rise to the facial artery, which provides branches to the cartilaginous part of the external nose; however, these branches are not specifically described in anatomical terminology [3].

The superior part of the nasal septum is nourished by the anterior and posterior ethmoidal arteries, as well as the sphenopalatine artery. The anterior septum and nasal floor receive blood from terminal branches of the superior labial artery. Kiesselbach's plexus, also known as the Little area, is located in the lower anterior third of the nasal septum and is formed by the convergence of the superior labial, anterior ethmoidal, greater palatine, and sphenopalatine arteries.

The venous drainage of the nose, which mirrors the arterial pattern, is carried out by the anterior facial vein, the sphenopalatine vein, and the ethmoid veins. Notably, these veins directly communicate with the cavernous sinus and lack valves, facilitating the potential intracranial spread of infections. Lymphatic drainage from the nose flows anteriorly through the upper lip lymphatics and posteriorly through the deep cervical and retropharyngeal lymph nodes [4].

4. Nasal cavity

The nasal cavity is a complex structure comprising two cavities separated by the nasal septum, with each cavity having distinct anatomical features including the anterior nares, choana, roof, floor, medial, and lateral walls. The anterior nares are bounded by the columella medially and the ala laterally, while the choana is framed by the vomer medially and the medial pterygoid plate laterally. The cribriform plate of the ethmoid bone forms the roof of the nasal cavity, while the hard palate makes up the floor.

The medial wall of the nasal cavity is formed by the nasal septum which is comprised of cartilaginous and bony parts; the quadrangular cartilage being antero-inferiorly, the perpendicular plate of the ethmoid bone being postero-superiorly and the vomer is postero-inferiorly.

The lateral wall features three turbinates (conchae) and four corresponding spaces (meatuses), with the inferior turbinate being the largest and highly vascular, located above the inferior meatus where the nasolacrimal duct opens. The middle turbinate is of moderate size and is located centrally within the nasal cavity. Below it lies the middle meatus, which contains several notable structures: the bulla ethmoidalis, the largest cell of the anterior ethmoid sinus; the hiatus semilunaris, a semilunar groove below the bulla; and the uncinate process, a bone shelf medial to the bulla. The osteomeatal complex (OMC), crucial for drainage of the anterior group of sinuses, is bounded laterally by the lamina papyracea and medially by the middle turbinate. The superior turbinate is the smallest and highest, with the superior meatus below it, where the posterior ethmoid sinus opens. The sphenoethmoidal recess, located above the superior turbinate, drains the sphenoid sinus.

The nasal valve, the narrowest part of the nasal cavity, is located at the junction of the lower and upper lateral cartilages near the anterior end of the inferior turbinates.

The lateral nasal cavity, with its intricate anatomy, will be explored further in this chapter.

5. Ethmoid bulla

The ethmoid bulla is a prominent anterior ethmoid air cell originating from the lamina papyracea along the medial wall of the orbit. Despite the intricate nature of ethmoid bone anatomy, the ethmoid bulla remains the most recognizable and dependable ethmoid cell. It is located medial to the lamina papyracea, anterior to the vertical basal lamella of the middle turbinate, posterior to the uncinate process and postero-inferior to the frontal recess. Roughly 92% of patients exhibit well-pneumatized ethmoid bullae, making it a reliable landmark for anterior ethmoid sinuses during endoscopic procedures. In the remaining 8% of patients, the ethmoid bulla is either minimally pneumatized or completely absent, a condition referred to as torus lateralis [5].

6. Uncinate process, ethmoid infundibulum, and hiatus semilunaris

Uncinate process, a hook-shaped projection, is located anterior and inferior to the ethmoid bulla. Starting superiorly, near the agger nasi region at ethmoidal crest of the maxilla, the uncinate process extends downward and backward ending in a free edge with no bony attachment. This curving will result in both vertical and horizontal portions to the uncinate process (**Figure 3**).

The way the vertical portion of the uncinate process attaches superiorly is significant, since it can influence the layout of the frontal recess and the outflow pathway of the frontal sinus. Not recognizing this alignment can result in insufficient drainage of the frontal sinus after surgery.

Most often, the uncinate process attaches laterally to the lamina papyracea. In this alignment, the frontal sinus drains medially to the uncinate process insertion and directly into the middle meatus. In uncommon instances, the uncinate process attaches to the upper section of the middle turbinate or directly to the skull base. In these scenarios, the frontal sinus drainage route will be lateral to the uncinate process, flowing into the ethmoid infundibulum [6].

Hiatus semilunaris, a two-dimensional space between the posterior free margin of the uncinate process and the ethmoid bulla that serves as an "entrance" to the three-dimensional ethmoid infundibulum.

The *ethmoid infundibulum is a channel that is* bordered anteromedially by the uncinate process, posteriorly by the ethmoid bulla, and laterally by the lamina papyracea.

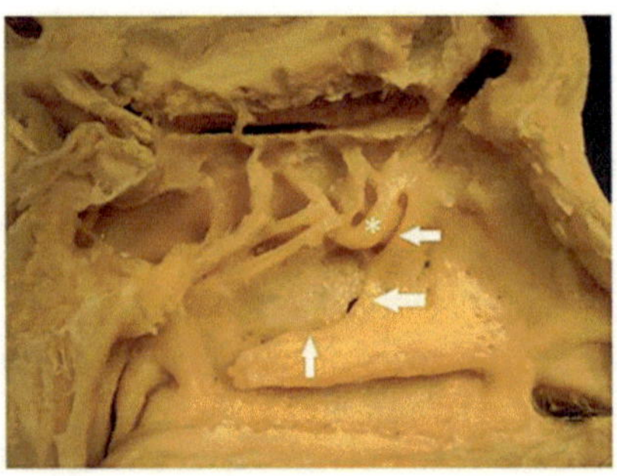

Figure 3.
Parasagittal cadaveric dissection of the lateral nasal wall. The arrows are demonstrating the hook-shaped uncinate process with its vertical and horizontal parts. (*) Ethmoid bulla [6].

This structure is important because it acts as an innate drainage channel for the maxillary sinuses and often the frontal sinuses.

7. Agger nasi region

Agger nasi is an area of ridge bone on the anterior region of lateral nasal wall, when pneumatized, it is identified as an agger nasi cells which are the most anterior ethmoid sinus cells. The agger nasi region is bordered superiorly by the frontal sinus, superiorly and posteriorly by the frontal recess, by nasal bones anteriorly and by anciante inferomedially. Recognizing and excising the agger nasi cell is essential in the surgical management of frontal sinus disease, as it often plays a major role in narrowing the frontal recess and sinus outflow tract [5].

8. Middle turbinate

The middle turbinate consists primarily of three sections, oriented in various planes and firmly attached to the lateral nasal wall, orbit, and skull base.

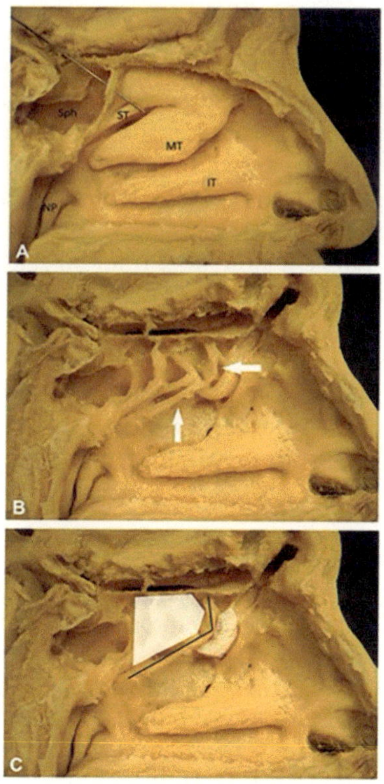

Figure 4.
Gross cadaveric dissections of the lateral nasal wall. (A) The three turbines are unbroken and in position. (IT) Inferior turbinate; (MT) middle turbinate; (ST) superior turbinate. (B) The vertical and horizontal portions of the MT basal lamella can be seen (arrows) after the parasagittal sections of the MT and ST have been excised. (C) The anterior ethmoid cavity is separated from the posterior ethmoid cavity by the vertical and horizontal parts of the MT basal lamella (demarcated in a black line) [6].

The anterior part of the middle turbinate, which is the first portion of the middle turbinate to be seen during anterior rhinoscopy or nasal endoscopy, is oriented in the sagittal plane. It is attached anterosuperiorly to the crista ethmoidalis of the maxilla, and base of the skull near the lateral lamella of cribriform plate of the ethmoid bone with a free edge anteroinferiorly.

The middle turbinate basal lamella has a vertical part which is aligned in the coronal plane and a horizontal part aligned in an axial plane (**Figure 4**).

The vertical part divides the anterior and posterior ethmoid complexes and attaches to the skull base superiorly and to the lamina papyracea laterally. The horizontal part, which forms posteriorly, can be used to locate the sphenopalatine foramen.

Pneumatization of the middle turbinate is possible and the nomenclature of these structures is determined by the location of the pneumatized portion. It is referred to as a concha bullosa if the anterior parasagittal-oriented portion is pneumatized and it is referred to as an interlamellar cell if the vertical section of the basal lamella is pneumatized [5].

9. Posterior ethmoid complex

The posterior ethmoid sinus is bounded anteriorly by the vertical portion of the middle turbinate basal lamella, posteriorly by an anterior wall of the sphenoid sinus, laterally by the laminia papyacea and superiorly by the skull base.

An Onodi cell, which is the term anatomical variant, is a highly pneumatized posterior ethmoid cell that has persisted in growing superolaterally in relation to the true sphenoid. Studies have shown that the incidence of sphenoethmoidal air cells in individuals ranges from 3.4 to 60%, a variation that is probably attributed to differences in definitions rather than actual rates of occurrence [7].

Since the term "sphenoethmoid cell" better describes the anatomy in this region, it is now preferred over "Onodi cell." sphenoethmoidal cell when present can be mistaken for a sphenoid sinus. Before sinus dissection, radiographic identification of sphenoethmoid cells is crucial to avoid incomplete dissection that may result from not identifying them beforehand.

The other primary reason for the significance of these air cells is that they are situated in close proximity to the internal carotid artery and optic nerve, with a mere 0.03 mm (median 0.08 mm) of bone dividing them [7].

10. Ethmoid roof and skull base

The anterior skull base is divided by the middle turbinate vertical lamella into the cribriform plate medially and the roof of the ethmoid laterally. The ethmoid roof is made up of two parts: a thinner vertical portion, named the lateral cribriform plate lamella (LCPL), and the thicker horizontal portion named fovea ethmoidalis which an extension of the orbital plate of the frontal bone.

The skull base is thicker laterally and as it gets medially it becomes thinner reaching only (0.2 mm thickness) along the cribriform plate which represents the thinnest point in the anterior skull base and the most frequent location for iatrogenic cerebrospinal fluid leaks during sinus surgery [6].

Furthermore, there is a positive correlation between the depth of the olfactory cleft and the chance of iatrogenic injury during surgery. The ethmoid roof is classified by Keros classification system into—shallow type I (1–3 mm), medium type II (4–7 mm), and deep type III (8–16 mm). As the depth increases the risk for complications increases during endoscopic endonasal surgery which can include CSF leak or intracranial hemorrhage [8].

10.1 Maxillary sinus

Out of all the paranasal sinuses, the maxillary sinus is the biggest and the first to develop. The maxillary sinus is positioned within the maxillary bone, it has a pyramid shape, with its base near the nasal cavity and its apex directed toward the zygomatic bone. The boundaries of the maxillary sinus include the floor, which is formed by the alveolar process of the maxilla housing the roots of the maxillary teeth; the roof is formed by the floor of the orbit; the medial wall is formed by the lateral nasal wall; and the lateral wall is facing the zygomatic process.

The usual opening of the maxillary sinus is through a single natural ostium positioned in the posterior part of the ethmoid infundibulum. Nonetheless, up to 23% of patients may develop accessory ostia through the anterior or posterior fontanelles. These fontanelles are situated in the lateral nasal wall and lack bone; in some instances, the mucosa and connective tissue covering them are insufficient, creating an alternative drainage route for the sinus [6].

The maxillary sinus can show different levels of pneumatization and occasionally, hypoplasticity. The orbital contents tend to fill a greater volume of the midface as the volume of the maxillary sinus decreases. Due to the increased orbital volume-to-maxillary sinus volume ratio in cases of maxillary sinus hypoplasia, paranasal sinus surgeons need to proceed with caution during surgery. The uncinate process is often shifted inferolaterally and positioned near the orbital wall in cases of the hypoplastic maxillary sinus. Furthermore, marked degrees of maxillary sinus hypoplasia may be associated with an underdeveloped uncinate process because of their shared developmental origins [5].

10.2 Sphenoid sinus

Positioned within the sphenoid bone at the central skull base, the sphenoid sinuses are a pair of large paranasal spaces found posterior to the ethmoid sinuses. The natural ostium of the sphenoid sinus connects the sphenoid sinus with the roof of the nasal cavity via the sphenoethmoidal recess.

The sphenoid ostium opens in the anterior superior wall of the sinus and it is situated approximately 1.0–1.5 cm above the superior edge of the posterior choana and it is typically situated approximately midway between the posterior insertion of the superior turbinate and the nasal septum [5].

Due to its variability, the sphenoid sinus is one of the most complex sinuses. Its proximity to critical neurovascular structures makes it challenging for endoscopic surgeons to navigate [9]. These relationships include the pituitary gland which is located posterior and superior to the sphenoid cavity. Above it lies the optic chiasm, with the optic nerves extending laterally toward the orbital apices. The internal carotid arteries are positioned lateral to the pituitary gland and clivus, often creating impressions in the sphenoid sinus walls and forming the opticocarotid recesses. The cavernous sinus, which houses oculomotor; trochlear; ophthalmic and maxillary divisions of trigeminal; and abducent cranial nerves, is situated lateral to the sphenoid sinus.

Additionally, the vidian nerve can be seen as an impression along the inferolateral aspect of the sinus floor. Understanding these anatomical relationships is vital to avoid damaging these structures during surgery.

The right and left sphenoid sinuses are separated by a sphenoid intersinus septum. It is common for these sinuses to develop unevenly, and demonstrate differences in size and pneumatization. When planning sphenoid sinus surgery, it is essential to thoroughly assess the sphenoid intersinus septum. This septum can deviate to one side and may attach near critical structures like the internal carotid artery or optic nerve [10].

Awareness of the anatomical layout and positioning of these neighboring structures can make interventions considerably safer, given the potentially devastating outcomes of damaging any of them.

10.3 Frontal sinus

The frontal sinus has a trapezoidal or triangular shape and it develops last among the paranasal sinuses. In newborns, the frontal sinus (FS) is barely noticeable, but it becomes detectable on radiographs around age 4. Its growth parallels that of the craniofacial area, attaining its maximum size approximately at age 18 [11].

The frontal sinus is bounded inferiorly by the orbital roof, anteriorly by the terminal frontal table of the frontal bone, and posteriorly by the posterior table dividing the sinus from the brain.

Despite the notable anatomical differences in the frontal sinus area, the right and left frontal sinuses are divided by a thin bone known as the frontal intersinus septum. The right and left frontal sinuses are often unequal in size, with the septum leaning toward one side. The frontal sinus is commonly bilateral; however, it might be unilateral in up to 15% of cases and absent in 8% of cases [12].

In some cases, the septum becomes pneumatized, resulting in an intersinus septal cell that can drain into the left or right frontal sinus. Surgeons may mistakenly think they have drained the opposite sinus by entering this cell.

Other pneumatized cells that should be taken into consideration in the frontal sinus region are the supraorbital ethmoid cells, which extend over the orbit, as they can also confuse without proper preoperative imaging.

Connecting the frontal sinus to the anterior ethmoid cells, the frontal recess is an inverted cone featuring a narrow upper end at the internal frontal ostium and a broader lower end that merges into the anterior ethmoid cells. Endoscopic surgeons need to have a comprehensive understanding of the frontal recess.

Despite that the drainage pathway can vary based on the superior insertion of the uncinate process, the frontal recess bony structures remain fairly the same. The frontal recess is bound medially by the anterior superior portion of the middle turbinate and laterally by the lamina papyracea. The frontal sinus forms its upper limit while the nasal cavity forms its lower limit. The agger nasi cells constitute the anterior border and the ethmoid bulla forms the posterior border [11].

Author details

Basma S. Almaadani
Independent Researcher, Amman, Jordan

*Address all correspondence to: basma.maadani@gmail.com

References

[1] Tu J, Inthavong K, Ahmadi G. The Human Respiratory System. 2012

[2] Sharma AR et al. Clinical outcome following septoplasty with or without inferior turbinate reduction. International Journal of Otorhinolaryngology and Head and Neck Surgery. 2020

[3] Pilsl U, Anderhuber F. The external nose: The nasal arteries and their course in relation to the nasolabial fold and groove. Plastic & Reconstructive Surgery. 2016;**138**(5):830e-835e. DOI: 10.1097/prs.0000000000002626

[4] Facs EWCMD. Nasal Anatomy: Embryology, Skin and Soft Tissues, Blood Supply and Lymphatics. n.d. Available from: https://emedicine.medscape.com/article/835134-overview

[5] Kennedy DW, Hwang PH. Sinonasal development and anatomy. In: Rhinology: Diseases of the Nose, Sinuses, and Skull Base. New York: Thieme; 2012c

[6] Villari RC, Wise SK. Surgical endoscopic anatomy of the nose, paranasal sinuses, and skull base. In: Devaiah AK, Marple BF, editors. Rhinology and Endoscopic Skull Base Surgery. Vol. 3. New York: Thieme; 2013

[7] KN, Gaillard F. Sphenoethmoidal Air Cell. Radiopaedia.org; 2008. DOI: 10.53347/rid-1776

[8] Fars ASM. Paranasal Sinus Anatomy: Overview, Gross Anatomy, Microscopic Anatomy. n.d. Available from: https://emedicine.medscape.com/article/1899145-overview

[9] Budu V, Mogoantă CA, Fănuţă B, Bulescu I. The anatomical relations of the sphenoid sinus and their implications in sphenoid endoscopic surgery. PubMed. 2013. Available from: https://pubmed.ncbi.nlm.nih.gov/23529304/

[10] Famurewa O, Ibitoye B, Ameye S, Asaleye C, Ayoola O, Onigbinde O. Sphenoid sinus pneumatization, septation, and the internal carotid artery: A computed tomography study. Nigerian Medical Journal/Nigerian Medical Journal. 2018;**59**(1):7. DOI: 10.4103/nmj.nmj_138_18

[11] Dassi CS, Demarco FR, Mangussi-Gomes J, Weber R, Balsalobre L, Stamm AC. The frontal sinus and frontal recess: Anatomical, radiological and surgical concepts. International Archives of Otorhinolaryngology. 2020;**24**(03):e364-e375. DOI: 10.1055/s-0040-1713923

[12] Ochs MW. Fractures of the Upper Facial and Midfacial Skeleton. Elsevier eBooks; 2008. pp. 905-933. DOI: 10.1016/b978-1-4160-2445-3.50097-2

Section 2

Medical Rhinology

Chapter 2

Nonallergic Rhinitis

Carlos Ríos-Deidán, Diana Villacrés-Silva, Daniela Saénz-Chávez and María Peña-Vásquez

Abstract

Non-allergic rhinitis (NAR) is a heterogeneous nasal disease with high global prevalence. While the specific factors contributing to the origin of NAR remain uncertain, there is indication that neurogenic factors play a significant role in the development of NAR. There are seven subtypes with overlapping presentations, including senile or geriatric rhinitis, gustatory rhinitis, drug-induced rhinitis, hormonal rhinitis, smokers' rhinitis, occupational rhinitis and idiopathic rhinitis. The approach to treatment is focused on alleviating symptoms and parallels the methods used for allergic rhinitis. Patients are advised to minimize exposure to identified triggers whenever feasible. Initial treatments involve the use of primary interventions such as intranasal corticosteroids, intranasal antihistamines, and intranasal ipratropium. Combination therapies may be considered if single interventions do not effectively manage symptoms. The surgery is considered in patients refractory to medical therapy, the reduction of inferior turbinate hypertrophy is a surgical procedure with an excellent outcome, besides the selective neurectomy of the vidian branches, has also proven to be effective.

Keywords: idiopathic rhinitis, endotypes, subtypes non allergic, intranasal corticosteroid, intranasal antihistamines, nasal surgical procedures

1. Introduction

Rhinitis constitutes an alteration of the nasal mucosa and stands out as one of the prevalent reasons for medical consultations globally. Its incidence has shown a rising trend over time. This condition is notably widespread in both Europe and America, with geographical variations possibly attributed to diverse environmental conditions, lifestyle factors, and host-related risk factors [1, 2].

The classification of rhinitis includes infectious rhinitis, allergic rhinitis, non-allergic rhinitis, and mixed rhinitis. Although non-allergic and allergic rhinitis exhibit certain overlapping symptoms, they diverge in terms of treatment approaches, affected demographic groups, and underlying pathophysiology [2]. **Figure 1** shows the phenotypes of chronic rhinitis.

Infectious rhinitis, as its name implies, is a condition triggered primarily by a virus, and to a lesser proportion, by bacteria, commonly recognized as the common cold. Symptoms encompass the presence of discolored nasal secretions, nasal obstruction or congestion, and the formation of crusts [3].

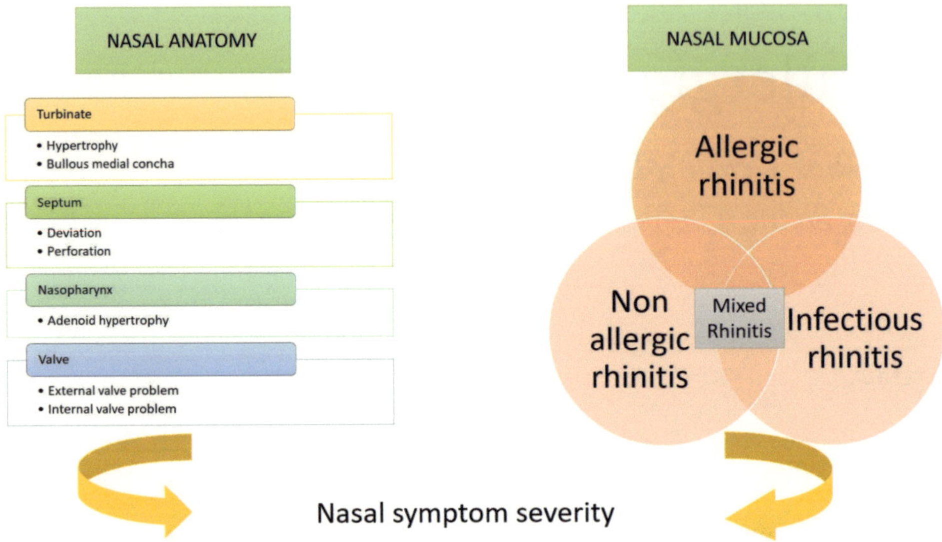

Figure 1.
Phenotypes of chronic rhinitis.

Allergic rhinitis is caused by IgE mediated reactions to inhaled allergens that trigger a type 2 immune reaction that starts in childhood and usually have seasonal exacerbation of symptoms. It is associated to asthma and conjunctivitis. Treatment included allergen avoidance, topic corticosteroid, antihistamines, and immunotherapy [2, 4].

Nonallergic rhinitis is a chronic condition that starts later in life which is defined as inflammation or dysfunction of the nasal mucosa present for at least 12 weeks per year without any indications of allergic sensitization. Diagnostic tests capable of detecting allergies include allergic skin tests, measurement of total immunoglobulin E (IgE), and specific serum IgE assays for certain allergens [1, 3, 5].

To establish a diagnosis of chronic rhinitis, it is necessary for two nasal symptoms to be consistently present for at least 1 hour each day. These symptoms include rhinorrhea (anterior or posterior), nasal obstruction or congestion, sneezing or nasal itch, cough, hyposmia, facial pressure, and eustachian tube dysfunction. These symptoms can be categorized as perennial/persistent, intermittent, or triggered by non-immunologic factors such as irritants, environmental humidity, or temperature variations [3, 5–7].

Mixed rhinitis has the characteristic of presenting several etiologic factors, whether known or not, that is why the diagnostic must be precise and a nasal endoscopy should be done to identify the presence of chronic rhinosinusitis with or without polyps and anatomic factors that may increase symptoms and are completely different pathologies [3].

2. Epidemiology

Limited data are available regarding the prevalence of general rhinitis, with even fewer statistics on Nonallergic rhinitis. However, estimates indicate that over 70 million individuals in the United States and 200 million globally are affected by chronic rhinitis. The European Community Respiratory Heath Survey (ECRHS), the most extensive international multicenter study on prevalence, reported a median self-reported nasal allergy prevalence of 20.9% in 1995, focusing on Western European countries. Notably, it did not delineate the prevalence of Nonallergic or allergic rhinitis. The absence of prevalence data for nonallergic rhinitis is attributed to the lack of a standardized definition and diagnostic criteria, as highlighted by Hellings. Nonetheless, Greiwe et al. proposed mechanistic theories involving autonomic dysfunction, resulting in altered sympathetic activity or parasympathetic overactivity. These mechanisms also encompass changes in the expression of receptor potential channels on sensory nerves in the nasal mucosa, serving as irritant sensors, and the involvement of neuropeptides leading to vasodilation and transudation [3, 7, 8].

Savoure et al. perform a study of worldwide prevalence of rhinitis and described unspecified rhinitis prevalence at 29.4% (1.1 to 63.3%), allergic rhinitis with 18.1% (1 to 54.5%) and 12% for nonallergic rhinitis (4 to 31.4%). They classified the median prevalence based on symptoms at 16.4% and based on IgE definition at 31.4% for nonallergic rhinitis. Bernstein et al., reported prevalence of 23, 34 and 43% for nonallergic rhinitis, mixed rhinitis, and allergic rhinitis respectively [7].

Between the prevalence reported Savoure et al. found studies from Europe, Africa, Asia, Oceania and America for non-specific rhinitis and allergic rhinitis but there were no data from America, Africa, and Oceania for nonallergic rhinitis. For Asia reported prevalence were ranging from 4 to 31.4% and for Europe 5.5 to 23.5% [9].

Savoure et al. in another study reported general prevalence of rhinitis were 53.4%, allergic rhinitis 36.5% and nonallergic rhinitis 16.9% and the onset of allergic rhinitis were 10 years earlier compared with nonallergic rhinitis which is consistent with Dykewicz that explain a study perform by Rondon et al. in which compared nonallergic and allergic rhinitis and demonstrated that patients with nonallergic rhinitis tend to be older and have severe symptoms and unlikely association to asthma. Savoure et al. reported in their study that moderate to severe rhinitis was found in higher percentage in nonallergic rhinitis (40%) compared with allergic rhinitis (24%) [1, 2].

Dykewicz et al. estimated nonallergic rhinitis affect 17–52% of adults in the United States and 34% may have a mixed rhinitis [1].

An important difference between nonallergic and allergic rhinitis is the seasonality of symptoms. Savoure et al. referred allergic rhinitis increased from March to June (spring) and nonallergic rhinitis increased during winter and decreases from May to September which is demonstrated in **Figure 2** [2].

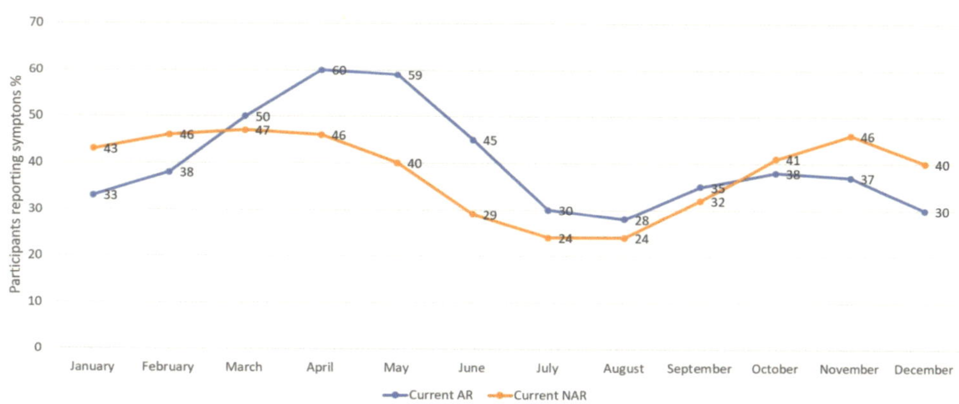

Figure 2.
Seasonality of symptoms of nonallergic rhinitis (NAR) and allergic rhinitis (AR) [2].

3. Histopathology of allergic rhinitis and non-allergic rhinitis

In the inflammatory process of rhinitis, the hair cells decrease and the number of goblet cells responsible for: rhinorrhea increased; capillary permeability causing edema and nasal obstruction [10], with stimulation of nerve endings responsible for sneezing and neuromodulators that recruit more inflammatory cells [11]. Chronic hypertrophic is characteristic with loss of cilia and a tendency to squamous metaplasia [12].

In their research, Ríos et al. discovered that histological examination of allergic rhinitis revealed a noteworthy increase in remodeling across all patterns, including epithelial markers and edema, except for fibrosis. Within the stromal marker, there was an elevated occurrence of subepithelial edema, and this was directly proportional to the presence of more than 10 eosinophils per field. This finding is clinically associated with heightened obstruction and severity of the condition [13].

Fibrosis was more frequently identified in the non-allergic variant, although this observation did not reach statistical significance. The collagen deposition in these cases is influenced by the upregulation of transforming growth factor beta [14].

3.1 Non-allergic rhinitis phenotypes

NAR needs to be differentiated from rhinitis patients experiencing an allergic response localized to the nasal mucosa, which is commonly referred to as local allergic rhinitis (LAR) [3]. NAR is characterized by rhinorrhea, blocked nose, sneezing, and/or itchy nose without clinical signs of infection or allergy. **Figure 3** shows the subtypes of non-allergic rhinitis [12].

Nonallergic noninfectious rhinitis (NANIR) encompasses a diverse set of patients experiencing rhinitis without apparent signs of infection, such as discolored secretions, and lacking systemic indications of allergic inflammation, such as allergen-specific IgE in the blood. Subcategories within NAR include drug-induced rhinitis, rhinitis associated with aging, hormonal rhinitis, including pregnancy-induced rhinitis, nonallergic occupational rhinitis, gustatory rhinitis, and idiopathic rhinitis [15–18]; its phenotyping is essential for choosing the best treatment option [19, 20].

3.1.1 Subgroups of NAR

3.1.1.1 Senile rhinitis or rhinitis

Senile rhinitis or rhinitis in elderly is characterized as rhinitis occurring in individuals aged 65 years and older, constituting an often-overlooked condition that impacts up to 29.8% of the population in this age group. The diagnosis of senile rhinitis typically pertains to individuals experiencing bilateral watery nasal secretions in the absence of endonasal mucosal or anatomical abnormalities. The symptoms are attributed to neurogenic dysregulation and are considered to arise late in onset [9].

3.1.1.2 Gustatory rhinitis

Is marked by a runny nose following the consumption of hot and spicy foods. This phenomenon is thought to be triggered by a gustatory reflex linked to an overactive neural system that is nonadrenergic, noncholinergic, or peptidergic [9].

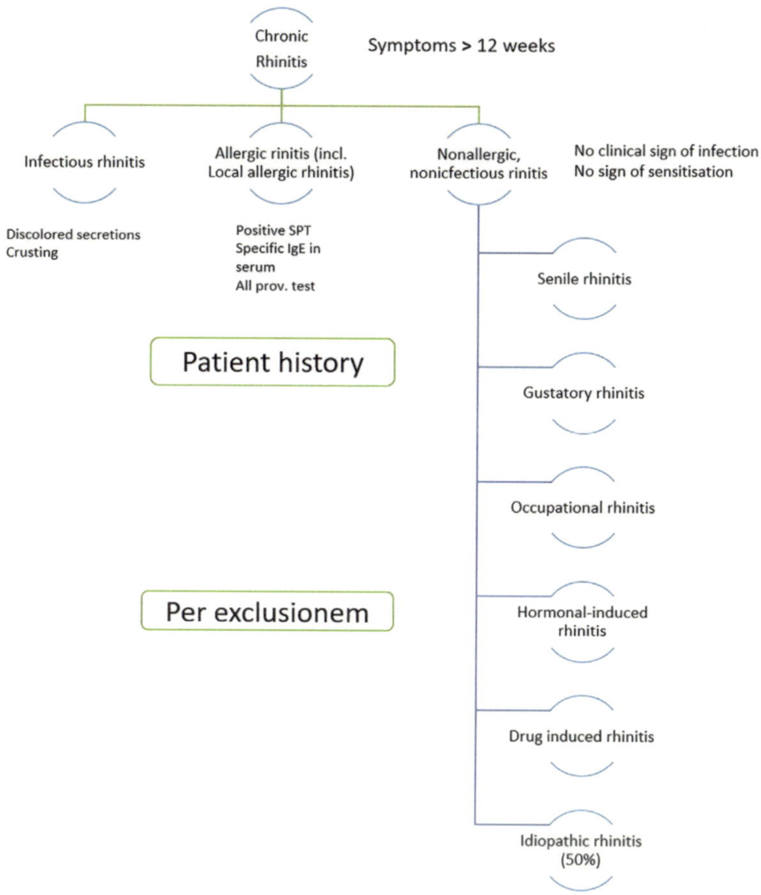

Figure 3.
Diagnosis of chronic rhinitis [3].

3.1.1.3 Occupational rhinitis

Is characterized as inflammation of the nasal mucosa resulting from exposure to a specific work environment and must be differentiated from "work-exacerbated" rhinitis [6]. Substances linked to occupational rhinitis include allergens of both high and low molecular weight (HMW and LMW) as well as irritants, according to reports. HMW agents could trigger a typical allergic inflammation mediated by IgE. The mechanisms behind chronic inflammation caused by LMW molecules are not clearly understood [9].

Prolonged exposure to occupational agents may lead to the development of asthma in patients. Recognizing occupational rhinitis is crucial for preventing occupational asthma. Rhinitis can emerge anew following prolonged exposure to airborne irritant chemicals in the occupational or environmental setting [20].

Nasal hyperreactivity to workplace triggers that are not specifically identified can also exist, with only about one-third of individuals experiencing relief during weekends or vacations. Individuals reported that their nasal issues were linked to sensitivity to nonspecific workplace triggers like air conditioning, dust and dry air [9, 12].

3.1.1.4 Hormonal rhinitis

Hormonal imbalances occurring during various stages such as the menstrual cycle, puberty, pregnancy, menopause, and specific endocrine disorders like hypothyroidism and acromegaly are frequently linked to NAR. Estrogens, induce nasal vascular engorgement, potentially leading to nasal obstruction and/or increased nasal secretions. Both beta-estradiol and progesterone contribute to an elevated expression of histamine H1-receptors on the nasal epithelium. Although hormonal changes are believed to have a notable impact, especially in conditions such as gestational rhinitis or pregnancy-induced rhinitis, the exact pathophysiology of hormonal rhinitis remains uncertain.

Smoking emerges as the sole universally acknowledged risk factor in pregnancy-induced rhinitis [12]. Regarding instances of new-onset allergic rhinitis and asthma after puberty, studies suggest that girls undergoing a delayed onset of menarche (first menstruation) are less susceptible to developing allergic rhinitis compared to those experiencing menarche at the average age.

Late menarche (occurring after 13 years of age) is statistically inversely correlated with the development of allergic rhinitis [9].

Additionally, the use of hormonal contraceptives is inversely associated with the onset of allergic rhinitis, suggesting that, aside from endogenous hormones, these contraceptives might provide protection against allergies and asthma in young women post-puberty [20].

3.1.1.5 Drug-induced rhinitis

A variety of drugs may cause nasal symptoms, primarily nasal obstruction. 11 Nasal obstruction and post-nasal drip being the most prominent complaints. Drug-induced rhinitis can be divided into two subgroups:

- *adverse events of systemic treatment*: includes prolonged oral intake of aspirin, ibuprofen, and other NSAID, beta-blockers, sedatives, antidepressants, oral contraceptives, or drugs used to treat erectile dysfunction. Peptidergic drugs activate human mast cells through a G-protein coupled receptor, and this interaction could be responsible for some forms of drug-induced rhinitis

- *abuse of decongestive nasal* therapy, most commonly recognized as rhinitis medicamentosa (RM), this condition arises from the extended use of powerful decongestant sprays, and it is advised to discontinue the use of these sprays suddenly [14].

Rhinitis medicamentosa (RM) is a result of a particular medication with localized action, operating through a distinct pathophysiological mechanism and showing connections with psychiatric conditions like anxiety or opioid use disorders. Mehuys et al. have revealed concerning rates of nasal decongestant (ND) abuse. Approximately half of individuals with persistent rhinitis, obtaining over-the-counter medication for their nasal issues, were found to be overusing ND, despite being informed about the recommended duration of use [21].

3.1.1.6 Smokers' rhinitis

Current smokers exhibited a significantly higher likelihood of having NAR compared to non-smokers (chi-square p = .034, OR 1.7 (1.04–2.8); RR 1.6 (1.004–2.6)). Among the NAR group, current smokers had a notably higher number of pack-years of smoking in comparison to individuals who were currently smoking but did not have NAR (32 ± 29 vs. 14 ± 14; t-test p = .014; MWU test p = .04) [20].

In contrast, former smokers had a comparable likelihood of having NAR when compared to those who had never smoked. The pack/year of smoking among former smokers did not differ between those with NAR and the control group [20].

3.1.1.7 Idiopathic rhinitis

Approximately half of individuals with NAR lack a clearly defined cause for their symptoms and are referred to as idiopathic rhinitis patients. The defining characteristic of these patients is the presence of Nasal Hyperreactivity. Recently, research has demonstrated an upregulation of the nociceptive TRPV1-substance P (SP) signaling pathway in idiopathic rhinitis patients, suggesting its probable involvement in the pathophysiology [2].

In her research, Klementina et al. affirms that Idiopathic Rhinitis (IR) represented the highest prevalence among the phenotypes, accounting for 39% of the NAR group. IR is characterized as a purely neurogenic phenotype, and identifying the disease with NHR as its most prominent symptom holds significant implications for its treatment [2].

Additionally, Klementina notes that 20% of the Non-Allergic Rhinitis group defied classification into any of the defined phenotypes. In this subset, a discernible pattern explaining the nature of nasal complaints eluded identification. Without NHR as a defining criterion for IR, these individuals might have been categorized under IR. This subgroup potentially comprises participants with anatomical factors contributing to nasal symptoms, such as septal deviation and/or inferior turbinate hypertrophy, and/or phenotypes that are yet to be recognized [20].

3.2 Diagnosis

The diagnosis of NAR relies on a comprehensive medical history, which includes the exclusion of clinically relevant sensitization to airborne allergens and the absence of clinical signs of rhinosinusitis [22]. In addition to the medical history, anterior rhinoscopy is employed to examine for signs of infection, endonasal crust formation, and/or significant anatomical deformities.

Nasal endoscopy is strongly recommended, as it enables a comprehensive evaluation of the entire endonasal cavity, including the ostiomeatal complex. Emphasizing the significance of nasal endoscopy in diagnosing prolonged cases of rhinitis is crucial, as it can unveil the presence of chronic rhinosinusitis without nasal polyps (CRSsNP) or with nasal polyps (CRSwNP) [21, 22].

The following diagnostic tests are not recommended in NAR: allergen provocation testing, microbiological analysis of the nasal content, nasal cytology or biopsies, measurement of total IgE or allergen-specific IgE in nasal secretions, CT scans of the sinonasal cavities, measurement of nasal hyperreactivity, measurement of markers of cerebrospinal fluid leakage (b2- transferrin or b-trace [21]. **Figure 4** shows an algorithm for diagnosis between the subtypes [20].

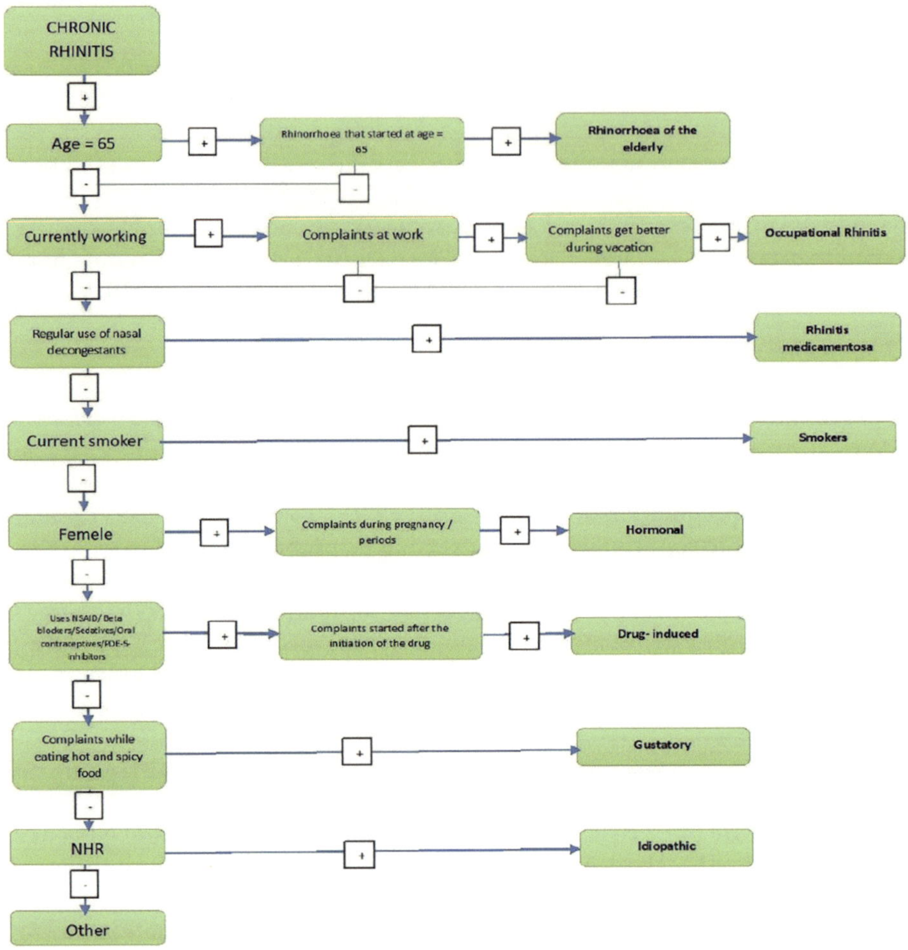

Figure 4.
Definitions of NAR phenotypes. NHR – Nasal hyper reactivity [20].

4. Endotypes of NAR

Varieties exist with diverse pathophysiological mechanisms, broadly categorized into a traditional inflammatory pathway, a neurogenic pathway, and other pathways that are predominantly unknown (**Figure 5**) [20].

The inflammatory pathway is present in a subset of NAR individuals. A Th2 cytokine inflammatory pattern, observed in patients with Allergic Rhinitis (AR) and those with work-related allergic rhinitis triggered by high molecular weight (HMW) allergens, characterizes this subgroup. Managing these patients typically poses no significant therapeutic challenges, as they generally respond favorably to nasal corticosteroid treatment, akin to individuals with Local Allergic Rhinitis (LAR) [20, 22].

Nevertheless, a portion of Non-Allergic Rhinitis (NAR) patients does not exhibit an infiltration of inflammatory cells in the nasal mucosa, suggesting involvement of a

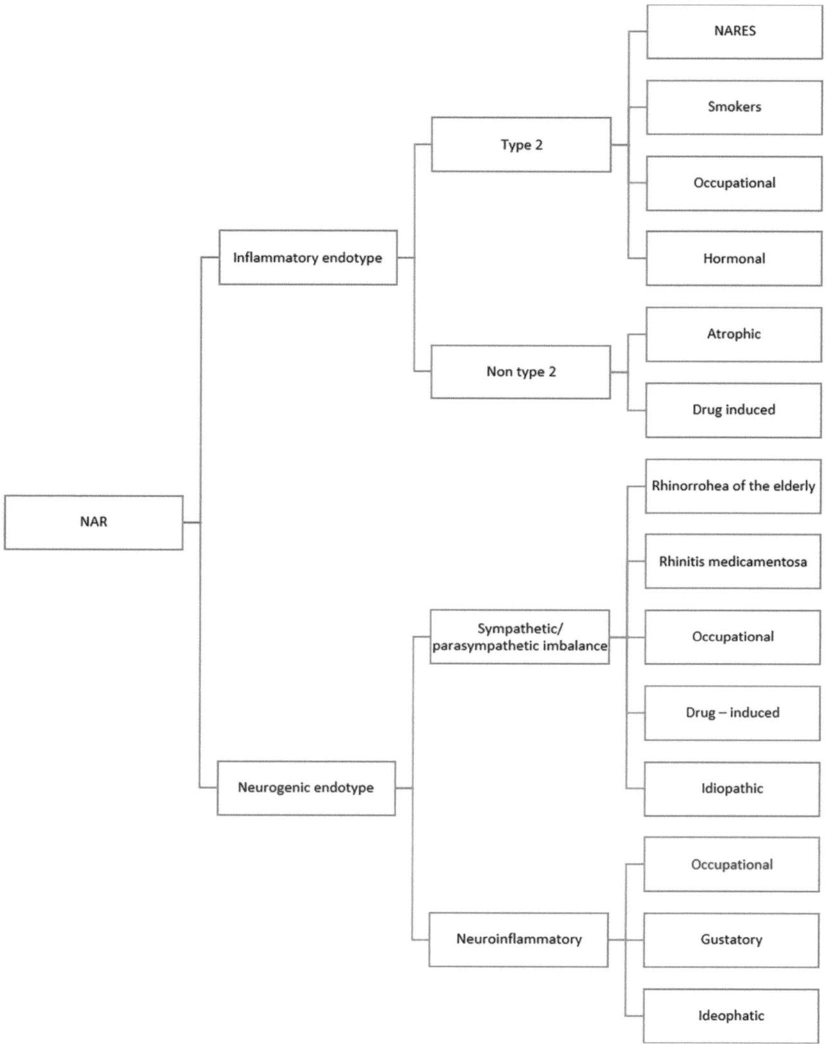

Figure 5.
Endotype and phenotype of NAR [20].

neurogenic mechanism. This includes conditions such as rhinitis of the elderly, gustatory rhinitis, certain types of occupational rhinitis, certain drug-induced rhinitis, and Idiopathic Rhinitis (IR). The intricate neural regulation of the upper airways involves various interacting nervous systems, where sensory, parasympathetic, and sympathetic nerves play roles in regulating processes in the nasal mucosa, including epithelial, vascular, and glandular functions [20–22].

Certain phenotypes appear to stem from a relatively straightforward regulatory disorder. For instance, rhinitis of the elderly primarily appears to involve a dysregulation in the balance between the parasympathetic and sympathetic neural systems. This can be addressed with the anticholinergic drug ipratropium bromide. Another example is rhinitis medicamentosa, where dysregulation of adrenergic receptors in the nasal mucosa results in a relative increase in parasympathetic drive, causing significant rhinorrhea and nasal obstruction [20].

Specialized epithelial chemosensors known as solitary chemosensory cells in the nasal cavity react to irritants via the canonical taste transduction cascade, activating peptidergic trigeminal nociceptive (pain) nerve fibers. The activation of these nasal cells can initiate comparable local inflammatory responses, including mast cell degranulation and plasma leakage. This response mirrors the effects of directly exciting trigeminal pain fibers using capsaicin. Importantly, this reaction is solely attributable to cholinergic neurotransmission and neural activity, without involving the release of local inflammatory mediators.

Idiopathic Rhinitis is believed to be a disorder affecting the nonadrenergic noncholinergic (NANC) or peptidergic neural system. Perivascular and intraepithelial nonadrenergic noncholinergic (NANC) sensory nerve fibers contain neuropeptides, including VIP, substance P (SP), and calcitonin gene-related peptide (CGRP). These neuropeptides are locally released from peptidergic neurons, mainly unmyelinated sensory C-fibers, in the nasal mucosa following activation by nonspecific stimuli. They can be responsible for the symptoms of IR. Stimulation may occur due to inflammatory mediators like histamine and bradykinin, as well as various inhaled irritants such as nicotine, chlorine, formaldehyde, and capsaicin, predominantly acting via the TRPA1 and TRPV1 receptors. Recent findings indicate an upregulation of the nociceptive TRPV1-SP signaling pathway in individuals with IR [18–20].

5. Treatment of NAR

Patients diagnosed with NAR are often offered different treatment modalities ranging from nasal topical medication to surgical [23].

The treatment modalities includes trigger avoidance, topical and systemic medications, and surgery [24–27].

There is little systematization in the protocols and there is little high-grade evidence regarding treatment, which causes difficult management with an impact on the quality of life and low patient satisfaction [23].

Intranasal corticosteroids are the first choice for NAR. Intranasal antihistamines have also been used in combination with the intranasal corticosteroids when monotherapy could not adequately control the symptoms of NAR [21].

5.1 Treatment modality

5.1.1 Intranasal corticosteroids

Act locally on the nasal mucosa, eliciting anti-inflammatory and immunosuppressant effects, modifying and reducing inflammation through suppression of the synthesis of pro-inflammatory cytokines and proinflammatory enzymes, inhibiting lymphocyte proliferation and chemotaxis [24, 25]. Avoid the systemic side effects of corticosteroids [25].

First-generation intranasal corticosteroids:

- Beclomethasone dipropionate

- Triamcinolone acetonide

- Flunisolide
- Budesonide

Second-generation intranasal corticosteroids:

- Fluticasone furoate
- Fluticasone propionate
- Mometasone furoate
- Betamethasone sodium phosphate
- Ciclesonide

Fluticasone propionate and beclomethasone dipropionate are the only topical nasal corticosteroids approved by the Food and Drug Administration for treatment of NAR [26, 27].

The adverse effects are [25]:

- Epistaxis (5–10%)
- Nasal irritation (5–10%, including dryness, burning and stinging)
- Headache
- Nasal septal perforation (<1%)
- Candida infection of the nose and pharynx
- Impaired wound healing after recent nasal surgery or trauma.

Intranasal corticosteroids are likely to work better for the inflammatory endotypes of non-allergic rhinitis [25].

The certainty of the evidence for most outcomes in this review was low or very low. Not all studies have shown these agents to be effective [25, 26].

It is unclear whether intranasal corticosteroids reduce patient-reported disease severity in non- allergic rhinitis patients compared with placebo when measured at up to 3 months. There is a lack of evidence comparing intranasal corticosteroids with other pharmacological treatments [23, 27].

It is unclear which is the best intranasal corticosteroid to use with respect to type, concentration, vehicle and how often to use it [25, 27].

5.1.2 Combination of intranasal corticosteroids antihistamines

Combination therapy of fluticasone propionate and azelastine hydrocholoride (AZE) – first-line therapies in seasonal allergic rhinitis – has been shown to be more effective compared against monotherapy with either class or placebo in seasonal allergic rhinitis [25, 27].

Combination therapy can be achieved either by using 2 separate nasal sprays or a combination product [27].

5.1.2.1 Topical antihistamine

Unclear mechanism(s) of action in since this is a non immunoglobulin E pathophysiology.

To diminish eosinophil activation and adhesion molecule expression and suppression of inflammatory cytokine generation [27].

Two topical antihistamine sprays have been studied in NAR, azelastine and olopatadine [23, 27].

Olopatadine and azelastine (0.1%) were compared for the treatment of NAR in a multicenter, randomized, both reduced congestion, rhinorrhea, postnasal drip, and sneezing; There were no statistically significant differences between their effects [23, 27].

In the review by Khoueir et al., all studies used Azelastine Hydrochloride (HCl) (137 mcg/aerosol) vs. olopatadine HCl 0.6% (665 mcg/spray). Both groups showed an importance decrease in TNSS after 14 days of treatment ($p < 0.001$) [23].

A statistically significant adverse effect reported was the sensation of bitter taste in patients who used azelastine. Another adverse effect reported in 3 studies was minor nasal bleeding [23].

5.1.2.2 Ipratropium bromide

The ipratropium bromide is the only topical anticholinergic approved in the treatment of NAR [24, 26]. In a study of 253 patients with perennial NAR the ipratropium nasal spray significantly reduced rhinorrhea versus placebo [16, 19]; dryness and epistaxis were uncommon [25].

5.1.2.3 Topical capsaicin

Reduced the density of the innervation of the nasal mucosa and the TRPV1-SP signaling pathway, without affecting the integrity and function of nasal epithelial cells or mast cells [24].

5.1.2.4 Oral decongestants

There is a lack of good evidence of effect, but theoretically it may be helpful for nasal stuffiness only [25].

5.1.2.5 Oral antihistamines

Limited data suggest that the newer nonsedating H1 antihistamines are not as effective in NAR, compared with allergic rhinitis.

5.1.2.6 Nasal saline lavage

The role is undefined. A 2007 Cochrane database review noted that nasal saline alone has not been demonstrated to be beneficial in CRS or more effective than an intranasal corticosteroid [26].

Intranasal saline sprays have been found effective in relieving postnasal drip, sneezing, and congestion [27].

5.2 Subgroups of NAR

- Senile rhinitis or rhinitis in elderly: ipratropium bromide is considered effective in reducing the severity and duration of the senile rhinitis (**Figure 6**) [24, 25, 27].

- Gustatory rhinitis: Ipratropium bromide is effective only for rhinorrhea.

  It is particular benefit in gustatory rhinitis [25, 27].

- Occupational rhinitis: The treatment is to avoid the trigger [27].

- Hormonal rhinitis: Nasal corticosteroids can be used to improve symptoms [27].

- Drug-induced rhinitis: Treatment is usually focused on cessation of the affecting agent, as well as support with intranasal corticosteroids [25].

- Smokers' rhinitis: the mainstay of treatment is triggering avoidance [25, 26].

- Idiopathic rhinitis: Intranasal capsaicin (8-methyl-N- vanillyl-6-nonenamide), the active component of chili peppers, appears to have a therapeutic effect in idiopathic rhinitis with long-lasting relief of symptoms from 6 to 9 month in 80% of well-selected IR patients ranging, based on several randomized controlled trials (**Figure 6**) [24, 25].

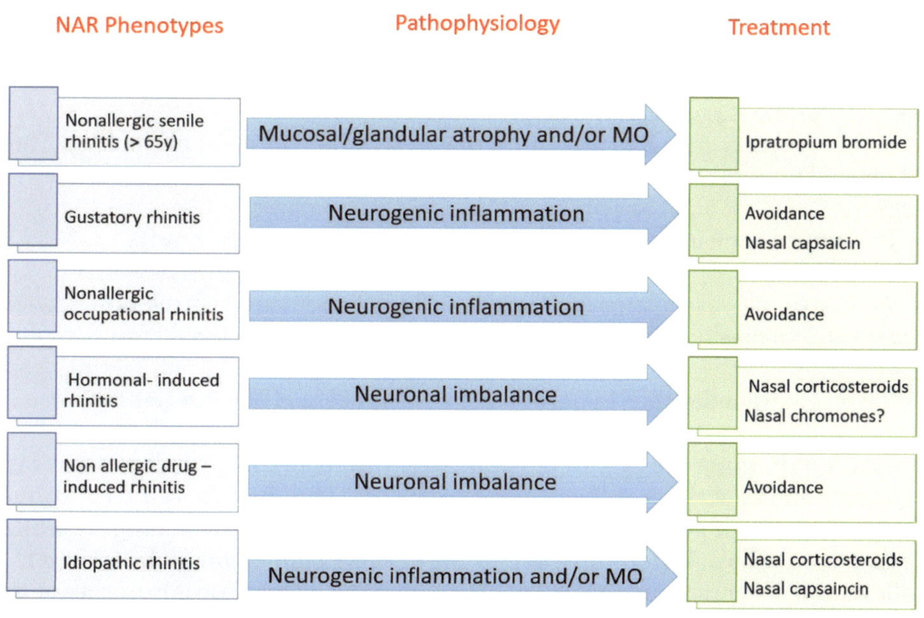

Figure 6.
Therapeutic strategy of nonallergic rhinitis.

5.3 Nonpharmacologic treatment

Surgical interventions are contemplated for patients with allergic rhinitis and non-allergic rhinitis (NAR) who do not respond adequately to medical therapy. A surgical procedure known for its favorable outcomes involves reducing the hypertrophy of the inferior turbinates. This intervention proves to be highly effective for individuals experiencing nasal obstruction that persists despite the clinical treatments described earlier [26].

The surgical techniques for inferior turbinate can be grouped into five groups [28]:

1. Turbinectomy is a procedure that removed turbinate: total, partial or by submucosal resection using cold steel instruments [SMR]

2. Microdebrider-assisted turbinoplasty [MAIT] which was defined by submucosal resection of the inferior turbinate with microdebrider.

3. Radiofrequency ablation [RFA] relying on the use of radio waves to create heat to cause tissue fibrosis and shrinkage.

4. Electrocautery techniques (monopolar/bipolar probe, diathermy electrode, straighttip electrode) which make use of an electrical current to destroy tissue.

5. Laser-assisted turbinoplasty (carbon dioxide, diode, potassium titanyl phosphate) that uses light energy to ablate tissue.

Irrespective of the method employed and the equipment utilized, the primary objectives remain to enhance air passage and minimize complications [29]. The selection of the most suitable technique depends on individual cases, taking into consideration factors such as the anatomy of the inferior turbinate (whether hypertrophy is more associated with bone or mucosa), the extent of hypertrophy (whether it is predominantly anterior or posterior), the availability of equipment, and the proficiency of the surgeon [29].

Another aspect to take into account is the preservation of nasal physiology. The fundamental physiological principles for this approach are [30]:

- The critical zone is the nasal valve region (inferior turbinate head);

- Small area increase = large increase in airflow (exponential correlation between area and airflow);

- Nasal sensation depends in part on the nasal mucosa of the inferior turbinate.

Hence, diminishing the volume of the inferior turbinate could result in a decrease in the quantity and/or activity of the responsive mucosa. This, in turn, could contribute to the amelioration of symptoms such as rhinorrhea, sneezing, and itching, all while maintaining the physiological functions of the mucosal epithelium [31]. Moreover, the alleviation of symptoms may be associated with the disruption of branches of the posterior nasal nerve, key in the mechanisms of sneezing and hypersecretion [31].

In a Systematic Review by Zhang et al. indicates that all techniques described above, significantly improved nasal obstruction post-operatively based on patient-reported outcome measures. No statistical significance was observed when stratifying outcomes by AR and NAR groups, improving is similar for both populations [28], RFA and MAIT improved rhinorrhea significantly compared to baseline [28]. A systematic review by Acevedo et al. assessing outcomes on VAS nasal obstruction found similar results when comparing RFA and MAIT [32], in contrast, Mirza et al. concluded that MAIT resulted better [33].

In relation to alternative techniques, RFA demonstrated comparable enhancements in nasal obstruction compared to electrocautery and turbinectomy/Septoplasty with Mucosal Reduction (SMR) [28]. Improvements in all physiological indicators of nasal patency, encompassing active anterior rhinomanometry (nasal resistance), acoustic rhinometry (nasal cavity volume), and nasal airflow, were observed with RFA, MAIT and laser interventions [28, 32, 33].

RFA and MAIT initially improved, but after 1-year, VAS obstruction scores worsened when compared to 3–6 months postoperatively [28], these results suggest that peak effect following inferior turbinate reduction is achieved within the first year; however patients are still significantly improved relative to baseline; with improvement a favor MAIT at 5 years.

An alternative technique outlined is the endoscopic vidian neurectomy (EVN) and posterior nasal neurectomy (PNN), involving both surgical (SPNN) and cryoablative (CPNN) methods. The systematic review indicates that EVN, SPNN, and CPNN serve as effective and safe surgical options for patients with non-allergic rhinitis (NAR) unresponsive to medical management. Notably, SPNN and CPNN are linked to lower rates of complications, including dry eye and palatal/cheek numbness, in comparison to EVN [34]. Another systematic review reveals that both selective vidian neurectomy (SVN) and PNN result in significantly reduced symptoms in both short-term and long-term postoperative periods when compared to the preoperative phase [35].

Another noteworthy aspect to emphasize pertains to the choice between septoplasty alone or septoplasty combined with turbinoplasty, it is observed that patients undergoing the combined procedure experience more pronounced symptomatic improvement [36].

In summary, the surgical treatment of inferior turbinate hypertrophy should be personalized based on the patient's clinical condition, with a primary focus on enhancing air space and nasal function [29]. In this context, the importance of experience-based medicine appears to outweigh the reliance on evidence.

6. Conclusions

NAR is a diverse nasal condition characterized by a high global prevalence. Its etiology is related to neurogenic factors. NAR comprises seven subtypes with overlapping presentations. Treatment strategies should prioritize the avoidance of known triggers, and initial therapeutic approaches involve the use of intranasal corticosteroids, intranasal antihistamines and intranasal ipratropium. Combinations of these therapies may be considered if monotherapy fails to adequately manage symptoms. Surgical interventions are contemplated for patients resistant to medical therapy, with effective outcomes noted in procedures such as the reduction of inferior turbinate hypertrophy and selective neurectomy of the vidian branches.

Author details

Carlos Ríos-Deidán[1,2,3]*, Diana Villacrés-Silva[3], Daniela Saénz-Chávez[3] and María Peña-Vásquez[4]

1 Otorhinolaryngology Department of Medical Sciences Faculty, Central University of Ecuador, Quito, Ecuador

2 Otorhinolaryngology Unit, Carlos Andrade Marin Specialties Hospital, Quito, Ecuador

3 Otorhinolaryngology Unit, DAME Clinic, Quito, Ecuador

4 Internal Medicine Unit, Enrique Garces Hospital, Quito, Ecuador

*Address all correspondence to: deidancar@hotmail.com

References

[1] Dykewicz MS, Wallace DV, Amrol DJ, Baroody FM, Bernstein JA, Craig TJ, et al. Rhinitis 2020: A practice parameter update. Journal of Allergy and Clinical Immunology. 2020;**146**(4):721-767

[2] Savouré M, Bousquet J, Leynaert B, Renuy A, Siroux V, Goldberg M, et al. Rhinitis phenotypes and multimorbidities in the general population: The CONSTANCES cohort. European Respiratory Journal. 2023;**61**(2):2200943

[3] Hellings PW, Klimek L, Cingi C, Agache I, Akdis C, Bachert C, et al. Non-allergic rhinitis: Position paper of the European academy of allergy and clinical immunology. Allergy. 2017;**72**(11):1657-1665

[4] Bousquet J, Anto JM, Bachert C, Baiardini I, Bosnic-Anticevich S, Walter Canonica G, et al. Allergic rhinitis. Nature Reviews Disease Primers. 2020;**6**(1):95

[5] Rombaux P, Hellings PW. Rinitis crónica no alérgica. EMC - Otorrinolaringología. 2014;**43**(3):1-10

[6] Baroody FM. Nonallergic rhinitis: Mechanism of action. In: Immunology and Allergy Clinics of North America. Vol. 36. W.B. Saunders; 2016. pp. 279-287. DOI: 10.1016/j.iac.2015.12.005

[7] Bernstein JA, Singh U. Neural abnormalities in nonallergic rhinitis. Current Allergy and Asthma Reports. 2015;**15**(4):18

[8] Greiwe J, Bernstein JA. Nonallergic rhinitis: Diagnosis. In: Immunology and Allergy Clinics of North America. Vol. 36. W.B. Saunders; 2016. pp. 289-303. DOI: 10.1016/j.iac.2015.12.006

[9] Savouré M, Bousquet J, Jaakkola JJK, Jaakkola MS, Jacquemin B, Nadif R. Worldwide prevalence of rhinitis in adults: A review of definitions and temporal evolution. In: Clinical and Translational Allergy. Vol. 12. John Wiley and Sons Inc; 2022. DOI: 10.1002/clt2.12130

[10] Wheatley LM, Togias A. Allergic rhinitis. The New England Journal of Medicine. 2015;**372**(5):456-463. DOI: 10.1056/NEJMcp1412282

[11] Campo P, Salas M, Blanca-López N, Rondón C. Local allergic rhinitis. Immunology and Allergy Clinics of North America. 2016;**36**(2):321-332. DOI: 10.1016/j.iac.2015.12.008

[12] Rios-Deidan C, Reyes M, Acosta-Castillo T, Escalante-Fiallos E, Villacres-Silva D. Histological characteristics of chronic allergic rhinitis versus non-allergy: Is there a difference in the remodeling? Indian Journal of Otolaryngology and Head & Neck Surgery. 2023;**75**(4):3364-3372. DOI: 10.1007/s12070-023-03922-y

[13] Kuhar HN, Tajudeen BA, Mahdavinia M, Gattuso P, Ghai R, Batra PS. Infammatory infltrate and mucosal remodeling in chronic rhinosinusitis with and without polyps: Structured histopathologic analysis: Infammation and mucosal remodeling in CRS. International Forum of Allergy & Rhinology. 2017;**7**(7):679-689. DOI: 10.1002/alr.21943

[14] Serpero L, Petecchia L, Sabatini F, Giuliani M, Silvestri M, Diblasi P, et al. The efect of transforming growth factor (Tgf)-β1 and (Tgf)-β2 on nasal polyp fbroblast activities involved upper airway remodeling: Modulation by futicasone propionate. Immunology Letters. 2006;**105**(1):61-67. DOI: 10.1016/j.imlet.2006.01.003

[15] Morais-Almeida M, Pite H, Pereira AM, et al. Prevalence and classification of rhinitis in the elderly: A nationwide survey in Portugal. Allergy. 2013;**68**(1150-1157):10

[16] Orban N, Maughan E, Bleach N. Pregnancy-induced rhinitis. Rhinology. 2013;**51**:111-119

[17] Hox V, Steelant B, Fokkens W, Nemery B, Hellings PW. Occupational upper airway disease: How work affects the nose. Allergy. 2014;**69**(282-291):12

[18] Jovancevic L, Georgalas C, Savovic S, Janjevic D. Gustatory rhinitis. Rhinology. 2010;**48**(7-10):13

[19] Van Gerven L, Alpizar YA, Wouters MM, et al. Capsaicin treatment reduces nasal hyperreactivity and transient receptor potential cation channel subfamily V, receptor 1 (TRPV1) overexpression in patients with idiopathic rhinitis. The Journal of Allergy and Clinical Immunology. 2014;**133**:1332-1339

[20] Avdeeva KS, Fokkens WJ, Segboer CL, Reitsma S. The prevalence of non-allergic rhinitis phenotypes in the general population: A cross-sectional study. Allergy. 2022;**77**:2163-2174. DOI: 10.1111/all.15223

[21] Gomez E, Campo P, Rondon C, et al. Role of the basophil activation test in the diagnosis of local allergic rhinitis. The Journal of Allergy and Clinical Immunology. 2013;**132**:975-976

[22] Rondon C, Campo P, Galindo L, et al. Prevalence and clinical relevance of local allergic rhinitis. Allergy. 2012;**67**:1282-1288

[23] Khoueir N, Khalaf MG, Assily R, Rassi S, Hamad WA. Intranasal antihistamines in the treatment of idiopathic non-allergic rhinitis: A systematic review and meta-analysis. Rhinology. 2023;**61**(4):290-296. DOI: 10.4193/Rhin21.380

[24] Meng Y, Wang C, Zhang L. Diagnosis and treatment of non-allergic rhinitis: Focus on immunologic mechanisms. Expert Review of Clinical Immunology. 2020;**17**:51-62

[25] Segboer C, Gevorgyan A, Avdeeva K, Chusakul S, Kanjanaumporn J, Aeumjaturapat S, et al. Intranasal corticosteroids for non-allergic rhinitis. Cochrane Database of Systematic Reviews. 2019;**2019**(11):CD010592

[26] Daramola O, Kern R. An update regarding the treatment of nonallergic rhinitis. Current Opinion in Otolaryngology & Head and Neck Surgery. 2016;**24**:10-14

[27] Lieberman PL, Smith P. Nonallergic rhinitis treatment. Immunology and Allergy Clinics of North America. 2016;**36**:305-319

[28] Zhang K, Pipaliya R, Miglani A, Nguyen S, Schlosser R. Systematic review of surgical interventions for inferior turbinate hypertrophy. American Journal of Rhinology & Allergy. 2023;**37**(1):110-122. DOI: 10.1177/19458924221134555

[29] Roithmann R. Inferior turbinectomy: What is the best technique? Brazilian Journal of Otorhinolaryngology. 2018;**84**:133-134

[30] Sozansky J, Houser SM. The physiological mechanism for sensing nasal airflow: A literature review. International Forum of Allergy & Rhinology. 2014;**4**:834-838

[31] Maniaci A, Di Luca M, La Mantia I, et al. Surgical treatment for

the refractory allergic rhinitis: State of the art. Allergie. 2021;**1**(1):48-62. DOI: 10.3390/allergies1010005

[32] Acevedo JL, Camacho M, Brietzke SE. Radiofrequency ablation turbinoplasty versus microdebrider-assisted turbinoplasty: A systematic review and meta-analysis. Otolaryngology–Head and Neck Surgery. 2015;**153**(6):951-956. DOI: 10.1177/0194599815607211.)

[33] Mirza A, Alandejani T, Shawli H, Alsamel M, Albakrei M, Abdulazeem H. Outcomes of microdebrider-assisted versus radiofrequency-assisted inferior turbinate reduction surgery: A systematic review and meta-analysis of interventional randomised studies. Rhinology. 2020;**58**(6):530-537. DOI: 10.4193/Rhin19.350

[34] Senanayake P, Wong E, McBride K, Singh N. Efficacy of vidian neurectomy and posterior nasal neurectomy in the management of nonallergic rhinitis: A systematic review. American Journal of Rhinology & Allergy. 2022;**36**(6):849-871. DOI: 10.1177/19458924221105933

[35] Niu X, Chen Y, Zhou T, Xiao H. Endoscopic vidian and vidian-branch neurectomy for refractory allergic rhinitis: A systematic review. International Forum of Allergy & Rhinology. 2023;**14**(3):679-694. DOI: 10.1002/alr.23259

[36] Samarei R, Mabarian S. A randomised trial comparing the subjective outcomes following septoplasty with or without inferior turbinoplasty European annals of otorhinolaryngology. Head and Neck Diseases. 2020;**137**:277-283. DOI: 10.1016/j.anorl.2020.01.024

Chapter 3

Current Medical Management of Chronic Rhinosinusitis in Adults

Moath Abdulrahman Alfaleh

Abstract

Rhinosinusitis is a prevalent disorder that places a heavy financial strain on society in terms of medical expenses and lost productivity. It is characterized by discomfort and pressure in the face, sinus and nasal lining irritation, nasal obstruction, rhinorrhea, and loss of smell. It occurs with or without nasal polyps. This chapter aims to evaluate and review each medical intervention for chronic rhinosinusitis to provide analysis and medical management recommendations. It is known that medical management of chronic sinusitis is a complex subject with many options available with different advantages and disadvantages, and so during our literature review, we focused on reaching to recommendations based on the latest and most accurate available studies from double-blinded randomized controlled clinical trial and meta-analyses using Ovid Medline, CINAHL, Scopus, and Cochrane. We focused on studies that compared the different types of medical management options to a placebo and in some instances to other drugs regarding how the patient's quality of life and disease burden improved. Based on our literature review, medications that showed benefit and improved patient quality of life were intranasal/systemic steroids, saline irrigation, biologics Anti-IL-5 and Anti-IL4/IL13, antihistamine, and montelukast. Other mentioned medical interventions need larger and higher-quality studies.

Keywords: chronic rhinosinusitis, nasal polyps, topical steroid, biologics, quality of life

1. Introduction

In much of the world, rhinosinusitis is a prevalent disorder that places a heavy financial strain on society in terms of medical expenses and lost productivity. It is characterized by discomfort and pressure in the face, sinus and nasal lining irritation, nasal obstruction, rhinorrhea, and loss of smell. It can occur with or without nasal polyps. This chapter aims to evaluate and review each medical intervention for chronic rhinosinusitis to provide analysis and medical management recommendations. As it is known that medical management of chronic sinusitis is a complex subject with many options available with different advantages and disadvantages, and so, during our literature review we focused on reaching to our recommendations based on the latest and most accurate available studies from double-blinded randomized controlled clinical trial and meta-analyses through using Ovid Medline, CINAHL, Scopus, and Cochrane, in which we focused on studies that compared the different types of medical management options to a placebo and in some instances to other drugs in regard to how the patient's quality of life disease burden improved. Based on our literature review, medications that showed benefit and improved patient quality of life were intranasal/systemic steroid, saline irrigation, biologics Anti-IL-5 and Anti-IL4/IL13, antihistamine, and montelukast. Other mentioned medical interventions need larger and higher-quality studies.

2. Short-term oral antibiotics for chronic rhinosinusitis (CRS)

In basic and secondary care settings, individuals with chronic rhinosinusitis (CRS) are frequently offered short-term regimens of antibiotics [1]. More proof is required to fully understand the function of antibiotics in the treatment of CRS. For this evaluation, 4 weeks or less is considered a short treatment period for antibiotics. Two published randomized clinical studies with a placebo control have examined the impact of short-term antibiotics in CRS [2, 3]. Thirty-two individuals with acute on chronic CRS exacerbations participated in a single-center, placebo-controlled trial [2]. In this study, an immediate worsening of sinonasal symptoms in the previous 4 weeks (nasal secretion, nasal obstruction/congestion, sense of smell, and/or face discomfort) in patients with underlying CRS was considered as an acute aggravation of the CRS condition. For 2 weeks, patients were randomly assigned to receive either a placebo or amoxicillin/clavulanate. Despite the fact that both groups' nasal secretions and nasal blockage significantly improved from baseline, on the Visual Analog Scale-Severity Scoring Assessment, there were no statistically significant changes between the groups [nasal secretion: mean difference (MD) -2 (−16.1, 12.1), $p = 0.44$; nasal obstruction: MD – 6.6(−10.6, 24), $p = 0.78$]. Nose endoscopic scores at day 14 did not differ between the groups (overall endoscopy score, $p = 0.88$; nasal polyps, $p = 0.58$; oedema, $p = 0.36$; nasal secretion, $p = 0.42$). After 3 months, the groups' improvements in quality of life (Sino-nasal Outcome Test, SNOT-22) were comparable [MD -2.7 (−20.36, 14), $p = 0.75$]. There was a notable observation of bacterial eradication in 29% of the amoxicillin/clavulanate group and 9% of the placebo group; nevertheless, there was no statistically significant difference seen between the groups ($p = 0.37$).

A double-blinded, placebo-controlled randomized controlled study (RCT) was carried out by Van Zele et al. [3] to evaluate the efficacy of doxycycline in the treatment of CRS with nasal polyps (CRSwNP) in 28 patients. After starting doxycycline therapy, there was a significant reduction in postnasal drip symptom ratings at week 2 ($p = 0.044$) and a tendency toward a decrease in rhinorrhea at week 8 ($p = 0.058$) (no adjustments were made for multiple testing). During the trial period, doxycycline did not significantly affect rhinorrhea, postnasal drip, or nasal congestion. It also did not significantly affect loss of smell. For 3 months, the doxycycline group showed a little but significant decrease in nasal polyp size (0,5 on an 8-point scale) in comparison to the placebo group ($p = 0.015$).

Ninety patients were included in the analysis of a single-blind, randomized, controlled trial that took place between 2018 and 2019 [4]. For 4 weeks, the patients in the clarithromycin group were given 500 mg pills twice a day. For 4 weeks, the second group took 500 mg pills of azithromycin every day. The group treated with azithromycin saw a considerably greater rate of complete symptom remission (71.1 vs. 24.4%, $P < 0.001$) than that treated with clarithromycin. There was no significant difference ($P = 0.120$) in the baseline Lund Mackay ratings between the groups. Nevertheless, both groups saw a substantial decrease in Lund Mackay scores following the intervention; however, the azithromycin group experienced a significantly larger improvement ($P < 0.001$).

In a randomized controlled trial, the authors measured the levels of two frequently given antibiotics, roxithromycin and doxycycline, in the sinonasal mucus, serum, and sinonasal tissues of patients with CRS after a one-week treatment [5]. Two groups were randomly assigned to twenty patients after they undergone functional endoscopic sinus surgery (FESS) for chronic rhinosis (CRS): (1) 100 mg of doxycycline every day for 7 days and 2) 300 mg of roxithromycin per day for 7 days. Liquid chromatography-tandem mass spectrometry was used to detect drug levels in steady state in blood, sinonasal tissues, and mucus. The amounts of doxycycline in the mucus were found to be substantially lower than those in the tissue (mean mucus/tissue ratio ¼ 0.18, $p < 0.0001$) and serum (mean mucus/serum ratio ¼ 0.16, $p < 0.001$). The concentration of roxithromycin in the mucus was likewise considerably lower than that in the tissue (mean mucus/tissue ratio ¼ 0.60, $p < 0.001$) and serum (mean mucus/serum ratio ¼ 0.37, p¼0.002).

Because of the complexity of bacterial biofilm antimicrobial tolerance, the effectiveness of doxycycline and roxithromycin in sinonasal mucus *in vivo* cannot be predicted solely from reported minimum inhibitory concentrations. However, these results suggest that low mucosal penetration of antibiotics may be one of the factors contributing to the limited efficacy of these agents in the treatment of CRS.

2.1 Conclusion

It is unclear whether or not the administration of a brief course of antibiotics has an effect on patient outcomes in people with acute exacerbations of CRS when compared with placebo because of the extremely low quality of the available data.

Furthermore, it is unclear whether or not the administration of a brief course of antibiotics has an effect on patient outcomes in adults with CRS when compared to placebo because of the extremely low quality of the data. Larger, higher-quality trials are required, particularly to assess the efficacy of brief courses of antibiotics in cases of acute exacerbations of CRS.

3. Long-term oral antibiotics for chronic rhinosinusitis (CRS)

Long-term antibiotic therapy was defined for the purposes of this systematic review as any course of treatment lasting more than 4 weeks. A few placebo-controlled trials have examined the effects of treating acute rhinosinusitis with antibiotics, despite the fact that this approach is often used. Two such studies assess the results of prolonged administration of macrolide antibiotics. The effects of long-term macrolide treatment on signs, symptoms, and patient-reported quality of life outcomes have been examined in two completed placebo-controlled trials [6, 7]. Both trials assessed a 12-week course of therapy; however, Wallwork et al. assessed roxithromycin at 150 mg daily, while Videler et al. assessed azithromycin (AZM) at 500 mg per week. In terms of nasal endoscopy, saccharin transit time, and SNOT-20 scores, Wallwork and colleagues demonstrated a notable improvement. In Wallwork and colleagues, groups were separated into low and high IgE levels (less than 200 μg/L against more than 200 μg/L); the low IgE group showed a substantial improvement in endoscopy, saccharin transit time, and nasal lavage of IL-8. Nevertheless, Videler et al.'s results did not improve, and since IgE was not assessed, a subgroup analysis was not feasible. By the time the study came to a close, Videler and colleagues had seen a nonsignificant improvement in 35% of the placebo group and 51% of patients on Azithromycin. Remarkably, 50% of the Azithromycin group claimed "improvement/cure" when the groups were reassessed by phone 12 weeks after the antibiotics ended, compared to 9% in the placebo group. This was statistically significant ($p = 0.017$). The response rate for Wallwork and colleagues was 67% in the roxithromycin group and 22% in the placebo group. Despite having comparable study numbers (60 vs. 64), the inclusion criteria of these research differed. Videler included individuals with and without nasal polyps, but Wallwork et al. exclusively included patients without nasal polyps. In the Wallwork et al. paper, patients were sub-analyzed according to their IgE levels, as previously mentioned. The majority of nonresponders were in the high IgE group. Another important difference was the dosing regimen, which in the study done by Videler et al. it was a once-weekly dosing compared to Wallwork et al.'s low-dose daily dosage.

A double-blind, randomized, placebo-controlled trial was conducted by de Oliveira et al. [8]. For 12 weeks, patients were given 500 mg of azithromycin orally three times a week. Sino-Nasal Outcome Test (SNOT-22), and nasal polyp biopsies were used to assess improvement. Data gathered 3 months after therapy and before treatment were compared. Evaluation of quality of life took place at the 1-year follow-up. For a duration of 12 weeks, treatment with 500 mg azithromycin three times a week resulted in a reduction in the amount of eosinophils in polyp biopsies, polyp staging, and a substantial clinical improvement based on subjective improvement as assessed by a quality-of-life questionnaire (the SNOT-22). On the other hand, the placebo group did not exhibit any noteworthy changes before or during the intervention. Furthermore, after the 1-year follow-up after therapy, the azithromycin group's improved quality of life persisted. Even though the current literature suggests that azithromycin therapy should be administered only to CRS patients with low polyposis and IgE, the outcomes that were seen in this particular patient group were unexpected.

Metanalysis conducted by Do Hyun Kim et al. to assess the efficacy of doxycycline (DOX) versus traditional therapy for individuals suffering from nasal polyps and refractory chronic rhinosinusitis (CRSwNP) [9, 10]. Up until September 2023, six databases were checked. Studies comparing improvements in refractory chronic sinusitis-related symptoms in groups treated with DOX and control groups were obtained. Lund-Kennedy (LK) score [−0.3670 (range - 0.6173; −0.1166); I2 = 92.8%], nasal polyposis score [−0.9484 (−1.2287; −0.6680); I2 = 92.5%], patient-reported sinonasal outcome test (SNOT) score [−0.3141 (−0.4622; −0.1660); I2 = 91.2%], and nasal obstruction score [−0.1813 (−0.3382; −0.0243); I2 = 86.2%] were all significantly decreased by DOX. The DOX group had considerably less nasal polyposis at the beginning of therapy, the completion of treatment, and 4 and 8 weeks later, according to subgroup analyses based on the measurement timepoints. Additionally, improvements were shown by the LK scores both during and after therapy. Over time, the therapy group's SNOT score tended to drop. The symptoms of nasal blockage improved both during and 4 weeks after therapy. In conclusion, by decreasing inflammation and recurring polyposis, DOX improves the postoperative endoscopic results for patients with CRSwNP who are resistant.

3.1 Risks of macrolide antibiotics

Patients having a history of congenital or documented acquired QT prolongation; ventricular cardiac arrhythmia, including torsades de pointe; or hypokalemia (risk of prolonging QT-time) should not be administered macrolides. Because of the increased risk of myopathy, including rhabdomyolysis, macrolides should not be taken concurrently with HMG-CoA reductase inhibitors (statins) that are substantially metabolized by CYP3A4 (lovastatin or simvastatin). Colchicine users should not take macrolides antibiotic, just like they should not take other potent CYP3A4 inhibitors. Patients with severe liver failure combined with renal impairment should not use macrolides.

3.2 Conclusion

In the studies shown previously, there was some promising positive effect for macrolides and doxycycline in improving the quality of life for patients with chronic rhinosinusitis without polyps and with polyps, respectively, but due to the low quality of evidence and the danger of these antibiotics adverse, we prefer to steer away from use of long-term macrolides. In the future, we are in need for a larger high quality trials.

4. Topical antibiotics for chronic rhinosinusitis (CRS)

The possible existence of bacterial biofilm in sinuses provides justification for using local antibiotics to treat patients with resistant, hard-to-treat rhinosinusitis. The sinus mucosa in these individuals stays inflamed even after washing with saline, anti-inflammatory corticosteroid therapy, and/or systemic antibiotics. These patients usually have undergone sinus surgery and so have wide open sinuses that are easily accessible for local treatment.

Pseudomonas aeruginosa or *Staphylococcus aureus* are frequently grown from these sinuses (producing biofilms) [11].

Nebulization of a solution of 150 mg tobramycin every day for 7 days resulted in the elimination of bacteria based on posttreatment culture, which was considerably better than placebo; all of this according to research conducted by Bonfils et al. in which they recruited 59 patients with CRSwNP who still had symptoms following sinus surgery [12] Day 0 and Day 10 of the bacteriological examination were conducted in accordance with Day et al.'s [13] investigation of the existence of pathogenic strains in culture and their sensitivity to antibiotics, together with a cytological analysis of the quantity and presence of leukocytes (at Day 0 and Day 10). The bacterial isolates from day 10 posttreatment and pretreatment were compared.

Leukocyte counts and antibiotic sensitivity tests were also conducted. But there was no discernible difference in the symptoms [12].

Victoria et al. conducted a prospective randomized controlled trial that is single-blinded (doctors only). Acute exacerbations of chronic rhinosinusitis and gram-positive bacteria on culture were seen in the postsurgical subjects. Subjects were divided into 3 groups (saline group, povidoen-iodine group, and mupirocin group), and for 30 days, they had twice-daily sinus irrigations [14]. The Mupirocin (14/20, 70%) group of the 62 participants under analysis had a greater posttreatment culture-negative povidone-iodine rate than the povidone-idodine (9/21, 43%) and Saline (9/19, 47%) groups, although this difference was not statistically significant ($p = 0.28$). The Lund-Kennedy endoscopic score (povidone-iodine −3.5 [−7, −0.5] vs. mupirocin −2 [−4, 2] vs. saline −3 [−5, 0]; $p = 0.35$) and the Sinonasal Outcome Test-20 score (povidone-iodine −0.3 [−0.6, 0.05] vs. mupirocin −0.3 [−0.7, 0.05] vs. saline −0.4 [−0.8, 0.05]; $p = 0.74$) did not differ significantly. When compared to saline and povidone-iodine, mupirocin sinus irrigations produced a greater posttreatment culture "control" rate in patients who had previously had sinus surgery and had acute exacerbations of CRS and gram-positive bacteria on culture.

4.1 Conclusion

Patients with CRS do not appear to benefit more from topical antibiotic treatment than a placebo in terms of symptom improvement.

It can, however, result in a clinically insignificant improvement in symptoms. We do not know for sure if topical antibiotic treatment affects patient outcomes in people with CRS since the evidence is of extremely low quality.

5. Intranasal corticosteroid for chronic rhinosinusitis (CRS)

In a double-blinded placebo-controlled trial conducted by Leopold et al. [15], 323 patients with CRSwNP 27% of whom previously had sinus surgery were randomized to receive either an exhalation delivery system (EDS)–placebo or twice-daily EDS-FLU (93, 186, or 372 mg) for a total of 24 weeks (16 double-blind plus 8 open-label when all got 372 mg). The coprimary end goals were the total bilateral polyp grade at 16 weeks and the change in nasal congestion/obstruction at 4 weeks. Functioning, polyp removal, and symptoms were secondary end goals. On both coprimary end points, EDS-FLU outperformed EDS-placebo ($P < .001$ across all dosages). The average polyp grade steadily increased through week 24 ($P < .009$, all comparisons), and at that point, 25% of patients had at least one side of the polyp removed, compared to 8.7% with the EDS-placebo ($P < .014$, all comparisons). Additionally, patients' SNOT scores increased considerably ($P < .05$ for all dosages) from those in patients receiving EDS-placebo (221.1 to 221.4 vs. 211.7 at week 16). All four of the defining illness symptoms were considerably improved by EDS-FLU (all dosages) at the end of the double-blind phase. 69% of the patients who received EDS-FLU experienced "much" or "very much" improvement. The percentage of patients who could have surgery dropped by 62–67%. The safety profile matched the information from earlier trials. In this study, they concluded that in individuals with chronic rhinosinusitis with nasal polyps, EDS-FLU results in a clinically and statistically significant improvement in all four diagnostically defined illness symptoms, polyp grade, and quality of life.

Japanese patients with chronic rhinosinusitis with nasal polyps have among them a common variant known as eosinophilic chronic rhinosinusitis (ECRS). Refractory eosinophilic airway inflammation, ECRS is strongly linked to asthma and necessitates comprehensive management as part of the unified airway concept [16, 17]. Kobayashi et al. in a double-blinded placebo-controlled trial reported a series of ECRS patients with asthma treated with fine-particle inhaled corticosteroid (ICS) exhalation through the nose (ETN). A double-blind randomization process was used to assign 23 patients with severe ECRS who were not responding to intranasal corticosteroid therapy to either the HFA-134a-beclomethasone dipropionate (HFA-BDP) metered-dose inhaler (MDI) ETN ($n = 11$) or the placebo MDI ETN ($n = 12$) over a 4-week period. Assessments were made of changes from the baseline in the nasal polyp score, smell test, computed tomographic (CT) score, and quality of life (QOL) score. Measurements of fractionated exhaled nitric oxide (FENO), a sign of eosinophilic airway inflammation, were made. Evaluation of the corticosteroid response was done both before and after therapy. Furthermore, a particle deposition model was used to illustrate the deposition of small particles. Purified eosinophils were co-incubated with BEAS-2B human bronchial epithelial cells to assess corticosteroid sensitivity and investigate the function of eosinophils on airway inflammation. Treatment with HFA-BDP MDI ETN increased corticosteroid sensitivity, and all evaluated clinical endpoints without impairing pulmonary function. HFA-BDP MDI ETN therapy decreased FENO and blood eosinophil count. According to the visualization investigation, fine particle deposition in the middle meatus, including the sinus ostia, was caused by ETN at expiratory flow rates of 10–30 L/min. Corticosteroid resistance was produced when BEAS-2B cells and eosinophils were co-incubated.

A metanalysis conducted by Bognanni et al. looked for randomized controlled trials comparing INCS administered in any way to placebo or other INCS administration types in studies stored in Medline, Embase, and Central between September 1, 2021 and the creation of the database [18]. Sixty-one randomized controlled trials with 7176 individuals and 8 treatments were examined. In comparison to a placebo, the quality of life linked to sinusitis may be enhanced by INCS rinse (mean difference [MD] −6.83, 95% confidence interval [CI] −11.94 to −1.71) and exhalation delivery system (EDS) (MD −7.86, 95% CI −14.64 to −1.08). In comparison to placebo, nasal obstruction symptoms are probably lessened after receiving INCS by stent/dressing (MD −0.31, 95% CI −0.54 to −0.08), spray (MD −0.51, 95% CI −0.61 to −0.41), and EDS (MD −0.35, 95% CI −0.51 to −0.18). The unfavorable effects of the various therapies did not significantly differ from one another (moderate certainty for INCS spray, extremely low to medium certainty for others). For CRSwNP, there are several effective INCS delivery methods that can enhance patient important results. The broadest possible range of evaluated outcomes seems to benefit from INCS using stent, spray, and EDS.

5.1 Conclusion

There is strong evidence that treating CRS patients with nasal corticosteroids over the long term is both safe and effective. They affect nasal symptoms and enhance one's quality of life.

6. Systemic corticosteroid for chronic rhinosinusitis (CRS)

For chronic rhinosinusitis with nasal polyps, short doses of systemic corticosteroids (7–21 days) are commonly administered in conjunction with local corticosteroids. In addition to lowering inflammation, systemic corticosteroids may also shrink nasal polyps. It is interesting to note that, in particular, olfactory mucosal inflammation is frequently decreased, leading to a quick improvement in olfaction (days) without any discernible change in polyp volume.

Shen et al. conducted a double-blind, placebo-controlled, randomized clinical study in which patient who underwent ESS with bilateral CRSwNP were randomly assigned to take oral prednisolone (30 mg/day) or a placebo for 2 weeks following operation. As the subjective outcomes, the visual analog scale (VAS) and Sino-Nasal Outcome Test 22 (SNOT-22) scores were selected and assessed at preoperative baseline, as well as 1, 3, and 6 months after surgery. The objective result was measured using Lund-Kennedy Endoscopic Scores (LKESs), which were assessed preoperatively, at 2 weeks, 2, 3, and 6 months after surgery [19, 20]. Out of the 100 patients that were included, only 82 with bilateral CRSwNP finished the 6-month follow-up. At every follow-up point, there was no discernible variation in the subjective results. Six months after surgery, the corticosteroid group reported an improvement in

LKESs (p = 0.05) among the objective results. Only patients with NECRSwNP (<10 eosinophils/HPF) showed a significant improvement in LKESs at 3 months post-operatively (p = 0.03), after classification by tissue eosinophils. In conclusion, oral corticosteroids administered after surgery did not result in further improvements in VAS or SNOT-22 scores; however, at 6 months after surgery, there was a tendency of improving LKES. At the three-month follow-up, this impact was significant only among NECRSwNP patients after stratification by tissue eosinophils.

In a meta-analysis conducted by Zhang et al. [21], 7 RCTs with a total of 414 participants were included in the meta-analysis out of the 337 pertinent papers that were found. In contrast to nonsteroid treatments or placebos, systemic corticosteroids significantly increased peak nasal inspiratory flow (PNIF) (SMD 42.39; 95% CI, 28.95 to 55.84; P < .00001), decreased the size of nasal polyps (SMD ~4.76; 95% CI, ~6.99 to ~2.52; P < .0001), and improved nasal obstruction scores (standardized mean difference; SMD ~2.81; 95% confidence interval [CI], ~4.68 to ~0.95; P¼.003). The low-dose subgroup (less than 50 mg/day prednisone) and high-dose (greater than or equal to 50 mg/day prednisone) had comparable advantages. But individuals receiving large dosages of prednisone reported experiencing sleeplessness and gastrointestinal problems more frequently. No discernible variation was seen in the frequency of any adverse events among the groupings. In conclusion for individuals with CRSwNP, systemic corticosteroids significantly reduce the amount of nasal polyps and significantly improve PNIF and nasal symptoms. In CRSwNP, prednisone dosages under 50 mg/day were advised when the benefits of oral corticosteroids were weighed against possible side effects.

6.1 Conclusion

Both the nasal polyp score and the overall symptom score are significantly reduced after a brief course of systemic corticosteroid therapy, either with or without local corticosteroid treatment. Up to 3 months after the commencement of treatment, the effect on the nasal polyp score is still significant, but by then, the effect on the symptom score has vanished.

7. Saline

In order to effectively manage CRS, nasal saline irrigation is thought to be crucial. By disrupting and eliminating antigens, biofilms, and inflammatory mediators as well as improving mucus clearance and ciliary beat activity, saline nasal irrigation may improve the function of the nasal mucosa through a number of physiological effects. It may also increase the hydration of the sol layer. Saline may also act as a carrier to provide the necessary volume to deliver medication into the sinuses. That being said, there is not much agreement on the optimal irrigation technique, the tonicity (concentration) of the saline solution, or the appropriateness of the devices, head position, volume, pressure, and frequency of washing.

In a non-blinded randomized controlled trial Glotakis et al. evaluated 174 patients with CRSwNP (whom 154 were posteopratively), they were divided to 3 groups [22]. The first group used 250 ml of 1175% Emser Salt solution (EmsSalt) (n = 59) twice daily for 1 year, the second group used 250 ml of isosmotic mineral salt mixture (IsoMix) (n = 58) twice daily for 1 year, and in the third group they were on no irrigation (n = 57) for one year. Outcomes were measured at 3, 6, 9, and 12 months including nasal symptoms, RQLQ, missed work days and postoperative condition of the mucosa. Results were significantly improved outcomes when comparing irrigation to no irrigation in terms of nasal symptoms and RQLQ. There is no discernible difference between mucosa and missed work days. There are no appreciable variations between the isosmotic mineral salt mixture and Emser Salt.

In a meta-analysis conducted by Lei Liu et al. [23, 24], in an effort to offer a guide for clinical nasal irrigation for the treatment of chronic rhinosinusitis, the authors compared the efficaciousness of hypertonic and isotonic saline in the management of rhinosinusitis. There were seven investigations in all. In four groupings, the effects of hypertonic saline on nasal symptoms were stronger. The patients with nasal secretion (SMD = 1.52; 95% CI: 1.04, 2.00; $p < 0.01$), congestion (SMD = 1.52; 95% CI: 1.04, 2.00; $p < 0.01$), headache (SMD = 0.82; 95% CI: 0.38, 1.26; $p < 0.01$), and overall symptomatic relief (SMD = 1.63; 95% CI: 0.83, 2.44; $p < 0.01$) were the first group of patients classified as (1). On the other hand, there was no difference in the improvement of radiologic scores (SMD = 2.44; 95% CI: −3.14, 8.02; $p < 0.01$) or scent (SMD = 0.47; 95% CI: −0.65, 1.59; $p = 0.41$). Furthermore, compared to the isotonic saline group, the hypertonic saline group had a higher improvement in mucociliary clearance time scores (SMD = 1.19; 95% CI: 0.78, 1.60; $p < 0.01$). Greater mild negative effects were observed with hypertonic saline. In conclusion, when treating chronic rhinosinusitis, hypertonic saline nasal irrigation is far more successful than isotonic saline and has less side effects. It also improves ciliary movement and nasal symptoms, but there is no discernible change in imaging results or improvement in smell.

7.1 Conclusion

Evidence provided showed that normal saline irrigation is effective in CRS in improving symptoms and quality of life compared to placebo but hypertonic saline is better than normal in regard to improving patients symptoms.

8. Anti-IgE

A substantial local IgE production that may contribute to chronic inflammation by persistently activating mast cells is a pathophysiological feature of CRSwNP [25]. Omalizumab is a recombinant humanized monoclonal antibody that has been studied for its potential use in the treatment of CRS in a randomized clinical study by Gevaert et al. [25]. Omalizumab works by selectively attaching to free circulating IgE, which inhibits the activation of effector cells such as mast cells, basophils, and dendritic cells and reduces the expression of IgE receptors on them. A subcutaneous dose of omalizumab (four to eight doses) or a placebo was administered to 24 individuals with CRSwNP with concomitant asthma for a period of more than 2 years. There were 20–700 kU/mL of total serum IgE. There was no discernible decrease in RSOM-31 or SF-36 following the subcutaneous delivery of four to eight doses of omalizumab. On the other hand, the omalizumab group showed a substantial improvement in the physical domain of the SF-36 and the Asthma Quality of Life Questionnaire (AQLQ), while the placebo group showed no meaningful changes. Comparing anti-IgE to baseline, there was a substantial reduction in the symptoms ratings for nasal congestion ($p = 0.003$), anterior rhinorrhoea ($p = 0.003$), loss of smell ($p = 0.004$), wheeze ($p = 0.02$), and dyspnea ($p = 0.02$). Anti-IgE medication did not help the patient's cough or spirometric findings. Over the course of the trial, a linear mixed model showed that the omalizumab group had a lower total nasal polyp score (NPS) than the placebo group ($p = 0.2$). The omalizumab group showed significantly higher Lund-MacKay scores on radiologic imaging ($p = 0.04$) when compared to the placebo group. Anti-IgE therapy was used, and the improvements in radiography and clinical outcomes were observed regardless of serum IgE levels. Patients who were allergic (−2.57; $p = 0.03$) and nonallergic (−2.75; $p = 0.06$) showed a decrease in total NPS after 16 weeks. Compared to nonallergic patients (20.66, $p = 0.75$), allergic patients showed an improvement in their Lund-Mackay CT scan scores (22.61, $p = 0.04$). On the other hand, the nonallergic group's total AQLQ score showed improvement (259.4, $p = 0.03$), but the allergic group's score (212.3, $p = 0.12$) did not. Among the patients covered, 22 out of 23 (95.7%) had at least one adverse event. In the omalizumab group, the most often reported adverse event was a common cold, which happened more frequently in the treatment group than in the control group ($p = 0.02$). In the omalizumab group, one patient died of lymphoblastic lymphoma a year after the research began. It is noteworthy that the analysis did not include four out of eight patients (or 50%) in the control group.

A randomized, double-blind, placebo-controlled study of anti-IgE for CRS was carried out by Pinto et al. [26] in 14 patients (12 of 14 with CRSwNP) who were not responding to conventional therapy. It was necessary for participants to have serum total IgE levels between 30 and 700 IU/ml. For 6 months, all patients received subcutaneous injections of either placebo or omalizumab at a rate of 0.016 mg/kg per IU every 2 to 4 weeks. The SNOT- 20 scores did not significantly differ between the treatments (median omalizumab −5.5, placebo −2.3, $p < 0.60$). Over the course of the study, the omalizumab group showed a clinically significant improvement (defined as at least 0.8) in their median change in SNOT-20 scores, while the placebo group showed no clinically significant change (−1.05 vs. -0.20, $p < 0.78$). All domains showed no significant differences between treatments, with the exception of vitality ($p < 0.05$; omalizumab 9.4, placebo 12.5). Comparing the omalizumab and placebo groups, there were no statistically significant differences in the sinus opacity in the CT scan, the median change in the percentage of eosinophils in nasal lavage, the median PNIF, the total nasal symptom scores, or the results of nasal endoscopy. Omalizumab-using patients utilized less antibiotics (0 vs. 1, $p < 0.32$) and fewer doses of steroids (median 0 vs. 1, $p < 0.043$) during the trial than placebo-using patients. During the trial, no negative outcomes or side effects happened [26].

In a meta-analysis study conducted by Qingwu Wu et al. [27, 28], randomized controlled studies comparing placebo and omalizumab in adult patients with CRSwNP, administered for a minimum of 16 weeks. Four randomized controlled trials with a total of 303 participants were found. Nasal Polyps Score (MD = −1.20; 95% CI −1.48 to −0.92), Nasal Congestion Score (MD = −0.67; 95% CI −0.86 to −0.48), Sino-Nasal Outcome Test-22 (MD = −15.62; 95% CI −19.79 to −11.45), Total Nasal Symptom Score (MD = −1.84; 95% CI −2.43 to −1.25), and decreased surgical requirement (risk ratio (RR) = 5.61; 95% CI 1.99 to 15.81) were among the significantly different outcomes when omalizumab was compared to placebo. The probability of adverse events (RR = 0.83; 95% CI 0.60 to 1.15), significant adverse events (RR = 1.40; 95% CI 0.29 to 6.80), and rescue systemic corticosteroid (RR = 0.52; 95% CI 0.17 to 1.61) did not vary either. In conclusion, omalizumab was shown to be safe and well-tolerated, and it improved endoscopic, clinical, and patient-reported outcomes in people with moderate to severe CRSwNP. Regarding adverse effects, current research on anti-immunoglobulin-E (IgE) therapy indicates a very low risk of anaphylaxis [27];

however, there is some evidence linking the drug to venous and arterial thromboembolic events that result in cardiovascular and cerebrovascular accidents [29].

8.1 Conclusion

There is a need for and an ongoing demand for larger population studies because the current studies' sample populations were too small. The literature recommends that Anti-IgE usage in CRSwNP cannot currently be recommended due to a lack of accessible data.

9. Anti-IL5

In the past, systemic and topical corticosteroids, long-term antibiotics, and surgery have all been used to treat CRS. Even with the application of optimal practices, some patients continue to have recalcitrant disease. Final development into an eosinophil requires interleukin 5 (IL-5), which also increases the mature cell's survival time within the tissue [30]. The care of patients with nonallergic asthma and CRSwNP may involve IL-5 as a target due to its observed elevation in nasal polyp tissue [31].

Bachert et al. conducted a double-blind, randomized, placebo-controlled experiment; patients with recurrent nasal polyposis who needed surgery were enrolled if they were between the ages of 18 and 70. In addition to daily topical corticosteroid treatment, patients received 750 mg intravenous mepolizumab or placebo every 4 weeks for a total of six treatments [26]. The number of patients who were no longer in need of surgery at Week 25 was the main outcome, which was determined by combining the endoscopic nasal polyp score with the visual analogue scale (VAS) score for nasal polyposis severity. Patient-reported outcomes (PRO), safety, improvement in individual VAS symptoms (rhinorrhea, mucus in the throat, nasal blockage, and sense of smell), and change in the nasal polyposis severity VAS score were among the secondary objectives. Mepolizumab ($n = 54$) or a placebo ($n = 51$) was administered to 105 patients. When compared to the placebo group, a considerably higher percentage of patients in the mepolizumab group (16[30%] vs. 5[10%], respectively, $P = 0.006$) no longer needed surgery. The nasal polyposis severity VAS score, endoscopic nasal polyp score, individual VAS symptom ratings, and SNOT-22 PRO score significantly improved in the mepolizumab group as compared to the placebo group. In terms of safety, mepolizumab was similar to a placebo. In conclusion, mepolizumab medication resulted in a higher improvement in symptoms and a reduction in the need for surgery compared to placebo in individuals with recurrent nasal polyposis on topical corticosteroids.

Gevaert et al. conducted a double-blind study conducted on thirty patients who had severe nasal polyposis (grade 3 or 4, or recurring after surgery) and were not responding to corticosteroid therapy. The patients were randomly assigned to receive either a placebo (n510) or two single intravenous injections of 750 mg of mepolizumab (n520) spaced 28 days apart [32]. Monthly assessments of the change from baseline in NPscore were conducted until week 8, 1 month following the last dose. At week eight, computed tomographic scans were also carried out. At week 8 relative to baseline, 12 out of 20 patients on mepolizumab had a substantially better NP score and computed tomographic scan score than 1 out of 10 patients on placebo. In conclusion, in 12 out of 20 individuals, mepolizumab produced a statistically significant decrease in NP size for at least 1 month following dosage. A possible new treatment strategy for those with severe eosinophilic nasal polyposis is IL-5 inhibition.

9.1 Conclusion

Because the evidence shown a considerable reduction in patients' need for surgery and an improvement in their symptoms, literature reviews recommend the use of mepolizumab in individuals with CRSwNP that is difficult to treat.

10. Anti-IL4/IL13

Strong type 2 immune mediators, IL-4 and IL-13, have different and overlapping roles. Both IL-13 and IL-4 play a significant role in IgE synthesis, eosinophil activation, mucus secretion, and airway remodeling. They also partially share receptors and signaling pathways. By starting T cell development toward the TH2 subtype, IL-4 is a key differentiating agent that drives a TH2 type response. Moreover, type 2-related chemokines and cytokines as TARC, fotaxin, IL-5, IL-9, and IL-13 are produced in response to IL-4. Additionally, the primary factors that cause B cells to flip their isotype class and generate Ig are IL-4 and IL-13. Dupilumab is a subcutaneous injection-based completely human monoclonal antibody targeting the IL-4 receptor subunit that blocks IL-4 and IL-13 signaling [28].

Bachert et al. conducted a 16-week follow-up randomized, double-blind, placebo-controlled parallel-group trial that involved 60 adults with chronic sinusitis and nasal polyposis who were not responding to intranasal corticosteroids. The study was carried out at 13 sites in the US and Europe between August 2013 and August 2014 [29]. Subcutaneous dupilumab (a 600 mg loading dose followed by 300 mg weekly; n = 30) or placebo (n = 30) plus mometasone furoate nasal spray for 16 weeks. Any changes in the primary end point were recorded, including endoscopic nasal polyp score at 16 weeks (range, 0–8; larger scores indicate worse condition). The 22-item SinoNasal Outcome Test score (range, 0–110; higher scores indicating worse quality of life; minimal clinically important difference = 8.90), the Lund-Mackay computed tomography (CT) score (range, 0–24; higher scores indicate worse status), the sense of smell as measured by the University of Pennsylvania Smell Identification Test (UPSIT) score (range, 0–40; higher scores indicate better status), symptoms, and safety were among the secondary end points. Fifty patients (mean [SD] age, 48.4 years [9.4 years], 34 men [56.7%], and 35 with concomitant asthma) who were randomly assigned were all enrolled in the trial; fifty-one of them finished it. The nasal polyp score exhibited a least squares (LS) mean change of −0.3 (95%CI, −1.0 to 0.4) when administered placebo and − 1.9 (95%CI, −2.5 to −1.2) when administered dupilumab (LS mean difference, −1.6 [95%CI, −2.4 to −0.7]; P < .001). For the Lund-Mackay CT total score, the LS mean difference between the two groups was −8.8 (95%CI, −11.1 to −6.6; P < .001). The 22-item SNOT showed significant improvements with dupilumab (LS mean difference between groups, −18.1 [95%CI, −25.6 to −10.6]; P < .001), as did the UPSIT, which measured sense of smell (LS mean difference, 14.8 [95%CI, 10.9 to 18.7]; P < .001). Nasopharyngitis (33% in the placebo group vs. 47% in the dupilumab group), injection site responses (7 vs. 40%, respectively), and headache (17 vs. 20%) were the most frequent side effects. In conclusion after 16 weeks, the endoscopic nasal polyp load in persons with symptomatic chronic sinusitis and nasal polyposis unresponsive to intranasal corticosteroids was lower when subcutaneous dupilumab was added to mometasone furoate nasal spray as opposed to mometasone alone [29].

Bachert et al. conducted a two international, multicenter, parallel-group, randomized, double-blind, and placebo-controlled trials evaluated the addition of dupilumab to the standard of therapy for people with severe CRSwNP. SINUS-52 was conducted in 117 centers across 14 nations, while SINUS-24 was conducted in 67 centers across 13 countries. Patients must have been 18 years of age or older, have bilateral CRSwNP, and still be experiencing symptoms even after using intranasal corticosteroids, getting systemic corticosteroids within the last 2 years, or having undergone sinonasal surgery [30]. For a duration of 24 weeks, patients in SINUS-24 were randomized (1:1) to receive subcutaneous dupilumab 300 mg or a placebo every 2 weeks. In SINUS-52, patients were randomized (1:1:1) to receive placebo every 2 weeks for 52 weeks, dupilumab 300 mg every 2 weeks for 24 weeks and then every 4 weeks for the final 28 weeks. Using a permuted block randomization schedule, all patients were assigned at random centrally. Randomization was stratified by nation, prior surgery at screening, and status of respiratory disease aggravated by nonsteroidal anti-inflammatory drugs or asthma at screening. Included were patients with and without co-occurring asthma. In an intention-to-treat population, coprimary outcomes included changes in nasal polyp score (NPS) from baseline to week 24, nasal congestion or obstruction, and sinus Lund-Mackay CT scores (a coprimary endpoint in Japan). A pooled population comprising the dupilumab group in SINUS-52 up to week 24, the dupilumab group in SINUS-24, and the placebo groups in both studies up to week 24 was used to evaluate safety. The results showed a significant and clinically meaningful decrease in MD was observed in the SNOT-22 score (scale 0–110) after 4–6 months (784 individuals; two studies; I2 = 0%). At 4–6 months, the rhinosinusitis disease severity (VAS) demonstrated a substantial and clinically meaningful decline of MD – 2.54 (95% CI −2.84–−2.23); two studies; 784 participants; I2 = 40%). A noteworthy and clinically meaningful decrease of MD – 0.86 (95% CI −0.98–−0.75); 784 individuals; two studies; I2 = 0%) was observed in the nasal congestion/obstruction score at 4–6 months. The UPSIT was utilized to assess smell. A significant and clinically meaningful drop of MD 10.83 (95% CI 9.59–12.08); 784 individuals; two studies; I2 = 0%) was observed in the UPSIT score at 4–6 months. In these investigations, the primary endpoint was the nasal polyp score. In these trials, the average nasal polyp score was approximately 6, which denotes a serious polyp illness. In two investigations with 784 patients, the nasal polyp score at 4–6 months shown a significant decrease of MD −1.79 (95% CI −2.01–−1.56); I2 = 65%. At 4–6 months, the Lund-Mackay score (scale 0-48) revealed a significant decrease in SMD −1.50 (95% CI −1.84–−1.16); there were 784 individuals in two investigations, with an I2 of 71%. Lastly, there was a notable effect on FEV1 and ACQ. Both studies demonstrated a significant improvement over placebo; however, the data could not be merged into a meta-analysis because ACQ5 and ACQ6 were employed. A meta-analysis of the FEV1 data revealed a substantial improvement in FEV1 (l) at 4–6 months of MD 0.21 (95% CI 0.20 – 0.22); there were 488 participants in two investigations, with an I2 of 0%. The most frequent side effects with placebo included headache, epistaxis, injection-site erythema, worsened nasal polyps and asthma, and rhinopharyngitis [30].

In a phase III trial study done by Anju et al. individuals with severe CRSwNP with or without concurrent AR were studied for the safety and effectiveness of dupilumab using this *post hoc* approach. This study used SINUS-24 weeks or SINUS-52 weeks protocol for dupilumab injections [33]. The analysis involved patients who were randomized every 2 weeks for 24 (SINUS-24) or 52 weeks (SINUS-52) to receive subcutaneous dupilumab 300 mg (n = 438) or placebo (n = 286). The first 24 weeks of treatment's pooled data are shown. The patient's self-reported history of allergic rhinitis (AR) status was used to examine changes from baseline in illness outcome measures and biomarker levels. In all, 46.7% (338 out of 724 patients) had AR. Patients with and without AR had largely comparable baseline characteristics. There was no discernible treatment difference between patients with and without AR at week 24, and dupilumab dramatically improved objective and patient-reported markers of CRSwNP, including loss of smell. It also lowered systemic and nasal biomarker levels compared placebo. Regardless of AR status, there was a substantial decrease in the use of systemic corticosteroids and/or sinonasal surgery with dupilumab compared to placebo ($p < 0.0029$). Dupilumab's safety profile was comparable in those with and without AR. In patients with severe CRSwNP, dupilumab showed statistically significant improvements in both clinical end goals and symptom scores when compared to placebo, regardless of concomitant AR status—a common patient subgroup that is frequently linked to worse CRSwNP outcomes [33].

In a meta-analysis conducted by Qingwu Wu [34], the included nine RCTs involved 1190 patients comparing three distinct biologics (omalizumab, mepolizumab, and dupilumab) with the placebo. Dupilumab demonstrated the highest level of efficacy for surface under the cumulative ranking curve (SUCRA) values of 0.900, 0.916, 1.000, and 0.807, respectively, for nasal polyp score (NPS), SNOT-22 score, University of Pennsylvania Smell Identification Test (UPSIT) score, and nasal congestion score (NCS). When it came to SUCRA values of 0.606, 0.500, and 0.693, omalizumab was the most effective drug in terms of SNOT-22, UPSIT, and NCS. For SUCRA values of 0.563, mepolizumab placed second in terms of NPS effectiveness, while for SUCRA values of 0.746, mepolizumab had the greatest risk of adverse events (AEs). In conclusion Omalizumab is the second-best option for CRSwNP, whereas dupilumab is the best option based on safety (AEs) and effectiveness (NPS). Mepolizumab had the highest risk of adverse events (AEs) while having the second-best effectiveness rating.

10.1 Conclusion

Because the evidence shown a considerable reduction in patients' need for surgery and an improvement in their symptoms, we recommend the use of dupilumab in individuals with CRSwNP that is difficult to treat.

11. Antihistamine

It is unknown how sensitization functions in CRSsNP. One may be tempted to hypothesize that an atopic person is more susceptible to developing CRS if they have allergic irritation in their nose. Nevertheless, there are contradictory studies assessing atopy as a risk factor for CRS. In a DBPCT, Haye et al. [35] randomly assigned 45 CRSwNP patients (16 of whom had allergies) to receive a placebo or 20 mg of cetirizine for a duration of 3 months. Cetirizine, according to the authors, decreased the number of days with a score for nasal sneeze and rhinorrhea that was less than one. At all time periods, the cetirizine-treated group experienced a score of 90–100% compared to the placebo group's 70–80% for rhinorrhea and 80–90% for sneezing. Nevertheless, information about individuals with and without allergies was not provided individually. The cetirizine had no influence on the size of nasal polyps.

11.1 Conclusion

When comparing antihistamines with placebos, the GRADE quality of the evidence was extremely low. Due to the limited number of studies and the absence of the most significant effectiveness measures, the evidence was downgraded. The effectiveness of regularly administering antihistamines to people with CRS is not established well enough. No meta-analysis is found in litertaure.

12. Anti-leukotrienes

Eosinophils and mast cells produce a family of inflammatory mediators called cysteineyl leukotrienes (CysLT) by breaking down arachidonic acid. CysLT are known to promote bronchoconstriction, mucus formation, edema, and neutrophil and eosinophil chemotaxis, and they may also have a role in the pathogenesis of asthma, rhinitis, and maybe CRSwNP.

Asthma, allergic rhinitis, and CRS with nasal polyposis (CRSwNP) have all been linked to the overproduction of CysLTs and overexpression of the receptor.

In a systemic review conducted by Wentzel et al. [36], studies evaluating the impact of leukotriene antagonists (LTAs) on clinical outcome markers of CRSwNP were included in a systematic review. Trials evaluating LTAs in CRS without nasal polyps or asthma symptoms alone met the exclusion criteria. The aim of this study is to evaluate the effects of LTA therapy on immunological markers, objective clinical outcomes, and nasal symptoms in CRSwNP, both alone and in combination with intranasal corticosteroids (INCSs). Twelve studies met the qualifying requirements: seven case series and five randomized control trials. Compared to placebo, LTAs significantly reduced CRSwNP symptoms; however, a meta-analysis of these randomized studies was not possible. The two studies that were included in the meta-analysis revealed that there was no difference in the treatment modalities, with a standardized mean difference of pooled total symptom scores between the LTA and INCS study arms of 0.02 (95% confidence range, −0.39-0.44). All trials reported improvements in immunological markers, clinical outcomes, and/or symptoms following LTA therapy; larger improvements were noted in a subset of symptoms compared to those seen with INCSs. Atopy, aspirin-exacerbated respiratory disorders, and concurrent asthma did not consistently or substantially alter these findings.

In a randomized controlled trial that evaluated adding montelukast to intransal spry therapy for chronic rhinosinusitis patients. In this study conducted by Suri et al. [37], 40 adult CRSwNP patients were randomly assigned to two groups. The subjects received budesonide nasal spray for 8 weeks, either with or without additional oral montelukast, and oral prednisolone for 14 days. After 8 weeks of therapy, subjects receiving supplemental oral montelukast reported a statistically significant improvement in their overall symptom score, sense of smell, and sneezing, which persisted for 4 weeks after the medication was stopped.

13. Conclusion

The data comparing placebo and montelukast was of extremely poor quality. The limited number of research and patients in the studies resulted in a downgrading of the evidence. The literature does not advise its usage unless a patient is unable to take nasal corticosteroids, based on the data currently available. Furthermore, the data contrasting montelukast with nasal corticosteroids is of low quality. The literature does not advise combining montelukast with nasal corticosteroids based on the available data. No meta-analysis is found.

Author details

Moath Abdulrahman Alfaleh
Department of Otolaryngology-Head and Neck Surgery, King Fahad Specialist Hospital, Dammam, Saudi Arabia

*Address all correspondence to: moathalfaleh@hotmail.com

References

[1] Lee LN, Bhattacharyya N. Regional and specialty variations in the treatment of chronic rhinosinusitis. The Laryngoscope. 2011;**121**(5):1092-1097

[2] Sabino HAC, Valera FCP, Aragon DC, Fantucci MZ, Titoneli CC, Martinez R, et al. Amoxicillin-clavulanate for patients with acute exacerbation of chronic rhinosinusitis: A prospective, double-blinded, placebo-controlled trial. International Forum of Allergy & Rhinology. 2017;**7**(2):135-142

[3] Van Zele T, Gevaert P, Holtappels G, Beule A, Wormald PJ, Mayr S, et al. Oral steroids and doxycycline: Two different approaches to treat nasal polyps. Journal of Allergy and Clinical Immunology. 2010;**125**(5):1069-1076

[4] Askari M, Khezri M, Zoghi G. Introducing a novel combination therapy with macrolides for the treatment of chronic rhinosinusitis: A randomized controlled trial. Iranian Journal of Medical Sciences. 2022;**47**(6):533

[5] Siu J, Klingler L, Wang Y, Hung CT, Jeong SH, Smith S, et al. Oral antibiotics used in the treatment of chronic rhinosinusitis have limited penetration into the sinonasal mucosa: A randomized trial. Xenobiotica. 2020;**50**(12):1443-1450

[6] Videler WJ, Badia L, Harvey RJ, Gane S, Georgalas C, Van Der Meulen FW, et al. Lack of efficacy of long-term, low-dose azithromycin in chronic rhinosinusitis: A randomized controlled trial. Allergy. 2011;**66**(11):1457-1468

[7] Wallwork B, Coman W, Mackay-Sim A, Greiff L, Cervin A. A double-blind, randomized, placebo-controlled trial of macrolide in the treatment of chronic rhinosinusitis. The Laryngoscope. 2006;**116**(2):189-193

[8] de Oliveira IS, Guimaraes AF, Arantes Pêgas GF, Machado CJ, Cassali GD, Tormin Borges Crosara PF, et al. Azitromycin for chronic eosihophilic rhinosinusitis with nasal polyp: A placebo-controlled trial. Rhinology. 2020;**58**(6):610-617

[9] Kim DH, Shin H, Stybayeva G, Hwang SH. A comparison of doxycycline and conventional treatments of refractory chronic rhinosinusitis with nasal polyps: A systematic review and meta-analysis. European Archives of Oto-Rhino-Laryngology. 2024;**281**:3949-3957

[10] Bonfils P, Escabasse V, Coste A, Gilain L, Louvrier C, Serrano E, et al. Efficacy of tobramycin aerosol in nasal polyposis. European Annals of Otorhinolaryngology, Head and Neck Diseases. 2015;**132**(3):119-123

[11] Day N, Mainardi JL, Malinvaud D, Bonfils P. Bacteriological study of ethmoid specimens from patients with nasal polyposis after ethmoidal surgery. Annales D'oto-laryngologie et de Chirurgie Cervico Faciale: Bulletin de la Societe D'oto-laryngologie des Hopitaux de Paris. 2009;**126**(4):196-202

[12] Lee VS, Pottinger PS, Davis GE. Tolerability and effectiveness of povidone-iodine or mupirocin versus saline sinus irrigations for chronic rhinosinusitis. American Journal of Otolaryngology. 2020;**41**(5):102604

[13] Leopold DA, Elkayam D, Messina JC, Kosik-Gonzalez C, Djupesland PG, Mahmoud RA. NAVIGATE II: Randomized, double-blind trial of

the exhalation delivery system with fluticasone for nasal polyposis. Journal of Allergy and Clinical Immunology. 2019;**143**(1):126-134

[14] Kobayashi Y, Yasuba H, Asako M, Yamamoto T, Takano H, Tomoda K, et al. HFA-BDP metered-dose inhaler exhaled through the nose improves eosinophilic chronic rhinosinusitis with bronchial asthma: A blinded, placebo-controlled study. Frontiers in Immunology. 2018;**9**:2192

[15] Tait S, Kallogjeri D, Suko J, Kukuljan S, Schneider J, Piccirillo JF. Effect of budesonide added to large-volume, low-pressure saline sinus irrigation for chronic rhinosinusitis: A randomized clinical trial. JAMA Otolaryngology–Head & Neck Surgery. 2018;**144**(7):605-612

[16] Bognanni A, Chu DK, Rank MA, Bernstein J, Ellis AK, Golden D, et al. Topical corticosteroids for chronic rhinosinusitis with nasal polyposis: GRADE systematic review and network meta-analysis. Journal of Allergy and Clinical Immunology. 2022;**150**(6):1447-1459

[17] Ecevit MC, Erdag TK, Dogan E, Sutay S. Effect of steroids for nasal polyposis surgery: A placebo-controlled, randomized, double-blind study. The Laryngoscope. 2015;**125**(9):2041-2045

[18] Giotakis AI, Karow EM, Scheithauer MO, Weber R, Riechelmann H. Saline irrigations following sinus surgery-a controlled, single blinded, randomized trial. Rhinology. 2016;**54**(4):302-310

[19] Zhang Y, Wang C, Huang Y, Lou H, Zhang L. Efficacy of short-term systemic corticosteroid therapy in chronic rhinosinusitis with nasal polyps: A meta-analysis of randomized controlled trials and systematic review. American Journal of Rhinology & Allergy. 2019;**33**(5):567-576

[20] Gevaert P, Calus L, Van Zele T, Blomme K, De Ruyck N, Bauters W, et al. Omalizumab is effective in allergic and nonallergic patients with nasal polyps and asthma. Journal of Allergy and Clinical Immunology. 2013;**131**(1):110-116

[21] Pinto JM, Mehta N, DeTineo M, Wang J, Baroody FM, Naclerio RM. A randomized, double-blind, placebo-controlled trial of anti-IgE for chronic rhinosinusitis. Rhinology. 2010;**48**(3):318

[22] Di Bona D, Fiorino I, Taurino M, Frisenda F, Minenna E, Pasculli C, et al. Long-term "real-life" safety of omalizumab in patients with severe uncontrolled asthma: A nine-year study. Respiratory Medicine. 2017;**130**:55-60

[23] Liu L, Pan M, Li Y, Tan G, Yang Y. Eficácia da irrigação nasal com solução salina hipertônica na rinossinusite crônica: revisão sistemática e metanálise. Brazilian Journal of Otorhinolaryngology. 2020;**86**:639-646

[24] Denburg JA, Sehmi R, Upham J. Regulation of IL-5 receptor on eosinophil progenitors in allergic inflammation: Role of retinoic acid. International Archives of Allergy and Immunology. 2001;**124**(1-3):246-248

[25] Johnson PS, Ohlsson B, von Buchwald C, Jannert M, Ahlner-Elmqvist M. A multi-centre study on quality of life and absenteeism in patients with CRS referred for endoscopic surgery. Rhinology. 2011;**49**(4):420-428

[26] Bachert C, Sousa AR, Lund VJ, Scadding GK, Gevaert P, Nasser S, et al.

Reduced need for surgery in severe nasal polyposis with mepolizumab: Randomized trial. Journal of Allergy and Clinical Immunology. 2017;**140**(4):1024-1031

[27] Wu Q, Yuan L, Qiu H, Wang X, Huang X, Zheng R, et al. Efficacy and safety of omalizumab in chronic rhinosinusitis with nasal polyps: A systematic review and meta-analysis of randomised controlled trials. BMJ Open. 2021;**11**(9):e047344

[28] Bagnasco D, Ferrando M, Varricchi G, Passalacqua G, Canonica GW. A critical evaluation of anti-IL-13 and anti-IL-4 strategies in severe asthma. International Archives of Allergy and Immunology. 2016;**170**(2):122-131

[29] Bachert C, Mannent L, Naclerio RM, Mullol J, Ferguson BJ, Gevaert P, et al. Effect of subcutaneous dupilumab on nasal polyp burden in patients with chronic sinusitis and nasal polyposis: A randomized clinical trial. JAMA. 2016;**315**(5):469-479

[30] Bachert C, Han JK, Desrosiers M, Hellings PW, Amin N, Lee SE, et al. Efficacy and safety of dupilumab in patients with severe chronic rhinosinusitis with nasal polyps (LIBERTY NP SINUS-24 and LIBERTY NP SINUS-52): Results from two multicentre, randomised, double-blind, placebo-controlled, parallel-group phase 3 trials. The Lancet. 2019;**394**(10209):1638-1650

[31] Peters AT, Wagenmann M, Bernstein JA, Khan AH, Nash S, Jacob-Nara JA, et al. Dupilumab efficacy in patients with chronic rhinosinusitis with nasal polyps with and without allergic rhinitis. In: Allergy and Asthma Proceedings. Vol. 44. issue no. 4. Ocean Side Publications, Inc.; 2023. pp. 265-274

[32] Gevaert P, Van Bruaene N, Cattaert T, Van Steen K, Van Zele T, Acke F, et al. Mepolizumab, a humanized anti–IL-5 mAb, as a treatment option for severe nasal polyposis. Journal of Allergy and Clinical Immunology. 2011;**128**(5):989-995

[33] Akinlade B, Guttman-Yassky E, de Bruin-Weller M, Simpson EL, Blauvelt A, Cork MJ, et al. Conjunctivitis in dupilumab clinical trials. British Journal of Dermatology. 2019;**181**(3):459-473

[34] Wu Q, Zhang Y, Kong W, Wang X, Yuan L, Zheng R, et al. Which is the best biologic for nasal polyps: Dupilumab, omalizumab, or mepolizumab? A network meta-analysis. International Archives of Allergy and Immunology. 2022;**183**(3):279-288

[35] Haye R, Aanesen JP, Burtin B, Donnelly F, Duby C. The effect of cetirizine on symptoms and signs of nasal polyposis. The Journal of Laryngology & Otology. 1998;**112**(11):1042-1046

[36] Wentzel JL, Soler ZM, DeYoung K, Nguyen SA, Lohia S, Schlosser RJ. Leukotriene antagonists in nasal polyposis: A meta-analysis and systematic review. American Journal of Rhinology & Allergy. 2013;**27**(6):482-489

[37] Suri A, Gupta R, Gupta N, Kotwal S. Montelukast as an adjunct to treatment of chronic rhinosinusitis with polyposis: A prospective randomized controlled trial. JK Science. 2015;**17**(2):92

Chapter 4

Psychological Impact of Rhinology Disorders

Zahra Ebrahim Soltani and Mohammad Elahi

Abstract

Rhinology disorders, including sinusitis, rhinitis, allergic diseases, and structural problems, often go unnoticed and undertreated. Yet, these seemingly localized issues can have far-reaching effects. Over time, problems in the rhino-sinus and upper airway systems can impact the nervous system, social interactions, and psychological well-being. Patients may experience anxiety, depression, and difficulties with learning, memory, and social behavior. These hidden complications are crucial for healthcare professionals to recognize, as addressing them can significantly improve patients' quality of life. This chapter will delve deeper into the intricate relationship between rhinology disorders and psychological impacts, exploring the possible underlying mechanisms. It will offer insights into effective management and treatment strategies, aiming to empower healthcare professionals to provide holistic care that encompasses the physical, psychological, and social aspects of their patients' lives.

Keywords: rhinitis, rhinosinusitis, olfactory bulb, depression, anxiety, cognition, behavior, mood, psychiatry, psychology

1. Introduction

Rhinology disorders, while predominantly recognized for their physical manifestations, conceal a profound psychological impact that is often overlooked in clinical practice. This chapter seeks to illuminate the intricate connections between rhinology disorders and mental health, offering healthcare professionals a comprehensive understanding of the emotional burdens these conditions impose on patients. By exploring the psychological ramifications of chronic rhinosinusitis, nasal polyps, allergic rhinitis, and other related disorders, we aim to reveal the extensive influence these ailments have on quality of life and mental well-being. Through this exploration, we will provide valuable insights and strategies to enhance patient care, emphasizing the importance of addressing both the physical and psychological aspects of rhinology disorders.

Understanding the psychological impact of rhinology diseases is crucial for several reasons, as these conditions often extend beyond physical symptoms to profoundly affect mental health and overall quality of life. Here are key reasons why this understanding is essential:

Comprehensive patient care: rhinology diseases often lead to persistent discomfort, pain, and functional impairments. These physical symptoms can contribute to significant psychological distress, including anxiety, depression, and sleep disturbances. Acknowledging the psychological dimensions of these conditions allows healthcare providers to offer more holistic and effective patient care, addressing both physical and mental health needs.

Improved quality of life: patients with rhinology disorders frequently experience a diminished quality of life due to ongoing symptoms such as nasal congestion, headaches, and reduced sense of smell. These symptoms can interfere with daily activities, social interactions, and work productivity. Understanding the psychological impact helps in developing targeted interventions that can alleviate these symptoms and improve patients' overall well-being.

Enhanced treatment outcomes: psychological factors can influence the course and treatment outcomes of rhinology diseases. For instance, stress and mental health conditions can exacerbate physical symptoms and hinder recovery. By integrating psychological assessment and support into the treatment plan, healthcare providers can enhance adherence to medical regimens, promote healthier coping strategies, and ultimately achieve better clinical outcomes.

Identification of comorbidities: patients with rhinology disorders often have comorbid psychological conditions that may go unrecognized and untreated. Understanding the psychological impact helps in early identification and management of these comorbidities, ensuring that patients receive comprehensive care that addresses all aspects of their health.

Patient-centered care: a patient-centered approach to healthcare emphasizes the importance of understanding the full spectrum of a patient's experience, including their psychological and emotional well-being. By recognizing and addressing the psychological impact of rhinology diseases, healthcare providers can build stronger therapeutic relationships, improve patient satisfaction, and empower patients to take an active role in their care.

Public health implications: rhinology disorders are prevalent and can lead to significant healthcare utilization and economic burden. By understanding the psychological impact, public health strategies can be developed to support mental health initiatives, improve access to comprehensive care, and reduce the overall burden of these diseases on the healthcare system.

Research and education: highlighting the psychological impact of rhinology diseases encourages further research into the biopsychosocial aspects of these conditions. It also underscores the need for interdisciplinary education, ensuring that healthcare professionals are equipped with the knowledge and skills to address both the physical and psychological needs of their patients.

2. Common rhinology disorders and their association with psychologic problems

2.1 Chronic rhinosinusitis

Chronic rhinosinusitis (CRS) significantly impacts patients' mental health, with a strong correlation between CRS and psychological disorders such as depression and anxiety. Depression in patients with CRS is often underdiagnosed, yet it profoundly affects treatment outcomes and healthcare utilization. Depression scores using the

Beck Depression Inventory (BDI) are higher in patients with CRS, even when controlling for other comorbid conditions such as asthma, allergies, and aspirin sensitivity. Notably, depression appears more prevalent in patients with CRS without nasal polyps (CRSsNP) compared to those with nasal polyps (CRSwNP). Nearly one-third of CRS patients screened positive for depression, indicating the need for routine mental health evaluations in this population [1, 2].

Patients suffering from both chronic sinusitis and nasal polyps are significantly more likely to experience depression and anxiety. The relationship between CRS and psychological disorders is bidirectional; while CRS exacerbates mental health issues, these psychological conditions can also worsen the symptoms and management of CRS. For example, patients with CRSwNP are often associated with more severe clinical disease, possibly because they are more likely to seek treatment early due to frequent associations with asthma and allergic rhinitis. This continuous medical attention may help mitigate some psychological impacts, although it does not eliminate them [3, 4].

Longitudinal studies reinforce the chronic nature of the psychological burden associated with CRS. In a large cohort study of over 48,000 participants, the incidence of depression and anxiety was significantly higher in the group with CRS compared to a control group over an 11-year follow-up period. Interestingly, the adjusted hazard ratio for developing depression and anxiety was higher in patients with CRSsNP than those with CRSwNP. This distinction highlights that while CRSwNP might lead to more severe physical symptoms and require frequent medical intervention, CRSsNP can have a more profound impact on mental health, possibly due to the chronic, lingering nature of the disease that may not prompt immediate clinical attention [4].

The relationship between CRS and psychological disorders is complex and multifaceted. For instance, while anxiety scores in CRS patients with asthma improved with pharmacological treatment, depression scores did not show the same improvement, suggesting that anxiety might be closely related to the severity of airway disease, whereas depression could be more independent of the physical symptoms of CRS. This finding underscores the necessity for separate mental health care pathways for managing depression in CRS patients. Furthermore, despite the high prevalence of mental health issues among CRS patients, the rates of depression did not differ significantly between those who underwent surgical treatment and those who did not, indicating that surgical interventions alone may not be sufficient to address the psychological impacts of CRS [3].

In conclusion, the interplay between chronic rhinosinusitis and psychological disorders like depression and anxiety necessitates an integrated approach to treatment that addresses both physical and mental health aspects. Routine screening for depression and anxiety in CRS patients should become a standard practice, and mental health interventions should be tailored to the unique needs of CRS patients. This holistic approach can potentially improve overall treatment outcomes and enhance the quality of life for those suffering from this chronic condition.

2.2 Nonallergic and allergic rhinitis

Rhinitis, encompassing both allergic rhinitis (AR) and nonallergic rhinitis (NAR), is significantly associated with psychological disorders, particularly depression and anxiety. Individuals with rhinitis are more prone to developing depression, with the risk being notably higher in patients with NAR compared to those with AR. The severity of rhinitis symptoms, rather than their persistence or seasonality, is the most critical factor influencing mental health outcomes. Severe symptoms are linked to a worse quality of life, poor sleep, and elevated levels of anxiety and depression. The continuous discomfort and impairment caused by rhinitis symptoms can lead to frustration and feelings of helplessness, further exacerbating psychological distress [5–7].

The psychological impact of rhinitis is further compounded by the presence of comorbid conditions. Patients with rhinitis are more likely to suffer from asthma, chronic bronchitis, and emphysema, contributing to their overall poorer health status. This burden of comorbidities correlates with a higher prevalence of significant depression, particularly among women. The physical strain and discomfort from these associated conditions can intensify mental health challenges, creating a cycle of worsening physical and psychological symptoms. Both AR and NAR are associated with increased frequencies of anxiety and depression, with NAR presenting a stronger association. The unpredictability and chronic nature of these conditions can lead to constant anxiety about managing symptoms and their impact on daily life [6, 8].

In addition to depression and anxiety, rhinitis has been linked to other serious mental health conditions. Adolescents with AR, for instance, have a significantly higher incidence of developing bipolar disorder later in life [9]. Furthermore, young patients with AR may experience difficulty in learning and attention, which can impact academic performance and social development [10]. These cognitive and attention deficits highlight the broader impact of rhinitis beyond immediate physical discomfort, affecting long-term educational and developmental outcomes. This highlights the potential long-term psychological effects of rhinitis, suggesting that early and comprehensive management is crucial in mitigating these impacts [7, 11].

Moreover, the impact of rhinitis on sleep is a critical factor linking it to psychological problems. Poor sleep quality and insomnia are common in individuals with severe rhinitis symptoms, leading to fatigue, irritability, and difficulty concentrating. Chronic sleep disturbances are well-known contributors to anxiety and depression, creating a feedback loop where psychological distress can further disrupt sleep. Addressing sleep issues through both pharmacological and behavioral interventions can significantly improve the overall mental health and quality of life for rhinitis patients. Effective management of sleep disturbances can break this cycle, providing a foundation for better psychological health [10, 12, 13].

These complex and varied associations underscore the necessity for healthcare professionals to adopt a holistic approach in managing rhinitis, considering both the physical symptoms and the profound psychological impacts. By understanding these connections, healthcare providers can develop more effective treatment strategies that address the full spectrum of patient needs, ensuring a higher quality of life and better mental health outcomes for those affected by rhinitis. Integrating mental health support into the care plan for rhinitis patients can facilitate early identification and treatment of psychological issues, promoting a more comprehensive approach to health and well-being.

2.3 Olfactory bulb dysfunction

Olfactory dysfunction, which includes anosmia (complete loss of smell), hyposmia (reduced smell), and hyperosmia (heightened smell), is prevalent in various psychological disorders such as anxiety, depression, schizophrenia, and bipolar disorder. These dysfunctions often manifest as impairments in odor detection, discrimination, and identification. The connection between olfactory deficits and mental health conditions is crucial as it impacts the diagnosis, treatment, and overall well-being of patients [14, 15].

In anxiety disorders, such as generalized anxiety disorder (GAD) and panic disorder, olfactory dysfunction is common. Patients with moderate to severe anxiety exhibit significant deficits in odor discrimination and identification, which correlate inversely with anxiety severity. This suggests that the poorer the olfactory function, the more severe the anxiety symptoms. Moreover, in panic disorder, a lower odor threshold is often found, indicating that these patients are more sensitive to odors, which correlates with the severity of their symptoms [14, 16–18].

Depression also significantly impacts olfactory function. Patients with depression commonly exhibit impairments in odor threshold, discrimination, and identification, with the olfactory threshold being the most notably affected. Olfactory dysfunction is a potential biomarker for depression, with successful treatment often reversing these deficits. Interestingly, olfactory training has been shown to improve depressive symptoms, although the results are somewhat controversial. These findings underscore the importance of considering olfactory assessments in the management of depression [19–24].

Schizophrenia is another condition where olfactory dysfunction is prevalent. Patients with schizophrenia, especially those in the early stages of psychosis or at high risk, often show impaired odor identification and discrimination. While some individuals with schizophrenia may experience olfactory hallucinations, these are not directly linked to the general olfactory deficits observed, suggesting different underlying neural mechanisms. The olfactory impairments in schizophrenia are associated with negative symptoms and reduced social and cognitive functioning, highlighting the need for comprehensive olfactory evaluations in these patients [25].

Bipolar disorder also shows a notable link with olfactory dysfunction. Patients, even during euthymic phases, often display impaired odor thresholds and reduced odor identification capabilities. However, olfactory discrimination does not seem to be significantly affected. The severity of clinical symptoms in bipolar disorder negatively correlates with odor sensitivity but not odor identification, suggesting a nuanced relationship between olfactory function and mood episodes. This relationship highlights the importance of olfactory assessments in understanding and managing bipolar disorder [20, 26].

The recent COVID-19 pandemic has brought additional focus to the link between olfactory dysfunction and psychological distress. COVID-19 often causes nasal inflammation leading to significant olfactory impairments, which can result in long-lasting neuropsychiatric sequelae. COVID-19-related anosmia can profoundly impact perceived quality of life and psychological well-being. This highlights the importance of olfactory function not just as a sensory ability but as a critical component of overall mental health. Additionally, other viruses, such as influenza and the common cold (caused by rhinoviruses and coronaviruses), have been shown to affect the olfactory bulb, leading to temporary or sometimes prolonged olfactory dysfunction. These viral infections can also trigger inflammatory responses in the nasal passages and olfactory pathways, contributing to sensory deficits and subsequent psychological impacts. These findings reinforce the need for integrated care approaches that address both olfactory and mental health in patients affected by olfactory dysfunction [27–29].

2.4 Structural nose problem

Structural nose problems, such as a deviated nasal septum, enlarged turbinates, and nasal polyps, can lead to significant nasal obstruction and congestion. These anatomical abnormalities can cause chronic nasal obstruction, which not only impairs physical health but also has profound psychological effects. The persistent nature of nasal obstruction often leads to frustration and can severely impact the quality of life. This chronic frustration can, in turn, contribute to the development of anxiety and depression [30].

One of the primary ways in which nasal obstruction impacts mental health is through its effect on sleep quality. Adequate sleep is essential for maintaining mood and overall mental health. Nasal obstruction can cause sleep disturbances such as sleep apnea or chronic snoring, which result in poor sleep quality [31]. Sleep deprivation has been closely linked to mood disorders, including anxiety and depression. Thus, individuals with chronic nasal obstruction are at a higher risk of developing these psychological conditions due to the compounded effects of disrupted sleep and the frustration caused by persistent breathing difficulties [32, 33].

Patients with unilateral or bilateral complete nasal obstruction exhibit higher psychiatric symptom scores across various personality traits. Although the scores for psychiatric symptoms were elevated in these patients, significant differences were noted particularly in dependent and antisocial personality traits compared to healthy controls [34]. This suggests that structural nose problems not only affect physical health but also influence personality characteristics and psychological well-being. The presence of higher scores for depression and anxiety, even if not significantly different from controls, highlights the mental health burden carried by patients with nasal obstruction [30, 34].

Nasal surgery, such as septoplasty or septorhinoplasty, has been shown to alleviate nasal symptoms and improve nasal obstruction. Interestingly, these surgical interventions also have positive effects on psychiatric symptoms. Patients with structural nasal problems and accompanying psychiatric symptoms can experience significant improvements in their mental health following corrective nasal surgery [35]. However, studies have suggested that isolated nasal surgery does not have a significant impact on apnea-hypopnea index in patients with obstructive sleep apnea [36]. This underscores the importance of a holistic approach in treating nasal obstruction, where both physical and psychological aspects are considered. Patients with personalities of dependent and antisocial personality disorders, in particular, may benefit from such surgical interventions, suggesting a close interplay between nasal structure and personality traits [34, 37].

Given the significant impact of structural nose problems on mental health, it is crucial for healthcare providers to adopt a multidisciplinary approach in managing these conditions. Evaluating patients from both an otorhinolaryngological and psychological perspective can ensure comprehensive care.

2.5 Granulomatosis with polyangiitis

Wegener's granulomatosis, now more commonly referred to as granulomatosis with polyangiitis (GPA), is a type of vasculitis that involves the inflammation of small and medium-sized blood vessels, often affecting the upper respiratory tract. Patients with GPA usually experience persistent nasal congestion and obstruction which are common due to inflammation and swelling of the nasal passages. While the physical manifestations of GPA are well-documented, the psychological impacts are equally significant but less frequently addressed. Research has revealed a high prevalence of psychiatric disorders among GPA patients, with paranoia and depression being particularly common. Additionally, there is a notable correlation between these psychiatric disorders and the neuroticism personality trait, suggesting that personality factors may influence the mental health outcomes of GPA patients [38].

Furthermore, psychosis has been reported as an initial manifestation in some cases of GPA, illustrating the severe psychological impact this disease can have. Some cases highlight the acute psychiatric symptoms that can arise in GPA [38, 39]. These cases underscore the potential for GPA to manifest with severe psychiatric symptoms, necessitating appropriate psychiatric intervention. Such cases, although rare, emphasize the importance of healthcare providers being vigilant about the psychiatric dimensions of GPA to ensure comprehensive treatment that addresses both the physical and mental health needs of the patient.

Depression, anxiety, sleep disturbances, and fatigue are common yet inadequately addressed comorbidities in GPA. These psychiatric conditions significantly impair the quality of life of GPA patients. Factors such as age, body mass index (BMI), and disease damage have been identified as predictors of depression in these patients. Targeting these potentially modifiable factors could help improve the psychological well-being of GPA patients [40].

The impact of GPA on daily life extends beyond the individual's physical health. There are substantial medical and functional morbidity among GPA patients. Many patients require long-term immunosuppressive treatment and experience significant reductions or constraints in their daily activities. These limitations negatively affect patients' normal daily living, employment, and income, as well as their family and close relationships. The substantial burden on quality of life highlights the need for a holistic approach in managing GPA, incorporating both medical treatment and psychosocial support [40, 41].

While depression is highly prevalent among patients with primary systemic vasculitis and associated with poorer outcomes, there is a need for further research specifically focusing on medium and large vessel vasculitis to better understand and address the psychological dimensions of these conditions. Overall, the psychological burden of GPA is profound, necessitating comprehensive care strategies that address both the physical and mental health aspects to improve patient outcomes and quality of life.

3. Mechanisms linking rhinology disorders and mental health

3.1 Inflammation and immune system mechanisms

The molecular mechanisms linking rhinological diseases to psychological disorders are intricately connected through pathways of inflammation and immune responses. Inflammatory responses in rhinological conditions have profound systemic effects, particularly influencing the central nervous system (CNS). CRS is predominantly associated with type 2 inflammation, where activated T helper (Th2) cells produce cytokines such as IL-4, IL-5, and IL-13. These cytokines facilitate the recruitment of eosinophils and the production of IgE antibodies, perpetuating a cycle of chronic inflammation and tissue remodeling [14, 42].

Elevated levels of IgE, frequently observed in AR, can cross the blood-brain barrier (BBB) and activate microglia, the brain's resident immune cells. This activation polarizes microglia to the M1 phenotype, which is associated with the production of pro-inflammatory cytokines such as IL-1β, IL-6, and TNF-α. These cytokines not only exacerbate neuroinflammation but also disrupt neuronal function, leading to symptoms of depression and anxiety. Additionally, IgE can activate astrocytes, further amplifying neuroinflammatory processes and potentially leading to neurodegeneration, thereby contributing to cognitive dysfunctions and psychological disturbances [43–45].

A significant pathway involved in this process is the activation of Fc epsilon RI alpha (FcεRIα) by IgE. FcεRIα is a high-affinity IgE receptor that, when activated by IgE, can upregulate its expression on immune cells, thereby enhancing the inflammatory response. This receptor is crucial in mediating allergic reactions and is found on various immune cells, including mast cells and basophils. Upon activation, FcεRIα increases BBB permeability and facilitates the infiltration of inflammatory cells into the CNS. The resulting neuroinflammation can lead to psychological disorders such as depression and anxiety. FcεRIα's role in increasing inflammatory cytokines in the CNS underscores its significance in linking peripheral allergic responses with central neuroinflammatory processes [46–48].

The role of innate immunity, particularly through the Toll-like receptor 4 (TLR4) and nuclear factor kappa-light-chain-enhancer of activated B cells (NF-κB) signaling pathways, is also critical in this context. TLR4 is a pattern recognition receptor that detects pathogen-associated molecular patterns (PAMPs) and damage-associated molecular patterns (DAMPs), leading to the activation of the NF-κB pathway. This activation triggers the production of various pro-inflammatory cytokines and chemokines, which are key mediators of the immune response. In rhinological conditions, the persistent activation of TLR4 and NF-κB can lead to chronic inflammation not only in the nasal passages but also in systemic circulation, impacting the CNS and contributing to psychological disorders such as behavioral and cognitive changes [49–51].

Chronic nasal obstruction, as seen in conditions like deviated nasal septum and enlarged turbinates, can lead to significant sleep disturbances, which are closely linked to mood disorders. Poor sleep quality and fragmented sleep, common in obstructive sleep apnea, result in intermittent hypoxia and oxidative stress, which can cause neuronal injury and cognitive impairments. The inflammatory mediators released during these episodes of intermittent hypoxia can further disrupt the BBB, allowing inflammatory cells and cytokines to infiltrate the CNS, thereby promoting neuroinflammation and psychological distress. The continuous cycle of poor sleep and heightened inflammation exacerbates symptoms of depression and anxiety, highlighting the interconnectedness of sleep, respiratory health, and mental well-being.

Olfactory dysfunction, which is common in conditions like CRS and AR, also has significant psychological implications. The loss of smell (anosmia) or reduced olfactory sensitivity (hyposmia) can lead to a diminished quality of life, contributing to depression and anxiety. Olfactory sensory neurons, when damaged by inflammation or infection (e.g., SARS-CoV-2), release pro-inflammatory cytokines such as IL-6 and TNF-α. These cytokines can travel to the brain and enhance neuroinflammation, affecting brain regions involved in mood regulation, such as the hippocampus and prefrontal cortex. This inflammatory cascade not only impacts mood but also impairs cognitive functions, leading to learning and memory dysfunctions [14, 52, 53].

Viral infections of the respiratory tract, such as influenza and COVID-19, further illustrate the link between rhinological diseases and psychological disorders. These infections trigger systemic inflammatory responses that extend to the CNS. For instance, the pro-inflammatory cytokines released during and after viral infections can lead to persistent neuroinflammation, contributing to long-term psychological effects such as depression and cognitive impairments. Influenza has been associated with an increased risk of depression, severe enough to require antidepressant treatment. Similarly, anosmia in COVID-19 patients has been linked to depression, which can persist for months after the resolution of acute symptoms. This highlights the role of viral-induced inflammation in the pathophysiology of psychological disorders [52–54].

In conclusion, the intricate relationship between rhinological diseases and psychological disorders is mediated through complex molecular mechanisms involving chronic inflammation, immune responses, and neuroinflammation. The activation of FcεRIα by IgE plays a crucial role in enhancing BBB permeability and promoting neuroinflammation, which links peripheral allergic responses to central psychological effects. Understanding these pathways provides insight into how conditions like CRS, AR, and olfactory dysfunction can lead to significant psychological distress and cognitive impairments. Additionally, the role of TLR4 and NF-κB in mediating innate immune responses highlights the broader implications of chronic inflammation in the pathophysiology of both rhinological and psychological conditions. This knowledge emphasizes the need for comprehensive management approaches that address both the respiratory and psychological aspects of these conditions, ultimately improving patient outcomes and quality of life.

3.2 Neurobiological mechanisms

The intricate relationship between rhinological diseases with psychological issues can be explained through neural mechanisms and signaling pathways. Rhinological diseases often lead to chronic inflammation, which not only affects the respiratory system but also has far-reaching effects on the CNS. The olfactory system, which begins with the olfactory neurons (ONs) in the olfactory epithelium (OE), plays a pivotal role in this connection. These neurons transmit odor information to the olfactory bulbs (OB), which then project to secondary olfactory structures such as the anterior olfactory nucleus (AON), piriform cortex, and olfactory tubercle, eventually reaching tertiary structures like the orbitofrontal cortex (OFC), insular cortex, and dorsal hippocampus. These pathways highlight how disruptions in the olfactory system can affect broader neural circuits involved in mood regulation and cognitive functions [55–57].

The amygdala and the orbitofrontal cortex (OFC) are key components of the neurocircuitry implicated in anxiety disorders. The primary olfactory cortex, which includes the cortical amygdala nuclei, transmits odor information to other parts of the amygdala [58, 59]. Dysfunction in the amygdala, a region crucial for emotional processing, may be a significant factor linking olfactory dysfunction and anxiety [16, 57, 59]. Odor discrimination, a central sensory process associated with the primary olfactory cortex, is correlated with anxiety symptom severity [60]. This suggests that central processing of olfactory information, rather than peripheral sensory thresholds, is more relevant to the development of anxiety symptoms in patients with olfactory dysfunction [56].

Inflammation in the anterior cingulate cortex (ACC) is another critical factor in the development of neuropsychiatric symptoms in conditions such as AR. Neuroinflammation in the ACC has been linked to anxiety and depression-like behaviors in AR. The anatomical proximity between the nasal cavity and the CNS facilitates the spread of inflammation and neuropeptides released by type C nociceptive nerves in the nasal cavity, which increase plasma extravasation and glandular secretion. Dysfunction of these nerves can lead to the release of proinflammatory cytokines, contributing to both AR and the development of psychological disorders [61].

CRS and other sinonasal diseases often result from viral or bacterial infections, which trigger an inflammatory response. Research has shown that sinus inflammation can alter brain activity, particularly in neural networks that regulate cognition, introspection, and responses to external stimuli. Functional MRI in CRS patients have revealed increased amplitude of low-frequency fluctuations (ALFF) in the left orbital superior frontal cortex and reduced connectivity in the right precuneus. These alterations in brain activity correlate with inflammation severity and psychological symptoms, as indicated by the positive correlation between ALFF values in the orbital superior frontal cortex and scores on the hospital anxiety and depression scale (HADS) [2].

The role of different types of neurons, particularly glutamatergic, GABAergic, and dopaminergic neurons, is critical in understanding how rhinological diseases impact brain function and contribute to psychological problems. Inflammation and neuroimmune responses can alter the balance of excitatory and inhibitory signaling in the brain [62]. Glutamatergic neurons, which are primarily excitatory, can be affected by increased levels of pro-inflammatory cytokines, leading to heightened neuronal excitability and potential excitotoxicity. This imbalance can result in cognitive impairments and mood disorders. Conversely, GABAergic neurons, which are inhibitory, may be downregulated or dysfunctional in inflammatory conditions, contributing to reduced inhibition and increased anxiety and depression [63, 64].

Dopaminergic neurons, which are involved in reward processing and motivation, can also be significantly impacted by inflammatory responses. Chronic inflammation can alter dopamine signaling pathways, leading to anhedonia, reduced motivation, and other depressive symptoms [62]. Dopaminergic signaling in regions such as the prefrontal cortex (PFC) and the ventral tegmental area (VTA) can be disrupted by inflammation [65]. These disruptions may further contribute to the psychological symptoms experienced by patients with rhinological diseases.

In summary, the neural mechanisms linking rhinological diseases to psychological problems involve complex interactions between inflammation, immune responses, and neural signaling pathways. The olfactory system's connection to key brain regions involved in mood regulation, such as the amygdala and OFC, plays a central role. Chronic inflammation, mediated by pathways such as TLR4/NF-κB and FcεRIα, leads to neuroinflammation and disruption of neural circuits, resulting in psychological disorders. Additionally, the balance of excitatory and inhibitory signaling through glutamatergic, GABAergic, and dopaminergic neurons is crucial in maintaining normal cognitive and emotional function. Disruptions in these neuronal pathways due to inflammation and immune responses underscore the importance of addressing both the respiratory and psychological aspects of rhinological diseases to improve patient outcomes and quality of life.

3.3 Social mechanisms

Social factors play a significant role in the interplay between rhinological diseases and psychological conditions. Social determinants of health, including socioeconomic status, access to healthcare, social support, and environmental conditions, can exacerbate both physical and mental health issues, creating a vicious cycle that is difficult to break.

Firstly, socioeconomic status profoundly impacts the prevalence and management of rhinological diseases. Individuals from lower socioeconomic backgrounds often face barriers to accessing quality healthcare, including diagnostic services and effective treatments for conditions like rhinitis and sinusitis. This lack of access can lead to the progression of these conditions into chronic states, which are associated with a higher risk of developing mental health problems such as depression and anxiety. The chronic discomfort and impaired functioning caused by untreated rhinological diseases can significantly diminish quality of life, leading to psychological distress.

Secondly, social support systems are crucial in managing both physical and mental health conditions. Individuals with strong social networks tend to have better health outcomes because they receive emotional support, practical help, and encouragement to seek medical care. Conversely, those who are socially isolated or have weak support networks are at greater risk of both rhinological diseases and psychological issues. The stress of dealing with health problems without adequate support can exacerbate symptoms of anxiety and depression, making it harder to manage chronic conditions like sleep apnea or structural nose problems.

Environmental conditions also play a vital role in this dynamic. Poor air quality, common in lower-income and densely populated urban areas, can aggravate conditions like rhinitis and sinusitis. Exposure to pollutants and allergens can lead to chronic inflammation of the nasal passages and respiratory system, which not only worsens physical symptoms but also contributes to cognitive impairments and mood disorders. Furthermore, individuals living in such environments often have limited access to clean, quiet, and safe spaces for rest, exacerbating sleep disorders like sleep apnea and increasing the risk of associated psychological problems.

Lastly, the stigma associated with certain rhinological conditions, particularly those that affect physical appearance or breathing, can lead to social anxiety and withdrawal. Conditions such as severe rhinitis, nasal polyps, or structural abnormalities can affect a person's self-esteem and social interactions. The fear of negative judgment or social embarrassment can discourage individuals from seeking help or participating in social activities, leading to increased feelings of isolation and depression. This social withdrawal further reinforces the negative impact on mental health, creating a feedback loop that is challenging to address.

In conclusion, the interplay between rhinological diseases and psychological problems is significantly influenced by social factors. Socioeconomic status, access to healthcare, social support, environmental conditions, and social stigma all contribute to the complexity of managing these intertwined health issues. Addressing these social determinants is crucial for developing comprehensive treatment strategies that not only focus on the medical aspects of rhinological diseases but also consider the broader social context to improve mental health outcomes and overall quality of life.

4. Diagnostic approaches

4.1 Assessing psychological symptoms in rhinology patients

Assessing psychological symptoms in rhinology patients involves a thorough understanding of both the physical and mental health aspects of the patient's condition. Rhinology patients often experience chronic discomfort, breathing difficulties, and sensory impairments such as loss of smell, which can significantly impact their psychological well-being. The assessment process should start with a comprehensive evaluation of the patient's medical history, including any pre-existing mental health conditions and current medications. Clinicians need to be vigilant for signs of depression, anxiety, cognitive dysfunction, and other psychological issues, as these can be both a consequence of and a contributor to chronic rhinological conditions [66–68].

4.2 Screening tools and questionnaires

Utilizing standardized screening tools and questionnaires is a crucial step in identifying psychiatric problems in rhinology patients. The Beck Depression Inventory (BDI) is one of the most widely used self-report measures for assessing the severity of depression. It consists of 21 multiple-choice questions that explore various aspects of depressive symptoms, such as mood, pessimism, sense of failure, self-dissatisfaction, and physical symptoms like fatigue and changes in sleep patterns. The BDI helps clinicians gauge the intensity of depression and monitor changes over time, providing valuable insights into the patient's mental health [69, 70]. The Generalized Anxiety Disorder 7 (GAD-7) scale is a brief, seven-item questionnaire specifically designed to screen for generalized anxiety disorder. Patients rate the frequency of anxiety-related symptoms over the past 2 weeks, including feelings of nervousness, inability to stop worrying, and physical symptoms such as restlessness and muscle tension. The GAD-7 is easy to administer and interpret, making it an effective tool for identifying anxiety disorders in rhinology patients [71]. The Patient Health Questionnaire-9 (PHQ-9) is a multipurpose instrument for screening, diagnosing, monitoring, and measuring the severity of depression. It consists of nine questions that reflect the diagnostic criteria for major depressive disorder in the DSM-5. Patients indicate how often they have been bothered by each symptom over the past 2 weeks. The PHQ-9 is highly effective in primary care settings and can be easily incorporated into routine rhinology assessments to identify patients in need of further mental health evaluation [72].

The Hospital Anxiety and Depression Scale (HADS) is another useful instrument for screening both anxiety and depression in rhinology patients. It consists of 14 items, seven for anxiety and seven for depression, and is designed to avoid reliance on somatic symptoms that might be confounded with physical illness. The HADS is particularly valuable in settings where patients may have overlapping physical and psychological symptoms [70].

The Sinonasal Outcome Test (SNOT-22) is specifically designed to assess the impact of sinonasal conditions on patients' quality of life. It includes questions about physical symptoms, emotional well-being, and functional limitations caused by rhinological issues. While not a direct measure of psychiatric symptoms, the SNOT-22 provides insight into how sinonasal problems affect mental health and daily functioning, allowing clinicians to tailor their interventions accordingly [72, 73].

The Brief Symptom Inventory (BSI) is a shorter version of the Symptom Checklist-90-Revised (SCL-90-R) and is used to evaluate psychological distress and psychiatric disorders. It includes 53 items and provides scores on nine primary symptom dimensions and three global indices of distress. The BSI can be useful for a broad assessment of psychological symptoms in rhinology patients, capturing a range of potential mental health issues [74].

For patients with suspected cognitive dysfunction, the Montreal Cognitive Assessment (MoCA) is a valuable tool. The MoCA assesses various cognitive domains, including attention, concentration, executive functions, memory, language, visuoconstructional skills, conceptual thinking, calculations, and orientation. This comprehensive screening tool can help identify cognitive impairments that may be contributing to or exacerbated by chronic rhinological conditions [68].

These tools allow clinicians to systematically evaluate the psychological state of the patient and provide a baseline for further investigation. Implementing these screenings in routine rhinology care can aid in early detection of mental health issues, facilitating timely intervention.

4.3 Clinical interviews

Clinical interviews remain a cornerstone of psychiatric assessment in rhinology patients. These interviews should be structured yet flexible to allow for a comprehensive understanding of the patient's psychological and physical health. During the interview, clinicians should explore the patient's mood, anxiety levels, sleep patterns, and cognitive functioning. It is also important to discuss the impact of rhinological symptoms on daily life and mental health. Open-ended questions and active listening can help patients feel comfortable sharing their experiences, which can reveal insights into their mental health status. Clinicians should be trained to recognize subtle signs of psychiatric distress and to differentiate between symptoms caused by rhinological issues and those stemming from psychological conditions [68, 75].

4.4 Integrating psychological assessment into routine care

Integrating psychological assessment into routine rhinology care is essential for holistic patient management. This integration can be achieved by developing multidisciplinary teams that include ENT specialists, psychologists, and psychiatrists. Regular training for rhinology clinicians on the importance of mental health and how to use screening tools effectively can enhance this process. Routine mental health screenings should be part of the standard care protocol for patients with chronic rhinological conditions. Creating a seamless referral pathway to mental health professionals for patients who need further evaluation and treatment is also crucial. By addressing both the physical and psychological aspects of rhinology patients' health, clinicians can improve overall treatment outcomes and patient well-being.

In conclusion, the diagnostic approach to psychiatric problems in rhinology patients involves a multi-faceted strategy that includes thorough assessment of psychological symptoms, utilization of standardized screening tools and questionnaires, comprehensive clinical interviews, and the integration of psychological assessments into routine care. By adopting these practices, clinicians can ensure a more holistic approach to patient care, addressing both the physical and mental health needs of rhinology patients.

5. Therapeutic strategies and interventions for psychiatric problems in rhinology patients

Addressing the psychiatric problems associated with rhinology diseases requires a multidisciplinary approach. Effective management integrates pharmacotherapy, surgical interventions, and psychological therapies to improve both physical and mental health outcomes. Here, we explore various therapeutic strategies and interventions tailored to rhinology patients experiencing psychological distress.

5.1 Pharmacotherapy

Pharmacotherapy plays a vital role in managing psychiatric symptoms in rhinology patients. Antidepressants, such as selective serotonin reuptake inhibitors (SSRIs) and serotonin-norepinephrine reuptake inhibitors (SNRIs), are commonly prescribed to alleviate symptoms of depression and anxiety [76]. These medications can help stabilize mood, reduce anxiety, and improve overall quality of life and it should prescribe under psychiatric. However, it is essential to prescribe these medications under psychiatric guidance, taking into account potential drug interactions. For instance, chlordiazepoxide, alprazolam zolpidem, amitriptyline, chlorpromazine, thioridazine, reserpine, and citalopram can induce rhinitis in some patients [77]. Notably, there is a lack of sufficient studies evaluating the pharmacotherapy effect on psychological symptoms in rhinology patients, highlighting the need for further research to improve this field and inform evidence-based treatment strategies.

Additionally, nasal corticosteroids and antihistamines can be beneficial in reducing the physical symptoms of rhinology diseases like AR, indirectly improving psychological well-being by alleviating discomfort and sleep issues. However, it is important to note that complete relief of rhinosinusitis symptoms may not fully alleviate the psychological aspects of these diseases. Patients with rhinitis can experience anxiety, depression, or cognitive problems even after successful treatment of physical symptoms or regardless of allergen seasonality. Therefore, ongoing therapy may be needed to address psychological symptoms even after physical symptoms have been managed [11, 78].

It is crucial to tailor pharmacotherapy to each patient's specific needs, taking into account any potential interactions between medications used for rhinological conditions and those prescribed for psychiatric symptoms. Consulting with a psychiatry specialist is important for selecting the appropriate treatment for these patients' unique conditions. This collaborative approach ensures that both the physical and mental health aspects of rhinology diseases are effectively managed.

5.2 Surgical interventions

Rhinoplasty significantly influences the psychological health of patients, as highlighted by various studies. Post-surgery, patients often report notable improvements in their social interactions and relationships. These enhancements are frequently linked to better self-perception and body image, leading to reduced body dissatisfaction and increased satisfaction with their nasal appearance. Such positive effects contribute to improved social functioning and are sustained over follow-up periods ranging from months to years [79].

The psychological benefits of rhinoplasty extend beyond physical changes, impacting several mental health areas. Research shows significant reductions in anxiety, depression, emotional distress, and body dysmorphia among rhinoplasty patients, as a study demonstrated a marked decrease in anxiety and depression levels at the 12-month follow-up, as measured by the Beck Depression Inventory [80]. Similarly, other study reported reductions in anxiety, social phobia, depression, and body dysmorphia [79]. These findings underscore the potential of rhinoplasty to alleviate psycho-social distress and enhance overall mental well-being.

Furthermore, rhinoplasty has been shown to improve patients' quality of life. Studies utilizing the Rhinoplasty Outcomes Evaluation (ROE) questionnaire, Body Image Inventory, and EuroQol 5D questionnaire have documented significant enhancements in quality of life post-surgery. These improvements are often attributed to better self-perception, enhanced body image, increased self-esteem, and greater social participation and confidence. However, some research indicates that while psychological improvements are significant, other quality of life aspects may remain unchanged. It is essential for surgeons to identify psychological risk factors during preoperative consultations and carefully select patients to prevent postoperative psychological complications. Utilizing screening tools like the Quality of

Life Scale Short Form (SF-36) questionnaire and general health questionnaire-28 (GHQ-28) can help identify at-risk patients, leading to better surgical outcomes [79, 81–83].

In patients with rhinosinusitis, surgical intervention may be useful in the early stages to prevent the development of psychological disorders. However, in the chronic and late phases of the disease, surgical procedures are generally less helpful, as neuroinflammation has a long-lasting disturbing effect on the central nervous system [4]. In cases where structural abnormalities in the nasal cavity contribute to both physical and psychological distress, surgical interventions can offer significant relief. Procedures such as septoplasty, turbinate reduction, and sinus surgery aim to correct anatomical issues that cause chronic nasal obstruction and sinusitis. By improving nasal airflow and reducing sinus infections, these surgeries can alleviate symptoms that contribute to poor sleep quality, fatigue, and subsequent psychological problems [35, 37].

Post-surgical improvements in breathing and reduction in chronic pain or discomfort can lead to significant enhancements in mental health, as patients experience fewer physical symptoms that exacerbate their psychological distress. It is important for surgeons to collaborate closely with mental health professionals to monitor and support the patient's psychological recovery post-surgery.

5.3 Psychological interventions

Psychological interventions are essential components of a comprehensive treatment plan for rhinology patients with psychiatric problems. These interventions address the mental health issues directly, helping patients develop coping strategies and resilience against the psychological impact of their physical conditions.

Cognitive-Behavioral Therapy (CBT) is a widely used and evidence-based psychological treatment for depression, anxiety, and other mental health issues. CBT helps patients identify and challenge negative thought patterns and behaviors, replacing them with more positive and constructive ones. For rhinology patients, CBT can be particularly effective in managing health-related anxiety, improving mood, and enhancing coping mechanisms for dealing with chronic symptoms and their impact on daily life [84].

5.4 Support groups and counseling

Support groups and counseling provide essential emotional and social support for rhinology patients experiencing psychological distress. Support groups offer a platform for patients to share their experiences, gain insights, and receive encouragement from others facing similar challenges. This peer support can alleviate feelings of isolation and promote a sense of community and understanding.

Individual or group counseling with a licensed therapist can provide a safe space for patients to explore their emotions, develop coping strategies, and address any underlying mental health issues. Counseling sessions can focus on specific areas of concern, such as coping with chronic illness, managing stress, and improving interpersonal relationships.

Integrating these therapeutic strategies and interventions into the care of rhinology patients requires a collaborative approach among healthcare providers. By addressing both the physical and psychological aspects of rhinology diseases, patients can achieve better health outcomes and an improved quality of life.

6. Summary

Addressing psychological issues in patients with rhinological disorders is crucial for improving their overall quality of life. Integrating psychological assessment and support into both medical and surgical management plans can lead to better patient outcomes, enhanced satisfaction, and reduced complications. This comprehensive approach acknowledges the significant interplay between physical symptoms and mental health, ultimately fostering holistic patient care.

Incorporating psychological assessments into the preoperative and postoperative phases of rhinological care ensures that patients' mental health needs are identified and addressed. Preoperative psychological evaluations can help identify patients at risk for anxiety, depression, and body dysmorphic disorder, which are common in individuals seeking rhinoplasty or other nasal surgeries. By recognizing these issues early, healthcare providers can offer appropriate interventions such as counseling, therapy, or medication, thus preparing patients mentally and emotionally for surgery. This proactive approach can mitigate postoperative dissatisfaction and the desire for unnecessary revision surgeries, thereby improving patient satisfaction and outcomes.

Integrating psychological support into the treatment plan for medical management of rhinological disorders, such as chronic rhinosinusitis or allergic rhinitis, can significantly enhance quality of life. Patients with chronic nasal conditions often experience a diminished quality of life due to persistent symptoms like nasal obstruction, facial pain, and impaired sense of smell, which can lead to anxiety and depression. Providing psychological support alongside medical treatments can help patients cope better with their symptoms, adhere more effectively to treatment regimens, and experience an overall improvement in their mental and physical well-being. This dual approach ensures that both the physical and psychological aspects of their condition are addressed, leading to a more comprehensive and effective treatment outcome.

Lastly, incorporating psychological care into the surgical management plan of rhinology patients can reduce perioperative stress and enhance recovery. Educating patients about the surgical process, setting realistic expectations, and providing postoperative psychological support can alleviate anxiety and promote a smoother recovery. Postoperatively, continued psychological support can help patients adapt to changes in their appearance and function, fostering a positive adjustment and long-term satisfaction with the surgical results.

In conclusion, addressing psychological issues in patients with rhinological disorders through integrated medical and surgical management plans can significantly improve their quality of life. This comprehensive approach not only enhances patient satisfaction and outcomes but also underscores the importance of treating the whole patient, not just their physical symptoms. By recognizing and addressing the psychological dimensions of rhinology disorders, healthcare providers can offer more effective, holistic care that leads to better overall health and well-being for their patients.

Author details

Zahra Ebrahim Soltani[1*] and Mohammad Elahi[2]

1 Experimental Medicine Research Center, Tehran University of Medical Science, Tehran, Iran

2 Center for Orthopedic Trans-disciplinary Applied Research, Tehran University of Medical Science, Tehran, Iran

*Address all correspondence to: zahra_esoltani@yahoo.com

References

[1] Schlosser RJ, Storck K, Cortese BM, Uhde TW, Rudmik L, Soler ZM. Depression in chronic rhinosinusitis: A controlled cohort study. American Journal of Rhinology & Allergy. 2016;**30**(2):128-133

[2] Lin S, Nie M, Wang B, Duan S, Huang Q, Wu N, et al. Intrinsic brain abnormalities in chronic rhinosinusitis associated with mood and cognitive function. Frontiers in Neuroscience. 2023;**17**:1131114

[3] Vogt F, Sahota J, Bidder T, Livingston R, Bellas H, Gane SB, et al. Chronic rhinosinusitis with and without nasal polyps and asthma: Omalizumab improves residual anxiety but not depression. Clinical and Translational Allergy. 2021;**11**(1):e12002

[4] Kim JY, Ko I, Kim MS, Yu MS, Cho BJ, Kim DK. Association of chronic rhinosinusitis with depression and anxiety in a nationwide insurance population. JAMA Otolaryngology Head & Neck Surgery. 2019;**145**(4):313-319

[5] Mou YK, Wang HR, Zhang WB, Zhang Y, Ren C, Song XC. Allergic rhinitis and depression: Profile and proposal. Frontiers in Psychiatry. 2021;**12**:820497

[6] Roxbury CR, Qiu M, Shargorodsky J, Woodard TD, Sindwani R, Lin SY. Association between rhinitis and depression in United States adults. The Journal of Allergy and Clinical Immunology in Practice. 2019;**7**(6):2013-2020

[7] Begemann MJH, Linszen MMJ, de Boer JN, Hovenga WD, Gangadin SS, Schutte MJL, et al. Atopy increases risk of psychotic experiences: A large population-based study. Frontiers in Psychiatry. 2019;**10**:453

[8] Bedolla-Barajas M, Morales-Romero J, Pulido-Guillén NA, Robles-Figueroa M, Plascencia-Domínguez BR. Rhinitis as an associated factor for anxiety and depression amongst adults. Brazilian Journal of Otorhinolaryngology. 2017;**83**(4):432-438

[9] Chen M-H, Lan W-H, Hsu J-W, Huang K-L, Chen Y-S, Li C-T, et al. Risk of bipolar disorder among adolescents with allergic rhinitis: A nationwide longitudinal study. Journal of Psychosomatic Research. 2015;**79**(6):533-536

[10] Simons FE. Learning impairment and allergic rhinitis. Allergy and Asthma Proceedings. 1996;**17**(4):185-189

[11] Blaiss MS, Hammerby E, Robinson S, Kennedy-Martin T, Buchs S. The burden of allergic rhinitis and allergic rhinoconjunctivitis on adolescents: A literature review. Annals of Allergy, Asthma & Immunology. 2018;**121**(1):43-52.e3

[12] Chirakalwasan N, Ruxrungtham K. The linkage of allergic rhinitis and obstructive sleep apnea. Asian Pacific Journal of Allergy and Immunology. 2014;**32**(4):276-286

[13] Liu J, Zhang X, Zhao Y, Wang Y. The association between allergic rhinitis and sleep: A systematic review and meta-analysis of observational studies. PLoS One. 2020;**15**(2):e0228533

[14] Marin C, Alobid I, Fuentes M, López-Chacón M, Mullol J. Olfactory dysfunction in mental illness. Current Allergy and Asthma Reports. 2023;**23**(3):153-164

[15] Lombion-Pouthier S, Vandel P, Nezelof S, Haffen E, Millot JL. Odor perception in patients with mood disorders. Journal of Affective Disorders. 2006;**90**(2-3):187-191

[16] Chen X, Guo W, Yu L, Luo D, Xie L, Xu J. Association between anxious symptom severity and olfactory impairment in young adults with generalized anxiety disorder: A case-control study. Neuropsychiatric Disease and Treatment. 2021;**17**:2877-2883

[17] Burón E, Bulbena A, Bulbena-Cabré A. Olfactory functioning in panic disorder. Journal of Affective Disorders. 2015;**175**:292-298

[18] Clepce M, Reich K, Gossler A, Kornhuber J, Thuerauf N. Olfactory abnormalities in anxiety disorders. Neuroscience Letters. 2012;**511**(1):43-46

[19] Negoias S, Croy I, Gerber J, Puschmann S, Petrowski K, Joraschky P, et al. Reduced olfactory bulb volume and olfactory sensitivity in patients with acute major depression. Neuroscience. 2010;**169**(1):415-421

[20] Kazour F, Richa S, Abi Char C, Surget A, Elhage W, Atanasova B. Olfactory markers for depression: Differences between bipolar and unipolar patients. PLoS One. 2020;**15**(8):e0237565

[21] Naudin M, El-Hage W, Gomes M, Gaillard P, Belzung C, Atanasova B. State and trait olfactory markers of major depression. PLoS One. 2012;7(10):e46938

[22] Zucco GM, Bollini F. Odour recognition memory and odour identification in patients with mild and severe major depressive disorders. Psychiatry Research. 2011;**190**(2-3):217-220

[23] Pabel LD, Hummel T, Weidner K, Croy I. The impact of severity, course and duration of depression on olfactory function. Journal of Affective Disorders. 2018;**238**:194-203

[24] Atanasova B, El-Hage W, Chabanet C, Gaillard P, Belzung C, Camus V. Olfactory anhedonia and negative olfactory alliesthesia in depressed patients. Psychiatry Research. 2010;**176**(2-3):190-196

[25] Chen B, Klarmann R, Israel M, Ning Y, Colle R, Hummel T. Difference of olfactory deficit in patients with acute episode of schizophrenia and major depressive episode. Schizophrenia Research. 2019;**212**:99-106

[26] Swiecicki L, Zatorski P, Bzinkowska D, Sienkiewicz-Jarosz H, Szyndler J, Scinska A. Gustatory and olfactory function in patients with unipolar and bipolar depression. Progress in Neuro-Psychopharmacology & Biological Psychiatry. 2009;**33**(5):827-834

[27] Hasegawa Y, Ma M, Sawa A, Lane AP, Kamiya A. Olfactory impairment in psychiatric disorders: Does nasal inflammation impact disease psychophysiology? Translational Psychiatry. 2022;**12**(1):314

[28] Bochicchio V, Mezzalira S, Maldonato NM, Cantone E, Scandurra C. Olfactory-related quality of life impacts psychological distress in people with COVID-19: The affective implications of olfactory dysfunctions. Journal of Affective Disorders. 2023;**323**:741-747

[29] Othman BA, Maulud SQ, Jalal PJ, Abdulkareem SM, Ahmed JQ, Dhawan M, et al. Olfactory dysfunction as a post-infectious symptom of SARS-CoV-2 infection. Annals of Medicine and Surgery (London). 2022;**75**:103352

[30] Fidan T, Fidan V, Ak M, Sütbeyaz Y. Neuropsychiatric symptoms, quality

of sleep and quality of life in patients diagnosed with nasal septal deviation. Kulak Burun Boğaz Ihtisas Dergisi. 2011;**21**(6):312-317

[31] Magliulo G, Iannella G, Ciofalo A, Polimeni A, De Vincentiis M, Pasquariello B, et al. Nasal pathologies in patients with obstructive sleep apnoea. Acta Otorhinolaryngologica Italica. 2019;**39**(4):250-256

[32] Rezaeitalab F, Moharrari F, Saberi S, Asadpour H, Rezaeetalab F. The correlation of anxiety and depression with obstructive sleep apnea syndrome. Journal of Research in Medical Sciences : The Official Journal of Isfahan University of Medical Sciences. 2014;**19**(3):205-210

[33] Su VY, Chen YT, Lin WC, Wu LA, Chang SC, Perng DW, et al. Sleep Apnea and risk of panic disorder. Annals of Family Medicine. 2015;**13**(4):325-330

[34] Yıldırım YS, Deveci E, Ozucer B, Kurt Y. Nasal obstruction in adults: How it affects psychological status? The Journal of Laryngology & Otology. 2024;**138**(2):184-187

[35] Katamanin O, Saini S, Jafferany M. Psychological implications and quality of life after cosmetic rhinoplasty: A systematic review. Discover Psychology. 2024;**4**(1):16

[36] Schoustra E, van Maanen P, den Haan C, Ravesloot MJL, de Vries N. The role of isolated nasal surgery in obstructive sleep Apnea therapy-A systematic review. Brain Sciences. 26 Oct 2022;**12**(11):1446

[37] Marcus P. Psychological aspects of cosmetic rhinoplasty. British Journal of Plastic Surgery. 1984;**37**(3):313-318

[38] Mousavi N, Molaei A, Alesaeidi S, Eftekhari Seas N, Effatpanah M. Prevalence of psychiatric disorders among patients with granulomatosis with polyangiitis (Wegener's) and the predictive role of personality traits. Clinical Practice and Epidemiology in Mental Health. 2024;**20**:e17450179276345

[39] Peeters L, Igodt P. Acute psychiatric symptomatology in Wegener's granulomatosis. Tijdschrift voor Psychiatrie. 2006;**48**(4):325-329

[40] Shrivastava A, Jain S, Damaraju V, Naidu G, Dhir V, Rathi M, et al. Severity and determinants of psychosocial comorbidities in granulomatosis with polyangiitis and their impact on quality of life. Rheumatology International. 2023;**43**(8):1467-1477

[41] Koutantji M, Pearce S, Harrold E. Psychological aspects of vasculitis. Rheumatology (Oxford, England). 2000;**39**(11):1173-1179

[42] Marin C, Alobid I, López-Chacón M, VanStrahlen CR, Mullol J. Type 2 and non-type 2 inflammation in the upper airways: Cellular and molecular alterations in olfactory neuroepithelium cell populations. Current Allergy and Asthma Reports. 2024;**24**(4):211-219

[43] Seals MR, Moran MM, Leavenworth JD, Leavenworth JW. Contribution of dysregulated B-cells and IgE antibody responses to multiple sclerosis. Frontiers in Immunology. 2022;**13**:900117

[44] Rosenwasser LJ. Mechanisms of IgE inflammation. Current Allergy and Asthma Reports. 2011;**11**(2):178-183

[45] Ratzlaff RE, Cavanaugh VJ, Miller GW, Oakes SG. Evidence of a neurogenic component during IgE-mediated inflammation in mouse skin. Journal of Neuroimmunology. 1992;**41**(1):89-96

[46] Cui H, Liu F, Fang Y, Wang T, Yuan B, Ma C. Neuronal FcεRIα directly mediates ocular itch via IgE-immune complex in a mouse model of allergic conjunctivitis. Journal of Neuroinflammation. 2022;**19**(1):55

[47] Liu F, Xu L, Chen N, Zhou M, Li C, Yang Q, et al. Neuronal Fc-epsilon receptor I contributes to antigen-evoked pruritus in a murine model of ocular allergy. Brain, Behavior, and Immunity. 2017;**61**:165-175

[48] Crosson T, Wang JC, Doyle B, Merrison H, Balood M, Parrin A, et al. FcεR1-expressing nociceptors trigger allergic airway inflammation. The Journal of Allergy and Clinical Immunology. 2021;**147**(6):2330-2342

[49] Ebrahim Soltani Z, Badripour A, Haddadi NS, Elahi M, Kazemi K, Afshari K, et al. Allergic rhinitis in BALB/c mice is associated with behavioral and hippocampus changes and neuroinflammation via the TLR4/NF-κB signaling pathway. International Immunopharmacology. 2022;**108**:108725

[50] Elahi M, Ebrahim Soltani Z, Afrooghe A, Ahmadi E, Dehpour AR. Sex dimorphism in pain threshold and neuroinflammatory response: The protective effect of female sexual hormones on behavior and seizures in an allergic rhinitis model. Journal of Neuroimmune Pharmacology. 2024;**19**(1):16

[51] Yeh CF, Huang WH, Lan MY, Hung W. Lipopolysaccharide-initiated rhinosinusitis causes neuroinflammation and olfactory dysfunction in mice. American Journal of Rhinology & Allergy. 2023;**37**(3):298-306

[52] Klein R, Soung A, Sissoko C, Nordvig A, Canoll P, Mariani M, et al. COVID-19 induces neuroinflammation and loss of hippocampal neurogenesis. Research Square. 29 Oct 2021:rs.3.rs-1031824 [Preprint]

[53] Vanderheiden A, Klein RS. Neuroinflammation and COVID-19. Current Opinion in Neurobiology. 2022;**76**:102608

[54] Bornand D, Toovey S, Jick SS, Meier CR. The risk of new onset depression in association with influenza–A population-based observational study. Brain, Behavior, and Immunity. 2016;**53**:131-137

[55] Su CY, Menuz K, Carlson JR. Olfactory perception: Receptors, cells, and circuits. Cell. 2009;**139**(1):45-59

[56] Imamura F, Ito A, LaFever BJ. Subpopulations of projection neurons in the olfactory bulb. Frontiers in Neural Circuits. 2020;**14**:561822

[57] Krusemark EA, Novak LR, Gitelman DR, Li W. When the sense of smell meets emotion: Anxiety-state-dependent olfactory processing and neural circuitry adaptation. The Journal of Neuroscience. 2013;**33**(39):15324-15332

[58] Ressler KJ. Amygdala activity, fear, and anxiety: Modulation by stress. Biological Psychiatry. 2010;**67**(12):1117-1119

[59] Milad MR, Rauch SL. The role of the orbitofrontal cortex in anxiety disorders. Annals of the New York Academy of Sciences. 2007;**1121**:546-561

[60] Nguyen AD, Shenton ME, Levitt JJ. Olfactory dysfunction in schizophrenia: A review of neuroanatomy and psychophysiological measurements. Harvard Review of Psychiatry. 2010;**18**(5):279-292

[61] Gao Z, Lv H, Wang Y, Xie Y, Guan M, Xu Y. TET2 deficiency promotes anxiety and depression-like behaviors by activating NLRP3/IL-1β pathway in microglia of allergic rhinitis mice. Molecular Medicine. 2023;**29**(1):160

[62] Rodrigues J, Rocha MI, Teixeira F, Resende B, Cardoso A, Sá SI, et al. Structural, functional and behavioral impact of allergic rhinitis on olfactory pathway and prefrontal cortex. Physiology & Behavior. 2023;**265**:114171

[63] Tian J, Kaufman DL. The GABA and GABA-receptor system in inflammation, anti-tumor immune responses, and COVID-19. Biomedicines. 18 Jan 2023;**11**(2):254

[64] Luscher B, Shen Q, Sahir N. The GABAergic deficit hypothesis of major depressive disorder. Molecular Psychiatry. 2011;**16**(4):383-406

[65] Bali V, Simmons SC, Manning CE, Doyle MA, Rodriguez M, Stark AR, et al. Characterization of proinflammatory markers in the ventral tegmental area across mouse models of chronic stress. Neuroscience. 2021;**461**:11-22

[66] Strazdins E, Nie YF, Ramli R, Palesy T, Christensen JM, Marcells GN, et al. Association of mental health status with perception of nasal function. Journal of the American Medical Association Facial Plastic Surgery. 2017;**19**(5):369-377

[67] Rodrigues J, Franco-Pego F, Sousa-Pinto B, Bousquet J, Raemdonck K, Vaz R. Anxiety and depression risk in patients with allergic rhinitis: A systematic review and meta-analysis. Rhinology. 2021;**59**(4):360-373

[68] Chang F, Hong J, Yuan F, Wu D. Association between cognition and olfaction-specific parameters in patients with chronic rhinosinusitis. European Archives of Oto-Rhino-Laryngology. 2023;**280**(7):3249-3258

[69] Upton J. Beck depression inventory (BDI). In: Gellman MD, Turner JR, editors. Encyclopedia of Behavioral Medicine. New York, NY: Springer New York; 2013. pp. 178-179

[70] Rodrigues J, Pinto JV, Alexandre PL, Sousa-Pinto B, Pereira AM, Raemdonck K, et al. Allergic rhinitis seasonality, severity, and disease control influence anxiety and depression. The Laryngoscope. 2023;**133**(6):1321-1327

[71] Xu K, Linton S, Sunavsky A, Garvey S, Botting H, Steacy LM, et al. Anxiety in adults with allergic rhinitis during the coronavirus disease 2019 pandemic: A Canadian perspective. Annals of Allergy, Asthma & Immunology. 2022;**129**(5):627-634

[72] Vandelaar LJ, Jiang ZY, Saini A, Yao WC, Luong AU, Citardi MJ. PHQ-9 and SNOT-22: Elucidating the prevalence of depression in chronic rhinosinusitis. Otolaryngology and Head and Neck Surgery. 2020;**162**(1):142-147

[73] Farhood Z, Schlosser RJ, Pearse ME, Storck KA, Nguyen SA, Soler ZM. Twenty-two-item sino-nasal outcome test in a control population: A cross-sectional study and systematic review. International Forum of Allergy & Rhinology. 2016;**6**(3):271-277

[74] Quintana GR, Ponce FP, Escudero-Pastén JI, Santibáñez-Palma JF, Nagy L, Koós M, et al. Cross-cultural validation and measurement invariance of anxiety and depression symptoms: A study of the brief symptom inventory (BSI) in 42 countries. Journal of Affective Disorders. 2024;**350**:991-1006

[75] Mou YK, Wang HR, Zhang WB, Zhang Y, Ren C, Song XC. Allergic rhinitis and depression: Profile and proposal. Frontiers in Psychiatry. 4 Jan 2022;**12**:820497

[76] Gorman JM, Kent JM. SSRIs and SNRIs: Broad spectrum of efficacy beyond major depression. The Journal of Clinical Psychiatry. 1999;**60**(Suppl. 4):33-38; discussion 9

[77] Alromaih S, Alsagaf L, Aloraini N, Alrasheed A, Alroqi A, Aloulah M, et al. Drug-induced rhinitis: Narrative review. Ear, Nose & Throat Journal. 15 Nov 2022:1455613221141214

[78] Muñoz-Cano R, Ribó P, Araujo G, Giralt E, Sanchez-Lopez J, Valero A. Severity of allergic rhinitis impacts sleep and anxiety: Results from a large Spanish cohort. Clinical and Translational Allergy. 2018;**8**:23

[79] Margraf J, Meyer AH, Lavallee KL. Well-being from the knife? Psychological effects of aesthetic surgery. Clinical Psychological Science. 2013;**1**(3):239-252

[80] Moss TP, Harris DL. Psychological change after aesthetic plastic surgery: A prospective controlled outcome study. Psychology, Health & Medicine. 2009;**14**(5):567-572

[81] Kucur C, Kuduban O, Ozturk A, Gozeler MS, Ozbay I, Deveci E, et al. Psychological evaluation of patients seeking rhinoplasty. The Eurasian Journal of Medicine. 2016;**48**(2):102-106

[82] Piromchai P, Suetrong S, Arunpongpaisal S. Psychological status in patients seeking rhinoplasty. Clinical Medicine Insights. Ear, Nose and Throat. 2011;**4**:31-35

[83] Mahmoudi S, Zadeh MR. Development and validation of the Rhinoplasty outcomes evaluation (ROE) questionnaire: An analytical study. World Journal of Plastic Surgery. 2022;**11**(2):68-74

[84] White CA. Cognitive behavioral principles in managing chronic disease. The Western Journal of Medicine. 2001;**175**(5):338-342

Chapter 5

Frontal Sinus Malignancy

Deepak Saharan, Col. Vikas Sharma, Sameer Beniwal and Mansha Poonia

Abstract

Frontal sinus malignancy, though relatively rare compared to malignancies in other anatomical regions, presents unique challenges in diagnosis, treatment and management. These tumours can arise *de novo* within the frontal sinus or extend from adjacent structures, and their proximity to vital anatomical landmarks, such as the orbit, cranial fossa and critical vascular structures, complicates surgical intervention. This chapter delves into the anatomical intricacies of the frontal sinus, the various classifications of frontal sinus cells and tumours, and the diagnostic modalities, including contemporary molecular and genetic markers that are crucial for effective treatment planning. It explores the historical evolution of surgical techniques, highlighting the shift from invasive procedures to advanced endoscopic methods that offer enhanced visualisation and precision. This chapter also describes different forms of adjuvant treatment, such as external beam radiotherapy, proton and carbon ion therapy, chemotherapy and targeted/immune therapy.

Keywords: primary frontal sinus carcinoma, sinonasal undifferentiated carcinoma, frontal sinus carcinoma T staging, IDH2 mutation, Riedel-Schenke's procedure, endoscopic craniofacial resection, intensity-modulated proton therapy, carbon ion therapy, tislelizumab

1. Introduction

Frontal sinus malignancy, though relatively rare compared to malignancies in other anatomical regions, presents unique challenges in diagnosis, treatment and management. These tumours can arise *de novo* within the frontal sinus or extend from adjacent structures, and their proximity to vital anatomical landmarks, such as the orbit, cranial fossa and critical vascular structures, complicates surgical intervention.

The frontal sinus is distinguished by its developmental timeline, not being present at birth and maturing through childhood into early adolescence. Its complex anatomy, including variations in pneumatization and drainage pathways, underscores the necessity for precise imaging and detailed anatomical knowledge before any surgical approach.

Historically, the management of frontal sinus disease, including malignancies, has evolved significantly. From the early radical approaches involving extensive resections and external access to the modern minimally invasive endoscopic techniques, each phase has contributed to the current understanding and strategies employed in treating frontal sinus tumours.

This chapter delves into the anatomical intricacies of the frontal sinus, the various classifications of frontal sinus cells and tumours, and the diagnostic modalities crucial for effective treatment planning. It also explores the historical evolution of surgical techniques, highlighting the shift from invasive procedures to advanced endoscopic methods that offer enhanced visualisation and precision.

By examining the anatomical, pathological and surgical aspects of frontal sinus tumours, this chapter aims to provide a comprehensive overview for clinicians, aiding in the accurate diagnosis and effective management of these challenging cases.

2. Anatomy

Anatomical variations in the sinuses are common and can obstruct ventilation and drainage, leading to medication-resistant osteomeatal complex disease. Preoperative CT (computed tomography) scans of the paranasal sinuses are essential for identifying these variations and their proximity to vital structures like the orbit, skull base and important blood vessels, helping prevent surgical injuries.

The frontal sinus is the only sinus not present at birth, starting its development around age 2. It reaches the orbital roof between ages 5 and 7 and attains adult size by age 12. Each frontal sinus, positioned in the frontal bone separated by a bony septum, develops independently, often resulting in asymmetrical pneumatization, with one sinus typically crossing the midline and overlapping the other.

The frontal sinus walls are known as the outer and inner tables. The outer table is robust, while the inner table is a thin bony plate separating the sinus from the cranial fossa. Venous channels, called the foramina of Breschet, are found in the inner table and can allow sinus infections to spread to the brain. These foramina also serve as sites where mucosa can invaginate into the bone, potentially leading to mucocoele if not entirely removed during surgery.

The drainage of the frontal sinus is managed through the ostium located at the posteromedial part of the sinus floor, featuring an hourglass shape. This ostium lies within the frontal beak, a projection of the maxilla. The ostium's size and patency depend on the thickness of this beak. A large agger nasi cell results in a smaller beak and a wider ostium, while a small agger nasi cell leads to a more prominent beak and a narrower ostium. Additionally, a large agger nasi cell can obstruct the drainage pathway at the frontal recess [1].

Frontal sinuses can be underdeveloped (hypoplastic) in about 4% of people or absent (aplasia) in about 5%. They may also contain air cells extending from the ethmoidal sinus, known as frontoethmoidal cells, classified as follows [2]:

Type 1 frontal cell: A single cell above the agger nasi but below the frontal ostium.

Type 2 frontal cells: Two or more cells above the agger nasi but below the frontal ostium.

Type 3 frontal cell: A single cell extending from above the agger nasi into the frontal sinus but not exceeding 50% of its vertical height.

Type 4 frontal cell: A single cell extending into the frontal sinus, exceeding 50% of its vertical height, or an isolated cell within the frontal sinus.

Another variation is the frontal bullar cell, which extends from the suprabullar region into the frontal sinus along the posterior wall of the frontal recess. Unlike the suprabullar cell, it extends into the frontal sinus. Its anterior wall relates to the frontal

Cell type	Cell name	Definition
Anterior cells (push the drainage pathway of the frontal sinus medial, posteriorly or posteromedially)	Agger nasi cell	Cell that either sits anterior to the origin of the middle turbinate or sits directly above the most anterior insertion of the middle turbinate into the lateral nasal wall.
	Supra agger cell	Anterior-lateral ethmoidal cell, located above the agger nasi cell (not pneumatizing into the frontal sinus).
	Supra agger frontal cell (SAFC)	Anterior-lateral ethmoidal cell that extends into the frontal sinus. A small SAFC will only extend into the floor of the frontal sinus, whereas a large SAFC may extend significantly into the frontal sinus and may even reach the roof of the frontal sinus.
Posterior cells (push the drainage pathway anteriorly)	Supra-bulla cell	Cell above the bulla ethmoidalis that does not enter the frontal sinus.
	Supra-bulla frontal cell	Cell that originates in the supra-bulla region and pneumatizes along the skull base into the posterior region of the frontal sinus. The skull base forms the posterior wall of the cell.
	Supraorbital ethmoid cell	An anterior ethmoid cell that pneumatizes around, anterior to or posterior to the anterior ethmoidal artery over the roof of the orbit. It often forms part of the posterior wall of an extensively pneumatized frontal sinus and may only be separated from the frontal sinus by a bony septation.

Table 1.
The International Frontal Sinus Anatomy Classification (IFAC) [3].

sinus, while the posterior wall is adjacent to the anterior cranial fossa, requiring caution during surgery to avoid damaging the anterior skull base. The frontal intersinus septal cell, a pneumatized intersinus septum, may communicate with one or both frontal sinuses or exist as an isolated air cell, potentially obstructing the frontal sinus ostium.

Recent new classification of frontoethmoidal air cells is known as the International Frontal Sinus Anatomy Classification (IFAC), which was published in 2016. It has also classified the extent of endoscopic frontal sinus surgery (**Table 1**) [3].

Lymphatics of frontal sinus drain through lymphatic channel that drains skin and nasal cavity. Therefore, spread to periparotid, perifacial and submandibular lymph nodes is common. In the absence of skin involvement, locoregional spread to lymphatics is rare (10–15%). Distant metastasis is approximately 10% at the time of presentation [4].

3. Aetiology

Predisposing factors for development of primary frontal sinus carcinoma are unclear. However, in most of the instances contiguous spread of ethmoid, nasal or maxillary carcinoma is the cause of frontal sinus involvement. Moreover, lining mucosa of paranasal sinus is continuing throughout all sinuses. Therefore, it is assumed that carcinogenic agents responsible for development of ethmoid, nasal or maxillary sinus carcinoma are also a causative factor for development of frontal sinus carcinoma [5–8]. Distant metastasis of primary from kidney, lung, breast or prostate can involve frontal or frontoethmoidal sinus but it is very rare [9].

4. Age & sex

Due to rarity of primary frontal sinus malignancies, it is difficult to estimate age and sex predilection. However, these malignancies occur in late adulthood or elderly. Males are affected 1.7 times more than females [10].

5. Symptoms & signs

Most common presentations are forehead or medial upper eyelid swelling and pain, which are seen in the majority of patients. Diplopia and proptosis are seen in one quarter to half of the patients. Physical examination should include thorough head and neck examination, with special attention to periparotid and cervical nodes (**Figure 1**). Neurological examination including cranial nerve examination should also be done.

6. Pathology

Primary frontal sinus malignancy is extremely rare [11, 12]. In most of the instances, frontal sinus is involved by contiguous spread from nasal cavity or ethmoidal malignancy. WHO's (World Health Organisation) latest pathology classification for sinonasal tumours was published in 2022 [13]. Sinonasal pathologies (benign and malignant), which can affect any of the paranasal sinuses and skull base, are divided into hamartomatous, respiratory epithelial carcinoma, mesenchymal and other tumours. Respiratory epithelial carcinoma is the most common type as far as malignant tumours are concerned and others are rare. Osteoma is a common benign tumour that frequently affects frontal sinus; however, WHO presently classifies this under osteogenic tumours (**Table 2**).

Due to very low incidence of sinonasal malignancies, it is difficult to comment upon the most common site of origin for sinonasal malignant pathologies, except for

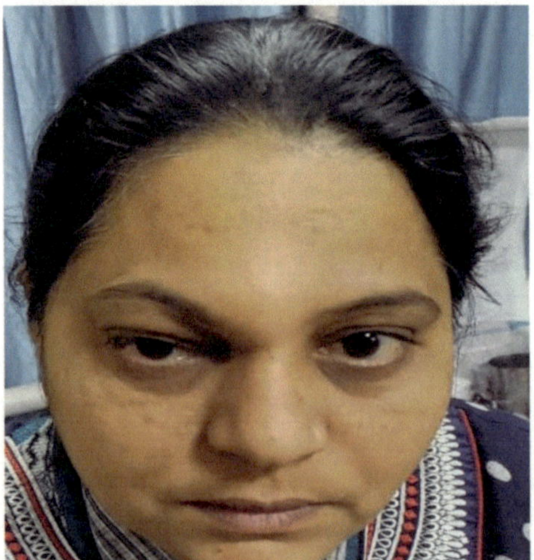

Figure 1.
Patient presented with swelling of forehead medial to right brow and medial canthus which is causing diplopia.

Diagnostic group	Category	Diagnostic entity section
Hamartomas		• Respiratory epithelial adenomatoid hamartoma • Seromucinous hamartoma • Nasal chondromesenchymal hamartoma
Respiratory epithelial lesions	Sinonasal papillomas	• Sinonasal papilloma, inverted type • Sinonasal papilloma, oncocytic type • Sinonasal papilloma, exophytic type
	Carcinomas	• Keratinizing squamous cell carcinoma • Non-keratinizing squamous cell carcinoma • NUT carcinoma • SWI/SNF complex-deficient sinonasal carcinoma • Sinonasal lymphoepithelial carcinoma • Sinonasal undifferentiated carcinoma • Teratocarcinosarcoma • HPV-related multiphenotypic sinonasal carcinoma
	Adenocarcinoma	• Intestinal-type adenocarcinoma of the sinonasal tract • Non-intestinal-type sinonasal adenocarcinoma
Mesenchymal tumours of sinonasal tract		• Sinonasal tract angiofibroma • Sinonasal glomangiopericytoma • Biphenotypic sinonasal sarcoma • Chordoma
Other tumours		• Sinonasal ameloblastoma • Adamantinomatous craniopharyngioma • Meningioma of sinonasal tract • Olfactory neuroblastoma

Abbreviations: NUT- nuclear protein in testes, SWI/SNF- SWItch/sucrose non-fermentable, HPV- human papilloma virus.

Table 2.
5th edition of the World Health Organisation (WHO) classification of tumours of the nasal cavity, paranasal sinuses, and skull base [13].

respiratory epithelial carcinoma that is most commonly originated from maxillary sinus or nasal cavity. Hence, primary involvement of frontal sinus by WHO-classified pathologies is extremely rare [10, 12, 14–16].

Bhojwani et al. conducted an analysis of all the frontal sinus malignancy cases reported between 1973 and 2012. They found that most common histologies involving frontal sinus were squamous cell carcinomas (SCCs) followed by mature B-cell NHL (non-Hodgkin's lymphoma), epithelial carcinomas NOS (not otherwise specified) and adenocarcinomas [10]. Another systematic review showed squamous cell carcinoma as the most common pathology, followed by mucosal melanoma and poorly differentiated carcinoma (neuroendocrine carcinoma or sinonasal undifferentiated carcinoma) [16].

7. Staging of frontal sinus carcinoma

The involvement of frontal sinus in primary carcinoma of maxillary, ethmoid or nasal cavity is classified as T4a according to AJCC (American Joint Committee on Cancer) 8th edition. Frontal sinus carcinoma has not been classified separately in AJCC TNM (tumour, node and metastasis) staging system. There are other classification systems that exist, such as the University of Florida classification system for sinonasal malignancies and classification for frontal inverted papillomas [15, 17].

University of Florida staging system for nasal cavity, frontal sinus and sphenoid sinus.

Stage 1- Tumours limited to the site of origin.

Stage 2- Extension to adjacent sites (orbit, paranasal sinuses, skin, nasopharynx, pterygomaxillary fossa).

Stage 3- Base of skull or pterygoid plates' erosion and/or intracranial extension.

Here, we are proposing a T classification for primary frontal sinus carcinoma. This might prove useful in clinical staging of these tumours. However, this needs validation with further studies.

Frontal sinus carcinoma T staging (our proposal)

T staging.

T1 – Limited to frontal sinus mucosa and/or inferior mucosal extension till frontal os with no bony involvement [Extenstion of tumor beyond frontal ostium will involve ethmoid sinus mucosa].

T2 - Involvement of frontal sinus mucosa with bony erosion of at least one (ant table, post table, orbital roof, interfrontal septum) or mucosal spread to anterior ethmoid air cells.

T3- Involvement of dura mater, periorbita, skin and subcutaneous tissue, nasal septum or nasal bone, posterior ethmoid air cells, lateral nasal wall, maxillary sinus, hard palate.

T4 Involvement of subdural space or brain, orbit, sphenoid sinus, nasopharynx, pterygoid plates and muscles, Internal Carotid Artery (ICA).

8. Pathological diagnosis

Histopathological diagnosis of sinonasal carcinoma is difficult, owing to rarity, overlapping features with other malignancies and a wide variety of pathological entities which affect the sinonasal region [13]. Preoperative pathology and postoperative histopathology often have discrepancies [18]. Immunohistochemistry is often required to narrow down the diagnosis (**Figure 2**).

9. Molecular diagnosis and genetic mutations

In recent years, there has been significant development in research as far as molecular diagnosis of sinonasal tumours is concerned. In the last decade, numerous studies have come up describing the role of genetic mutations in sinonasal carcinoma. With increasing molecular testing, new subtypes of tumours have been identified. These subtypes have been identified based on genetic alterations, such as specific gene mutations, chromosomal translocations, specific protein expression and infection by oncogenic virus (**Table 3**) [19–30].

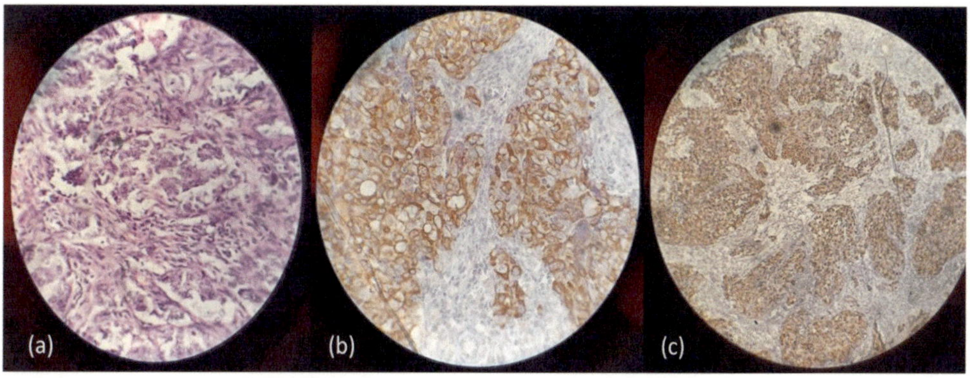

Figure 2.
(a) sheets of undifferentiated pleomorphic malignant epithelial cells on H&E 100X (b) CK7 positivity (c) CKAE1/CKAE3 positivity consistent with diagnosis of sinonasal undifferentiated carcinoma.

Sinonasal CA subtype	Biomarker	Type
Olfactory neuroblastoma	Methylation classifier	Diagnostic
Sinonasal melanoma	NRAS and KIT	Therapeutic
SNUC	IDH2 mutation	Diagnostic
NUT midline carcinoma	Translocation of gene encoding NUTM1	Diagnostic
Neuroendocrine carcinoma	Insulinoma-associated protein 1	Diagnostic
Sinonasal adenocarcinoma	KRAS	Therapeutic
Sinonasal SCC	EGFR, TrkB, TIL	Therapeutic and prognostic

Abbreviations: SNUC-sinonasal undifferentiated carcinoma, NUT- nuclear protein in testes, SCC- squamous cell carcinoma, NRAS- neuroblastoma ras viral oncogene homologue, KIT-tyrosine protein kinase, IDH-isocitrate dehydrogenase, KRAS- Kirsten rat sarcoma viral oncogene homologue, EGFR- epidermal growth factor receptor, TrkB-tropomyosin receptor kinase B, TIL-tumour-infiltrating lymphocytes, NUTM1- NUT Midline Carcinoma Family Member 1.

Table 3.
Recent biomarkers and therapeutic targets [19].

Taverna et al. have extensively studied the role of molecular markers in the causation of all types of sinonasal carcinoma and hinted upon the possibility of molecular classification for sinonasal carcinoma in the near future. Squamous cell carcinoma is the most common type followed by adenocarcinoma. Undifferentiated and poorly differentiated types of carcinoma are rare. A subset of squamous cell carcinoma associated with epidermal growth factor receptor (EGFR) alterations and rare variants with chromosomal translocations (DEK::AFF2; ETV6::NTRK and others) have been identified. The group of adenocarcinoma is very heterogeneous at molecular level; however, researchers identified potentially targetable genetic alterations with some success [22, 31]. EGFR exon 20 insertion mutation has also been studied in relation to inverted papilloma-associated squamous cell carcinoma [32]. TP53 mutation is seen in 80% of sinonasal squamous cell carcinoma and associated with poor survival [33].

Sinonasal undifferentiated carcinoma (SNUC) and poorly differentiated carcinoma are showing aggressive behaviour with poor prognosis. There have been attempts to understand tumour biology at molecular level. SNUC harbours isocitrate dehydrogenase 2 R172 gene mutation (IDH2R172) in >80% of cases. IDH2 mutated SNUC and large cell neuroendocrine carcinoma represent a phenotypic spectrum of the same entity, which is different from SMARCB1 (SWI/SNF-related, matrix-associated, actin-dependent regulator of chromatin, subfamily b, member 1)-deficient sinonasal and small cell neuroendocrine carcinoma. IDH2 mutated carcinomas are associated with better disease-free survival and lower propensity for distant pulmonary metastasis [25, 34]. However, IDH2 mutant sinonasal carcinomas are more likely to metastasize to liver and other rare organs. SMARCA4{SWI/SNF-related, matrix-associated, actin-dependent regulator of chromatin, subfamily a, member 4 (human)} deficiency was mostly associated with teratocarcinosarcoma [35].

10. Imaging

Historically, plain radiograph; the occipitofrontal (Caldwell view) for frontal sinus imaging was used. With the advent of new radiodiagnostic modalities, such as computed tomography scan (CT scan), etc., plain radiograph is no longer used.

10.1 Computed tomography (CT)

Computed tomography scan (CT scan) is a very useful technique for assessment of primary in frontal sinus malignancy patient. The CT scan should be acquired with slice thickness as less as possible (usually 0.75 to 0.6 mm with 128 slice machine).

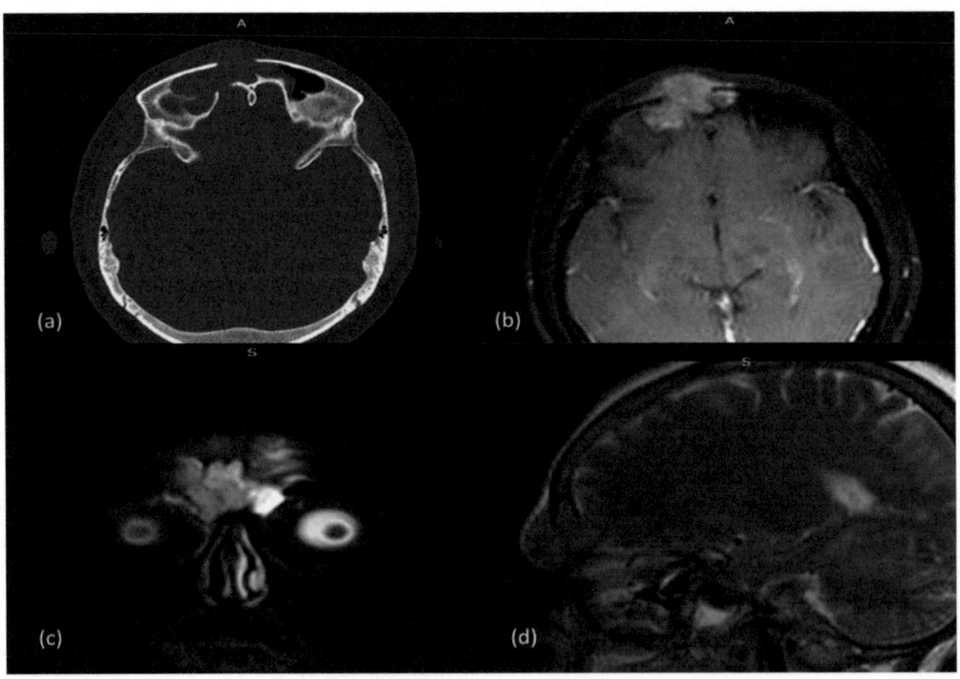

Figure 3.
(a) axial CECT bone window showing erosion of anterior and posterior table of frontal sinus (b) axial CEMRI T1 W showing enhancing tumor (c) coronal T2W MRI (d) saggital T2W MRI.

This can demonstrate involvement of the bone very well. Multiplanar reconstructed CT scan improves understanding in three dimensions [36]. It is important to consider synoptic reporting of CT scan and more important for an endoscopic surgeon to know anatomic variations of frontal sinus anatomy. Cone beam CT can also be used to study the anatomy of frontal sinus (**Figure 3a**) [37].

10.2 Magnetic resonance imaging (MRI)

Magnetic resonance imaging (MRI) and CT scan are complementary to each other in the assessment of frontal sinus malignancy cases where disease is extensive [38]. MRI demonstrates the involvement of soft tissues, such as dura or brain parenchyma more accurately. DW (diffusion-weighted) MRI can also be used to differentiate malignancy from benign lesions or inflammation (**Figure 3b–d**).

10.3 Positron emission tomography (PET)

Positron emission tomography (PET) CT is being increasingly used and few centres are using this modality routinely as baseline imaging and for preoperative assessment of distant metastasis in patients of sinonasal carcinoma [39]. ^{18}FDG (fluorodeoxyglucose) is the most commonly used tracer. The role of ^{68}Ga DOTA-TATE (dodecanetetraacetic acid-TATE or Tyr3-octreotate) has also been studied for neuroendocrine carcinoma of sinonasal carcinoma [40]. SUV (standardised uptake value) of ^{18}FDG may prove useful in treatment planning when the preoperative cytological or histological diagnosis is inconclusive [41].

11. Surgical approaches to frontal sinus

Various approaches have been described for frontal sinus surgery in the last two centuries. A brief description of all these approaches has been given in this section; however, some of them are no longer popular or not feasible in the management of frontal sinus carcinoma. Considerations in determining the surgical approach should include frontal sinus anatomy and its variations, tumour location and disease extension.

11.1 Frontal wall removal and obliteration of sinuses

11.1.1 First wave of open approaches

1800s - Schaeffer laid the groundwork for managing frontal sinusitis, sparking a wave of exploration in surgical techniques.

1884 – Ogston-Luc technique:

Anterior table trephination, nasofrontal duct dilatation, drainage tube insertion and frontal sinus mucosa treatment.

1895-Khunts technique:

Anterior table removal, manipulation of superior nasofrontal duct mucosa, nasofrontal duct stent placement and sinus obliteration using skin on the posterior table.

1898-Riedel-Schenke in 1898 proposed complete frontal sinus obliteration, which involved excising the anterior table and floor, stripping frontal sinus mucosa and utilising skin for sinus obliteration on the posterior table. Despite its radical approach, poor cosmesis led to its eventual abandonment; however, this procedure still holds importance in the surgical armamentarium of a Head & Neck Oncosurgeon (**Figure 4a–h**) [42].

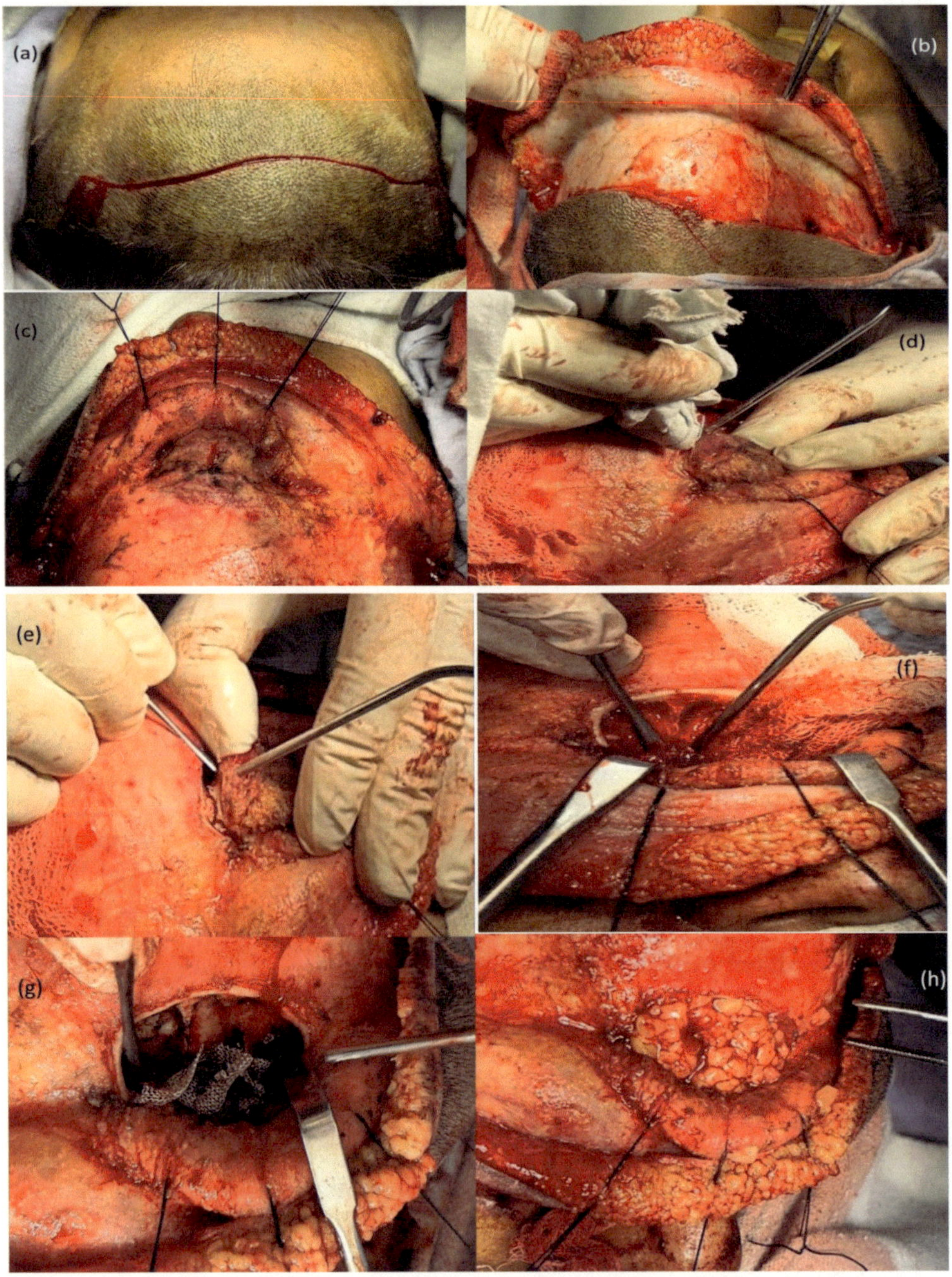

Figure 4.
Riedel's-schenke's procedure in a case of isolated frontal sinus sinonasal undifferentiated carcinoma through bicoronal approach (a)skin incision (b) subgaleal flap elevation (c) tumour delineation (d) drilling of bone with safety margin (e) removal of anterior bony table of frontal sinus with tumor (f) tumor removal from posterior wall (g) complete tumor removal done (h) fat obliteration.

11.1.2 Limitations of the procedures listed included

1. Poor cosmesis

2. Recurrent frontoethmoid outflow stenosis

11.2 Endonasal approaches

1915-Lothrop procedure.

Combined intranasal and extranasal techniques with widening of nasofrontal duct. Involved removing parts of the bilateral anterior ethmoids, the frontal sinus's anterior floor, as well as sections of the superior nasal septum and internasal septum.

11.2.1 Limitations of the procedure listed included

1. Technically demanding

2. Nasofrontal duct narrowing over time

3. Limited visibility

11.3 External approach

11.3.1 Second wave of external approaches

1921-Lynch &Howarth.

External ethmoidectomy technique. It involved a medial periorbital incision, with removal of ethmoidal air cells, excising the lamina papyracea, removal of the frontal sinus, curetting frontal sinus mucosa and placing a drainage tube for 5 days, followed by postoperative dilatation.

1958-Goodale and Montgomery.

Resurgence of osteoplastic flap technique [43].

11.3.2 Limitations of the procedure listed included

1. Restenosis

2. Cerebrospinal fluid (CSF) leaks

3. Frontal paraesthesia

4. Headaches

11.4 Endoscopic techniques

1970-Messerklinger, Kennedy & Stammberger.

Era of endoscopic sinus surgeries.

1990-Schaefer and Close.

Focused on advancing endoscopic approaches for frontal sinus disease.

1991-Draf.

Combined use of microscope and endoscope for frontoethmoidectomy.
Draf procedure is classified as follows-

1. **Draf I:** Simple drainage via anterior ethmoidectomy and opening of the naso-frontal duct

2. **Draf II:** Extended drainage through unilateral resection of the floor of the frontal sinus from the lamina papyracea to the nasal septum

   - **Draf IIa**: Unilateral approach removing a portion of the frontal sinus floor from the lamina papyracea to the middle turbinate.

   - **Draf IIb**: Unilateral approach removing a part of the frontal sinus floor from the lamina papyracea to the nasal septum.

   - **Draf IIc**: Proposed as an extension to IIb wherein frontal sinus is opened across the midline but not extended to the opposite frontal recess [44].

3. **Draf III**: Bilateral frontal sinus floor resection from the lamina papyracea of one orbit to the other and removal of the internisus septum [2].

11.4.1 Limitations of the procedure listed included

1. Limited two-dimensional (2D) visualisation

2. Steep learning curve

3. Difficult access especially in pneumatized frontal sinus

4. Increased risk of complications

5. Incomplete disease removal in case of disease lateralization

6. Postoperative complications (headaches, CSF leaks, infections)

11.5 Current practice

Endoscopic interventions have become predominant, with open procedures reserved for cases with distorted intranasal landmarks, failed multiple endoscopic surgeries, frontal sinusitis with intracranial or intraorbital complications and disease extending to the lateral boundaries. Combined approaches of external and endoscopic surgery may be employed in selected cases [45]. These developments reflect a progressive shift towards minimally invasive techniques and a greater reliance on endoscopic interventions in contemporary skull base and frontal sinus surgery [46].

This current practice of minimal invasive technique (endoscopic) may not be feasible alone in the management of malignant frontal sinus or skull base pathology.

Minimal involvement of frontal sinus in nasal or other paranasal sinus malignancy can be dealt with using endoscopic approach; however, involvement of frontal sinus with primary frontal sinus malignancy often requires external or combined approach. Riedel's procedure and osteoplastic flap technique can be employed to remove primary frontal sinus tumour *in toto*. The choice of procedure will be based on the involvement of anterior or posterior wall of the frontal sinus. Riedel's procedure can be used for tumour involving anterior wall or posterior wall or both while osteoplastic flap technique is feasible when anterior wall is not involved [42, 43]. Similarly, cranialization of the frontal sinus can be done when anterior wall is not involved [47].

11.6 Craniofacial resection

Required for skull base malignant lesions. Disease extension to frontal sinus involvement in skull base malignancy can be addressed by this approach. Conventional craniofacial resection is an open approach surgery that is not without facial scars and disfigurement eventually. Contemporary craniofacial resection includes purely endoscopic or endoscopy-assisted resection. Endoscopy has clear advantage of magnification and illumination. However, patient selection is of utmost importance [48, 49]. Subfrontal approach is an alternative when the disease necessitates the removal of superior orbit, glabella, nasal root and nasal bones.

12. Adjuvant treatment

Adjuvant treatment in the management of sinonasal tumours is often required and forms an important component of multimodal treatment plan. Given the complex three-dimensional (3D) anatomy of nose and paranasal sinuses, it is often impossible to get clear margins all around whatsoever definition of free margin taken into consideration. In most of the instances, one face of the tumour will always be close to or adherent to skull base structures, which makes the removal of tumour difficult with adequate free margin. Hence, all cases of malignant sinonasal tumours should be discussed in multidisciplinary tumour board preoperatively and postoperatively so as to make appropriate treatment plan tailored according to individual case need.

Below-described modalities are not necessarily to be used postoperatively. Cases where operative risks outweigh benefits, these modalities may be used upfront. Preoperative radiotherapy has a well-documented role in olfactory neuroblastoma and soft tissue sarcoma. Similarly, preoperative chemotherapy can be used for sinonasal undifferentiated carcinoma, neuroendocrine carcinoma, olfactory neuroblastoma and craniofacial sarcoma [50].

Various adjuvant treatment modalities have been described. These treatment modalities can be divided into conventional and newer.

Conventional

1. EBRT (External beam radiotherapy)- Radiation is delivered from a machine that aims at radiations at the diseased part. Most radiation therapy machines use photon beams. However, other source of radiation, such as protons and electrons, can also be used.

2. IMBT (Intensity-modulated brachytherapy)- Brachytherapy is a type of internal radiation therapy. This uses an applicator that delivers radiation dose. Intensity can be modulated using static or dynamic shielding.

3. Chemotherapy- It is a drug treatment that uses chemotherapeutic agents to kill tumour cells.

Newer

1. Hadron therapy (Particle therapy)- Hadron therapy is a type of radiation therapy that uses charged particles, such as proton, carbon ion, etc.., instead of X-rays (photons).

2. Targeted therapy/Immunotherapy (Monoclonal antibodies)- Targeted therapies are small molecule drugs or monoclonal antibodies. Small molecules can enter cells and attack the desired target. Monoclonal antibodies can directly stop or destroy cancer cells or they can mark cancer cells, so that they will be better seen and destroyed by the immune system.

12.1 EBRT

Postoperative radiotherapy has a well-established role in the management of sinonasal malignancies. External beam photon radiotherapy can be delivered as 2D, 3D, Intensity-modulated or volumetric arc radiotherapy. With the advent of intensity-modulated radiation therapy, it is now possible to deliver photon radiations to the tumour tissues precisely. This prevents normal tissue damage which is often a vital organ in a case of sinonasal malignancy, such as orbit, optic nerve, skull base or brain. Radiotherapy can be delivered alone or concomitantly with chemotherapy.

The role of radiotherapy has been studied for primary frontal sinus malignancies [51]. Surgery combined with postoperative radiotherapy might result in satisfactory efficacy. Slevin et al. carried out a retrospective multicentre study of outcomes for locally advanced squamous cell carcinoma that was treated with adjuvant or definitive IMRT. They found poor survival outcome and high locoregional treatment failure [52].

12.2 IMBT

Perioperative fractionated IMBT is useful for advanced or recurrent paranasal sinus SCC that is involving a skull base [53]. However, its role in frontal sinus carcinoma *per se* is not studied.

12.3 Chemotherapy

Chemotherapy plays a vital role in the management of sinonasal carcinoma. The use of multimodal approach (chemoradiotherapy or surgery plus chemoradiotherapy) has shown improved survival than surgery plus radiotherapy or radiotherapy alone [54]. Though we do not have robust data on chemotherapy for primary frontal sinus carcinoma, this seems true for frontal sinus as well and the results of studies on sinonasal carcinoma can be extrapolated. Undoubtedly, surgical resection is the preferred approach for treatment of sinonasal carcinoma like other primaries in the head and neck; however, proximity of paranasal sinuses to critical structures makes R0 resection difficult. In advanced sinonasal carcinoma, upfront surgery can lead to significant morbidity and loss of organ function. In such instances, induction or neoadjuvant chemotherapy can be very useful [55–57]. There have been few studies showing survival benefit with organ preservation in advanced sinonasal squamous cell carcinoma, which is the most common histopathology affecting paranasal sinuses [58]. Most commonly used chemotherapeutic regimens are TP (taxanes and platinum) or TP with 5 FU (fluorouracil) or ifosfamide [59]. Patients with complete response (CR) or partial response (PR) will be benefitted the most. Patients with stable disease or progressive diseases should be treated with surgery [60, 61].

12.4 Proton therapy

Proton therapy works on the principle of 'Bragg peak'. Protons preferentially deposit energy at the end of their path through the tissues. The desired depth to target malignant tissue can be achieved through computer programming. Lewis et al. selected 10 patients of sinonasal carcinoma who were treated with IMPT (intensity-modulated proton therapy). Dosimetrist generated VMAT (volumetric-modulated arc therapy) plans of all the patients for comparison. They concluded that VMAT had more homogeneous coverage of target volume and spares OAR (organ at risk) better than IMPT which were close to target volume; however, relative risk of secondary malignancy with VMAT compared to IMPT is 3.35. IMPT better spares nonadjacent OAR [62, 63]. Treatment of frontal sinus carcinoma with postoperative proton therapy has been reported [64]. Further studies are required with large number of patients to establish the role of proton therapy in the management of primary frontal sinus carcinoma.

12.5 Carbon ion therapy

The last decade has seen significant advancement in particle radiation treatment, including carbon ion radiation therapy. A recent meta-analysis done by Zhang and colleagues has shown significant benefit of using carbon ion radiation therapy in sinonasal malignancies [65]. The results of 2282 patients of sinonasal malignancies with different histologies from 44 observational studies were analysed. Patients were divided into three groups- 911 in carbon ion radiation therapy group, 599 in proton radiation therapy group and 772 in intensity-modulated photon radiation therapy group. Out of 11 frontal sinus malignancy patients, 9 patients received carbon ion therapy. Considering all subsites together, the overall survival was higher after carbon ion radiation therapy (75.1%) than proton radiation therapy (66.2%) and intensity-modulated photon radiotherapy (63.8%). Similarly, the local control rate was found to be significantly higher after carbon ion radiation therapy. Higher linear energy transfer (LET) and greater relative biological effectiveness (RBE) of carbon ion radiotherapy than proton and photon radiotherapy make carbon ion radiation therapy more useful for relatively radioresistant sinonasal histologies, such as adenoid cystic carcinoma, mucosal melanoma and sarcoma.

12.6 Targeted therapy/immunotherapy

In the era of targeted therapy, efforts are being made to identify specific molecular markers or genetic alterations or immune cells that can be targeted with novel biological agents [66]. This can pave the way for clinicians to provide a longer recurrence-free period and better survival to patients of sinonasal carcinoma. Discovery of different mutations such as IDH2 in sinonasal carcinoma may provide an opportunity for use of targeted therapy. Tislelizumab is a monoclonal antibody and found to be useful in a case of SMARCB1/INI-1-deficient sinonasal carcinoma [65].

Immunotherapy could have a potential role in the management of sinonasal undifferentiated carcinoma and NUT carcinoma, owing to the association of these tumours with immune system-related functional pathways [67–69]. Anti-PD-1 (programmed death ligand-1)-targeted antibodies, such as nivolumab and pembrolizumab, have been found to be useful in sinonasal squamous cell carcinoma, intestinal-type adenocarcinoma and sinonasal undifferentiated carcinoma [70–72]. No study is available on the treatment of isolated frontal sinus malignancy with targeted agents.

13. Conclusion

Frontal sinus is a very uncommon site to be affected by malignant neoplasm of both primary origin and distant secondary. Therefore, there are no robust data available regarding management and survival of frontal sinus malignancies. The management of frontal sinus malignancy poses a specific challenge for Head & Neck Oncosurgeon due to variable and complex anatomy, wide variety of pathologies affecting the frontal sinus and intimate relation with the skull base. Frontal sinus malignancy treatments require a multidisciplinary approach. The advent of endoscopic surgical techniques, newer radiation delivery technologies, particle treatment and targeted therapy has improved survival.

Conflict of interest

The authors declare no conflict of interest.

Author details

Deepak Saharan[1*], Col. Vikas Sharma[2], Sameer Beniwal[3] and Mansha Poonia[1]

1 Government Medical College, Kota, Rajasthan, India

2 Command Hospital, Kolkata, India

3 Government Medical College, Jhalawar, Rajasthan, India

*Address all correspondence to: drdeepu.ytl@gmail.com

References

[1] Wormald PJ. The agger nasi cell: The key to understanding the anatomy of the frontal recess. Otolaryngology and Head and Neck Surgery. 2003;**129**(5):497-507. DOI: 10.1016/S0194-59980301581-X

[2] Bent JP, Cuilty-Siller C, Kuhn FA. The frontal cell as a cause of frontal sinus obstruction. American Journal of Rhinology. 1994;**8**(04):185-191

[3] Wormald PJ, Hoseman W, Callejas C, Weber RK, Kennedy DW, Citardi MJ, et al. The international frontal sinus anatomy classification (IFAC) and classification of the extent of endoscopic frontal sinus surgery (EFSS). International Forum of Allergy & Rhinology. 2016;**6**(7):677-696. DOI: 10.1002/alr.21738. Epub 2016 Mar 14

[4] Osguthorpe JD, Richardson M. Frontal sinus malignancies. Otolaryngologic Clinics of North America. 2001;**34**(1):269-281. DOI: 10.1016/s0030-6665(05)70311-x

[5] Roush GC. Epidemiology of cancer of the nose and paranasal sinuses. Current concepts. Head and Neck Surgery. 1979;**2**:3-11

[6] Holmstrom M, Lund V. Malignant melanoma of the nasal cavity following occupational exposure to formaldehyde. British Journal of Industrial Medicine. 1991;**48**:9-11

[7] Lund VJ, Howard DJ, Battershill J. Olfactory neuroblastoma and occupation. In: Columbus F, editor. Focus on Neuroblastoma Research. Hauppauge, NY: Nova Science Publishers; 2007

[8] Cantor JO, Adelglass J, Cerreta JM, Kancherla PL, Samara M. Collision tumor of the frontal sinus: Evidence of prior intrasinus instillation of thorotrast. The Laryngoscope. 1981;**91**(5):798-803. DOI: 10.1288/00005537-198105000-00013

[9] Jiménez Oliver V, Lazarich Valdés A, Dávila Morillo A, Ruiz del Portal JM, Conde Jiménez M, Fernández Ruiz E, et al. Metástasis fronto-etmoidal de adenocarcinoma prostático. Presentación de un caso clínico y revisión de la literatura [Frontal ethmoid metastases of prostatic carcinoma. Report of one case and review of the literature]. Acta Otorrinolaringológica Española. 2001;**52**(2):151-154. Spanish. DOI: 10.1016/s0001-6519(01)78191-5

[10] Bhojwani A, Unsal A, Dubal PM, Echanique KA, Baredes S, Liu JK, et al. Frontal sinus malignancies: A population-based analysis of incidence and survival. Otolaryngology and Head and Neck Surgery. 2016;**154**(4):735-741. DOI: 10.1177/0194599815621878. Epub 2015 Dec 23

[11] Brownson RJ, Ogura JH. Primary carcinoma of the frontal sinus. The Laryngoscope. 1971;**81**(1):71-89. DOI: 10.1288/00005537-197101000-00007

[12] Jain R, Saharan D, Sukheeja D, Parashar G, Gill N, Meena V. Isolated frontal sinus undifferentiated carcinoma: First case report in the literature. Indian Journal of Otolaryngology and Head & Neck Surgery. 2024;**76**(1):1106-1109. DOI: 10.1007/s12070-023-04142-0. Epub 2023 Aug 26

[13] Thompson LDR, Bishop JA. Update from the 5th edition of the World Health Organization classification of head and neck tumors: Nasal cavity, paranasal sinuses and skull base. Head and Neck Pathology. 2022;**16**(1):1-18. DOI: 10.1007/s12105-021-01406-5. Epub 2022 Mar 21

[14] Athre R, Ducic Y. Frontal sinus hamartomas. American Journal of Otolaryngology. 2005;**26**(6):419-421. DOI: 10.1016/j.amjoto.2005.05.007

[15] Sham CL, van Hasselt CA, Chow SMW, Lee DLY, Cho RHW, Woo JKS, et al. Frontal inverted papillomas: A 25-year study. The Laryngoscope. 2020;**130**(7):1622-1628. DOI: 10.1002/lary.28245. Epub 2019 Aug 16

[16] Sileo G, Valentini M, Gravante G, Monti G, Arosio AD, Bignami M, et al. Sinonasal malignancies involving the frontal sinus: A mono-institutional experience of 84 cases and systematic literature review. Journal of Clinical Medicine. 2023;**12**(9):3186. DOI: 10.3390/jcm12093186

[17] Katz TS, Mendenhall WM, Morris CG, Amdur RJ, Hinerman RW, Villaret DB. Malignant tumors of the nasal cavity and paranasal sinuses. Head & Neck. 2002;**24**(9):821-829. DOI: 10.1002/hed.10143

[18] Ganti A, Tajudeen BA, Plitt MA, Rossi I, Gattuso P, Batra PS. Discordance in preoperative and postoperative histopathology of sinonasal tumors. American Journal of Rhinology & Allergy. 2018;**32**(2):101-105. DOI: 10.1177/1945892418762845

[19] Lechner M, Liu J, Lund VJ. Novel biomarkers in sinonasal cancers: From bench to bedside. Current Oncology Reports. 2020;**22**(10):106. DOI: 10.1007/s11912-020-00947-2

[20] Capper D, Engel NW, Stichel D, et al. DNA methylation-based reclassification of olfactory neuroblastoma. Acta Neuropathologica. 2018;**136**:255-271

[21] Chraybi M, Alsamad IA, Copie-Bergman C, Baia M, André J, Dumaz N, et al. Oncogene abnormalities in a series of primary melanomas of the sinonasal tract: NRAS mutations and cyclin D1 amplification are more frequent than KIT or BRAF mutations. Human Pathology. 2013;**44**:1902-1911

[22] Turri-Zanoni M, Medicina D, Lombardi D, Ungari M, Balzarini P, Rossini C, et al. Sinonasal mucosal melanoma: Molecular profile and therapeutic implications from a series of 32 cases. Head & Neck. 2013;**35**:1066-1077

[23] Wroblewska JP, Mull J, Wu CL, FujimotoM OT, Marszalek A, et al. SF3B1, NRAS, KIT, and BRAF mutation; CD117 and cMYC expression; and tumoral pigmentation in sinonasal melanomas: An analysis with newly found molecular alterations and some population-based molecular differences. The American Journal of Surgical Pathology. 2019;**43**:168-177

[24] Jo VY, Chau NG, Hornick JL, Krane JF, Sholl LM. Recurrent IDH2 R172X mutations in sinonasal undifferentiated carcinoma. Modern Pathology. 2017;**30**:650-659

[25] Dogan S, Chute DJ, Xu B, Ptashkin RN, Chandramohan R, Casanova-Murphy J, et al. Frequent IDH2 R172 mutations in undifferentiated and poorly-differentiated sinonasal carcinomas. The Journal of Pathology. 2017;**242**:400-408

[26] Takahashi Y, Gleber-Netto FO, Bell D, et al. Identification of markers predictive for response to induction chemotherapy in patients with sinonasal undifferentiated carcinoma. Oral Oncology. 2019;**97**:56-61

[27] Rooper LM, Bishop JA, Westra WH. INSM1 is a sensitive and specific marker of neuroendocrine differentiation in head and neck tumors. The American

Journal of Surgical Pathology. 2018;**42**:665-671

[28] Udager AM, Rolland DCM, McHugh JB, et al. High-frequency targetable EGFR mutations in sinonasal squamous cell carcinomas arising from inverted sinonasal papilloma. Cancer Research. 2015;**75**:2600-2606

[29] Li L, Zhu L. Expression and clinical significance of TrkB in sinonasal squamous cell carcinoma: A pilot study. International Journal of Oral and Maxillofacial Surgery. 2017;**46**:144-150

[30] Quan H, Yan L, Wang S, Wang S. Clinical relevance and significance of programmed death-ligand 1 expression, tumor-infiltrating lymphocytes, and p16 status in sinonasal squamous cell carcinoma. Cancer Management and Research. 2019;**11**:4335-4345

[31] Taverna C, Agaimy A, Franchi A. Towards a molecular classification of sinonasal carcinomas: Clinical implications and opportunities. Cancers (Basel). 2022;**14**(6):1463. DOI: 10.3390/cancers14061463

[32] Pacini L, Cabal VN, Hermsen MA, Huang PH. EGFR exon 20 insertion mutations in sinonasal squamous cell carcinoma. Cancers (Basel). 2022;**14**(2):394. DOI: 10.3390/cancers14020394

[33] Doescher J, Piontek G, Wirth M, Bettstetter M, Schlegel J, Haller B, et al. Epstein-Barr virus infection is strictly associated with the metastatic spread of sinonasal squamous-cell carcinomas. Oral Oncology. 2015;**51**(10):929-934. DOI: 10.1016/j.oraloncology.2015.07.008. Epub 2015 Aug 10

[34] Bell D, Hanna EY, Agaimy A, Weissferdt A. Reappraisal of sinonasal undifferentiated carcinoma: SMARCB1 (INI1)-deficient sinonasal carcinoma: A single-institution experience. Virchows Archiv. 2015;**467**(6):649-656. DOI: 10.1007/s00428-015-1853-1. Epub 2015 Sep 25

[35] Vuong HG, Le T, Le TTB, Le HT, El-Rassi ET, McKinney KA, et al. Clinicopathological features and prognostic outcomes of molecularly defined entities in the new edition of the WHO classification of sinonasal carcinoma. Frontiers in Oncology. 2023;**13**:1117865. DOI: 10.3389/fonc.2023.1117865

[36] Kew J, Rees GL, Close D, Sdralis T, Sebben RA, Wormald PJ. Multiplanar reconstructed computed tomography images improves depiction and understanding of the anatomy of the frontal sinus and recess. American Journal of Rhinology. 2002;**16**(2):119-123

[37] Štoković N, Trkulja V, Čuković-Bagić I, Lauc T, Grgurević L. Anatomical variations of the frontal sinus and its relationship with the orbital cavity. Clinical Anatomy. 2018;**31**(4):576-582. DOI: 10.1002/ca.22999. Epub 2017 Oct 30

[38] Ferrari M, Bossi P, Mattavelli D, Ardighieri L, Nicolai P. Management of sinonasal adenocarcinomas with anterior skull base extension. Journal of Neuro-Oncology. 2020;**150**(3):405-417. DOI: 10.1007/s11060-019-03385-8. Epub 2020 Jan 3

[39] Dondi F, Pasinetti N, Guerini A, Piazza C, Mattavelli D, Bossi P, et al. Prognostic role of baseline ^{18}F-FDG pet/CT in squamous cell carcinoma of the paranasal sinuses. Head & Neck. 2022;**44**(11):2395-2406. DOI: 10.1002/hed.27145. Epub 2022 Jul 12

[40] Liu KY, Goldrich DY, Ninan SJ, Filimonov A, Lam H, Govindaraj S, et al. The value of ^{68}Gallium-DOTATATE

PET/CT in sinonasal neuroendocrine tumor management: A case series. Head & Neck. 2021;**43**(6):E30-E40. DOI: 10.1002/hed.26695. Epub 2021 Mar 31

[41] Elkhatib AH, Soldatova L, Carrau RL, Hachem RA, Ditzel L, Campbell R, et al. Role of ^{18}F-FDG PET/CT differentiating olfactory neuroblastoma from sinonasal undifferentiated carcinoma. The Laryngoscope. 2017;**127**(2):321-324. DOI: 10.1002/lary.26194. Epub 2016 Aug 2

[42] Raghavan U, Jones NS. The place of Riedel's procedure in contemporary sinus surgery. The Journal of Laryngology and Otology. 2004;**118**(9):700-705. DOI: 10.1258/0022215042244831

[43] Weber R, Draf W, Keerl R, Kahle G, Schinzel S, Thomann S, et al. Osteoplastic frontal sinus surgery with fat obliteration: Technique and long-term results using magnetic resonance imaging in 82 operations. The Laryngoscope. 2000;**110**(6):1037-1044. DOI: 10.1097/00005537-200006000-00028

[44] Al Komser MK, Goldberg AN. Unilateral transnasal endoscopic approach to frontal sinuses: Draf IIc. Allergy Rhinology (Providence). 2013;**4**(2):e82-e87. DOI: 10.2500/ar.2013.4.0058

[45] Ketcham AS, Wilkins RH, Van Buren JM. A combined intracranial facial approach to the paranasal sinuses. American Journal of Surgery. 1963;**106**:698-703

[46] Lund VJ, Stammberger H, Nicolai P, Castelnuovo P, Beal T, Beham A, et al. European position paper on endoscopic management of tumours of the nose, paranasal sinuses and skull base. Rhinology. Supplement. 2010;**22**:1-143

[47] Horowitz G, Amit M, Ben-Ari O, Gil Z, Abergel A, Margalit N, et al. Cranialization of the frontal sinus for secondary mucocele prevention following open surgery for benign frontal lesions. PLoS One. 2013;**8**(12):e83820. DOI: 10.1371/journal.pone.0083820

[48] Hanna E, DeMonte F, Ibrahim S, Roberts D, Levine N, Kupferman M. Endoscopic resection of sinonasal cancers with and without craniotomy: Oncologic results. Archives of Otolaryngology – Head & Neck Surgery. 2009;**135**(12):1219-1224. DOI: 10.1001/archoto.2009.173

[49] Kim BJ, Kim DW, Kim SW, Han DH, Kim DY, Rhee CS, et al. Endoscopic versus traditional craniofacial resection for patients with sinonasal tumors involving the anterior skull base. Clinical and Experimental Otorhinolaryngology. 2008;**1**(3):148-153. DOI: 10.3342/ceo.2008.1.3.148. Epub 2008 Sep 30

[50] König M, Osnes T, Bruland Ø, Sundby Hall K, Bratland Å, Meling TR. The role of adjuvant treatment in craniofacial malignancy: A critical review. Frontiers in Oncology. 2020;**7**(10):1402. DOI: 10.3389/fonc.2020.01402

[51] Wang ZK, Yi JL, Zhang Y, Wu RY, Zhang JH, Wang JB, et al. Clinical characteristics and long-term outcomes following surgery combined with adjuvant radiotherapy for patients with frontal sinus malignancies. Journal of Neurological Surgery Part B: Skull Base. 2022;**84**(6):609-615. DOI: 10.1055/a-1924-9319

[52] Slevin F, Pan S, Mistry H, Denholm M, Shor D, Oong Z, et al. A multicentre UK study of outcomes for locally advanced sinonasal squamous cell carcinoma treated with adjuvant or definitive intensity-modulated

radiotherapy. Clinical Oncology (Royal College of Radiologists). 2021;**33**(10):e450-e461. DOI: 10.1016/j.clon.2021.05.012. Epub 2021 Jun 17

[53] Strege RJ, Kovács G, Maune S, Holland D, Niehoff P, Eichmann T, et al. Feasibility of combined operation and perioperative intensity-modulated brachytherapy of advanced/recurrent malignancies involving the skull base. Strahlentherapie und Onkologie. 2005;**181**(2):97-107. DOI: 10.1007/s00066-005-1274-1

[54] Kuo P, Manes RP, Schwam ZG, Judson BL. Survival outcomes for combined modality therapy for sinonasal undifferentiated carcinoma. Otolaryngology and Head and Neck Surgery. 2017;**156**(1):132-136. DOI: 10.1177/0194599816670146. Epub 2016 Oct 5

[55] Lilly G, Geltzeiler M. Induction chemotherapy for orbit preservation in sinonasal squamous cell carcinoma. Journal of Neurological Surgery Reports. 2021;**82**(3):e36-e37. DOI: 10.1055/s-0041-1736160

[56] Licitra L, Resteghini C, Bossi P. The evolving role of systemic therapy in the primary treatment of sinonasal cancer. Advances in Oto-Rhino-Laryngology. 2020;**84**:78-86. DOI: 10.1159/000457927. Epub 2020 Jul 30

[57] Melder KL, Geltzeiler M. Induction chemotherapy for locoregionally advanced sinonasal squamous cell carcinoma and sinonasal undifferentiated carcinoma: A comprehensive review. Cancers (Basel). 2023;**15**(15):3798. DOI: 10.3390/cancers15153798

[58] Abdelmeguid AS, Teeramatwanich W, Roberts DB, Amit M, Ferraroto R, Glisson BS, et al. Neoadjuvant chemotherapy for locoregionally advanced squamous cell carcinoma of the paranasal sinuses. Cancer. 2021;**127**(11):1788-1795. DOI: 10.1002/cncr.33452. Epub 2021 Feb 10

[59] Bossi P, Saba NF, Vermorken JB, Strojan P, Pala L, de Bree R, et al. The role of systemic therapy in the management of sinonasal cancer: A critical review. Cancer Treatment Reviews. 2015;**41**(10):836-843. DOI: 10.1016/j.ctrv.2015.07.004. Epub 2015 Jul 31

[60] Papazian MR, Gordon AJ, Chow M, Patel A, Pacione D, Lieberman S, et al. Sinonasal undifferentiated carcinoma with failed response to induction chemotherapy. Journal of Neurological Surgery Repors. 2022;**83**(3):e83-e86. DOI: 10.1055/a-1871-3586

[61] Amit M, Abdelmeguid AS, Watcherporn T, Takahashi H, Tam S, Bell D, et al. Induction chemotherapy response as a guide for treatment optimization in sinonasal undifferentiated carcinoma. Journal of Clinical Oncology. 2019;**37**(6): 504-512. DOI: 10.1200/JCO.18.00353. Epub 2019 Jan 7

[62] Lewis L, Kreinbrink P, Richardson M, Westerfield M, Doberstein M, Zhang Y, et al. Intensity modulated proton therapy better spares non-adjacent organs and reduces the risk of secondary malignant neoplasms in the treatment of sinonasal cancers. Medical Dosimetry. 2022;**47**(2):117-122. DOI: 10.1016/j.meddos.2021.11.002. Epub 2021 Dec 21

[63] Russo AL, Adams JA, Weyman EA, Busse PM, Goldberg SI, Varvares M, et al. Long-term outcomes after proton beam therapy for sinonasal squamous cell carcinoma. International Journal of Radiation Oncology, Biology, Physics.

2016;**95**(1):368-376. DOI: 10.1016/j.ijrobp.2016.02.042. Epub 2016 Feb 18

[64] Tirelli G, Capriotti V, Sartori G, Tofanelli M, Marcuzzo AV. Primary adenoid cystic carcinoma of the frontal sinus: Case description of a previously unreported entity and literature review. Ear, Nose, & Throat Journal. 2019;**98**(4):E8-E12. DOI: 10.1177/0145561319837881. Epub 2019 Mar 28

[65] Zhang W, Hu W, Hu J, Gao J, Yang J, Kong L, et al. Carbon ion radiation therapy for sinonasal malignancies: Promising results from 2282 cases from the real world. Cancer Science. 2020;**111**(12):4465-4479. DOI: 10.1111/cas.14650. Epub 2020 Oct 16

[66] Lehtinen N, Suhonen J, Rice K, Välimäki E, Toriseva M, Routila J, et al. Assessment of targeted therapy opportunities in sinonasal cancers using patient-derived functional tumor models. Translational Oncology. 2024;**44**:101935. DOI: 10.1016/j.tranon.2024.101935. Epub 2024 Mar 23

[67] De Cecco L, Serafini MS, Facco C, Granata R, Orlandi E, Fallai C, et al. A functional gene expression analysis in epithelial sinonasal cancer: Biology and clinical relevance behind three histological subtypes. Oral Oncology. 2019;**90**:94-101. DOI: 10.1016/j.oraloncology.2019.02.003. Epub 2019 Feb 11

[68] Stelow EB, Bellizzi AM, Taneja K, Mills SE, Legallo RD, Kutok JL, et al. NUT rearrangement in undifferentiated carcinomas of the upper aerodigestive tract. The American Journal of Surgical Pathology. 2008;**32**:828-834

[69] Evans AG, French CA, Cameron MJ, Fletcher CD, Jackman DM, Lathan CS, et al. Pathologic characteristics of NUT midline carcinoma arising in the mediastinum. The American Journal of Surgical Pathology. 2012;**36**:1222-1227

[70] Xu CC, Dziegielewski PT, McGaw WT, et al. Sinonasal undifferentiated carcinoma (SNUC): The Alberta experience and literature review. Journal of Otolaryngology - Head & Neck Surgery. 2013;**42**:2

[71] García-Marín R, Reda S, Riobello C, et al. CD8+ tumour infiltrating lymphocytes and tumour microenvironment immune types as biomarkers for immunotherapy in sinonasal intestinal-type adenocarcinoma. Vaccine. 2020;**8**:202

[72] Denaro N, Merlano M, Numico G, Garrone O, Bossi P. Complete response to immunotherapy in sinonasal undifferentiated carcinoma. Tumori. 2021 Dec;**107**(6):NP101-NP104. DOI: 10.1177/03008916211026971. Epub 2021 Jul 6

Section 3

Surgical Rhinology

Chapter 6

Extended Sinus Surgery: Indications, Anatomy, and Step-by-Step Techniques

Ghassan Alokby

Abstract

This chapter provides a comprehensive overview of advancements and methodologies in extended sinus surgeries. It elucidates surgical techniques, key anatomical landmarks, indications, and postoperative care necessary for effectively managing refractory chronic rhinosinusitis and other complex sinonasal conditions. A detailed examination of various approaches, such as Endoscopic Maxillary Mega-Antrostomy (EMMA), Modified Medial Maxillectomy, Denker's approach, and Draf IIb and III procedures, is discussed, highlighting the anatomical considerations, surgical steps, and postoperative protocols to optimize patient outcomes and minimize complications. Extended sinus surgeries offer enhanced access to difficult-to-reach areas, allow better management of extensive disease processes, and provide improved pathways for drainage and disease resolution, utilizing advanced endoscopic techniques that preserve critical structures and promote faster recovery compared to traditional methods. The evolution of endoscopic sinus surgery has significantly transformed the management of sinus diseases, making extended sinus surgeries crucial for patients who have failed conventional treatments and presenting a viable option for addressing severe and complex pathologies. By understanding the detailed surgical techniques and postoperative care outlined in this chapter, surgeons can achieve better outcomes and improve the quality of life for their patients.

Keywords: extended sinus surgery, chronic rhinosinusitis, endoscopic sinus surgery, maxillary sinus, frontal sinus, sphenoid sinus, postoperative care

1. Introduction

The management of chronic sinus disease has evolved significantly over the last decade. With the advent of endoscopic sinus surgery (ESS) and continuous advancements in understanding sinonasal anatomy, technology, and surgical techniques, the field of rhinology and skull base surgery has undergone transformative changes. Modern nasal endoscopes, with various angles, high-definition cameras, and navigation systems, have paved the way for these advancements, enabling more precise and effective interventions. These technological improvements have allowed surgeons to perform intricate procedures with better visualization and control, significantly reducing the risks associated with traditional open surgeries [1, 2].

Sinus surgery today is not only aimed at managing chronic sinus disease but also at addressing recalcitrant sinus disease and providing corridors to access various areas of the skull base. To fulfill these objectives, the limited approaches used in standard sinus surgery have been expanded, allowing for more comprehensive management of complex sinus conditions and improving surgical outcomes. Extended sinus surgeries have become a crucial aspect of modern rhinology, offering solutions for conditions that were previously deemed inoperable or required invasive open surgical techniques [3].

Extended sinus surgeries are indicated for several reasons, including the need to create gravity-dependent drainage pathways, as seen in cystic fibrosis patients. It is also indicated to ensure safe entry into extensively diseased sinuses where the natural ostium is obstructed by remodeling, often requiring access through the contralateral side. These procedures are crucial for patients who have failed appropriate medical therapy and traditional endoscopic sinus surgery. By expanding the surgical approach, surgeons can address more extensive disease processes, reduce the risk of recurrence, and improve overall patient outcomes [4].

In this chapter, we will delve into the various extended approaches in sinus surgery, focusing on their indications, key anatomical landmarks, and surgical techniques. By understanding these advanced procedures, we aim to provide a comprehensive guide for the effective management of refractory chronic rhinosinusitis and other complex sinonasal conditions. This knowledge will equip surgeons with the necessary skills to handle challenging cases and enhance the quality of care provided to patients.

2. Extended maxillary sinus procedures

A number of different procedures have been described to access the maxillary sinus, varying in their indications and conservativeness. Generally, extended sinus surgeries are indicated for the need for gravity-dependent drainage, as in cases of ciliary dyskinesia or thick secretions in cystic fibrosis. Other indications include the management of benign or malignant tumors, inverted papilloma, odontogenic sinusitis, mucocele, or fungus ball. These procedures are also used in cases of recalcitrant sinus disease or extensive osteogenesis [4].

3. Key anatomical landmarks

Identifying and understanding the key anatomical landmarks is crucial for the success of extended maxillary sinus procedures. These landmarks provide reference points that guide the surgeon during the procedure, ensuring precision and minimizing the risk of complications.

- *Inferior meatus wall:* Forms the lower boundary of the nasal cavity, crucial for procedures like inferior meatal antrostomy.

- *Medial floor of the orbit:* Serves as the superior boundary during maxillary sinus procedures. Careful identification and preservation of this structure are necessary to prevent injury to the orbital contents, which can result in serious complications such as vision loss or diplopia.

- *Nasolacrimal duct:* Located lateral to the lacrimal bone and the frontal process of the maxillary bone, it must be identified during maxillary sinus procedures to prevent postoperative complications like nasolacrimal duct injury, which can lead to chronic tearing and recurrent dacryocystitis.

- *Pyriform aperture:* The bony opening of the nasal cavity, serving as a landmark for accessing the anterior maxillary wall during procedures like Denker's approach. Understanding the anatomy of the pyriform aperture helps in the precise removal of bone to access the anterior and lateral walls of the maxillary sinus.

- *Hasner's valve:* Located at the distal end of the nasolacrimal duct, it must be preserved during inferior meatal antrostomy. Preserving this valve is crucial to maintain the normal function of the nasolacrimal duct and prevent postoperative obstruction.

4. Extended approaches to the maxillary sinus

4.1 Endoscopic maxillary mega-antrostomy (EMMA)

Endoscopic Maxillary Mega-Antrostomy (EMMA) is appropriate for patients with chronic recalcitrant maxillary sinusitis that has not responded to standard therapies and for those who benefit from gravity-dependent drainage. In our experience, a 30-degree nasal endoscope is used for this procedure, as it allows the surgeon to have better visualization of the operating field. The process begins with a partial inferior turbinectomy, where the anterior one-third of the inferior turbinate is preserved. This initial step is essential for maintaining some of the turbinate's physiological functions while allowing better access to the maxillary sinus.

The core of the procedure involves significantly widening the maxillary antrostomy down to the nasal cavity floor. This expansion facilitates gravity-assisted drainage, which is particularly beneficial for patients with impaired ciliary function or thick secretions, such as those with cystic fibrosis. The removal of a large portion of the medial wall of the maxillary sinus improves ventilation and drainage, reducing the likelihood of recurrent infections and inflammation.

Postoperatively, the large antrostomy enables effective irrigation and topical medication delivery directly into the sinus cavity, ensuring better disease management and reducing the likelihood of recurrence. Long-term outcomes have shown significant improvements in symptom resolution, highlighting the procedure's efficacy. Patients often report improved quality of life due to reduced symptoms and fewer sinus infections [5].

4.2 Extended maxillary sinusotomy with preservation of the inferior turbinate

This procedure differs from EMMA by preserving the inferior turbinate. Patients undergoing this procedure typically have failed a previous standard maxillary antrostomy. The procedure begins by revising the middle meatus antrostomy, ensuring a patent and wide middle meatus antrostomy with the removal of any residual uncinate process, and incorporating the natural ostium.

The inferior turbinate is temporarily outfractured and possibly reduced while preserving the medial mucosa, facilitating clear inspection of the lateral wall of the inferior meatus. An inferior meatus antrostomy is then performed, removing the lateral wall of the inferior meatus down to the floor of the nose. It is crucial to identify and preserve Hasner's valve (**Figure 1**), as this procedure does not involve removing the nasolacrimal duct. The middle meatus antrostomy and inferior meatus antrostomy are then combined into one common antrostomy, with the inferior turbinate preserved and attached anteriorly and posteriorly like a hammock. This approach maintains the physiological functions of the inferior turbinate while providing extensive access for sinus drainage and disease management (**Figure 2**).

4.3 Endoscopic medial maxillectomy

Endoscopic Modified Medial Maxillectomy is primarily indicated for managing benign sinonasal neoplasms and extensive maxillary sinus disease. The procedure starts with the complete removal of the inferior turbinate, which is essential to provide maximum access to the maxillary sinus. This step is crucial, as it allows for a more extensive view and access to the sinus cavity, enabling thorough removal of diseased tissue.

The surgical approach involves expanding the maxillary antrostomy to its anatomical limits, extending inferiorly to the nasal cavity floor, superiorly to the medial orbital floor, posteriorly to the posterior wall of the maxillary sinus, and anteriorly

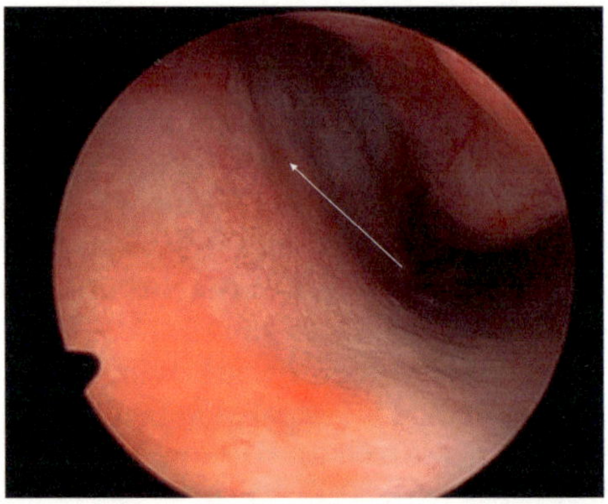

Figure 1.
The right inferior meatus ins visualized using a 30-degree endoscope and Hesner's valve is seen.

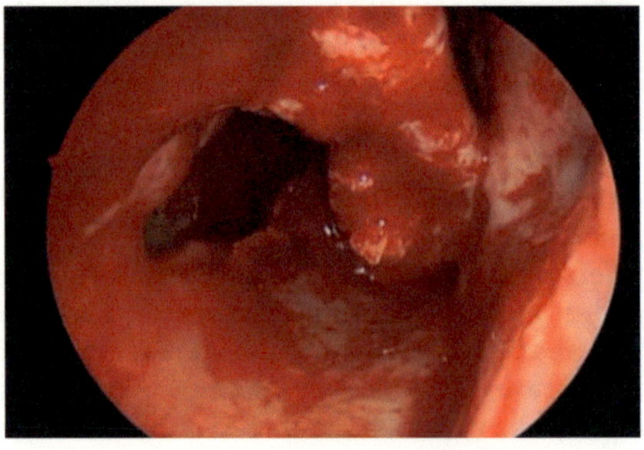

Figure 2.
Extended maxillary antrosotmy. This figure shows the inferior meatus antrostomy that is communicating with the middle meatus antrostomy while preserving the inferior turbinate.

to the anterior wall of the maxillary sinus. Complementary procedures include a complete ethmoidectomy and sphenoid sinusotomy. This comprehensive approach ensures that all diseased tissues are addressed, reducing the risk of recurrence [6].

This approach allows access to the medial wall of the orbit, which can be removed depending on its involvement with the disease, and the nasolacrimal duct, which is resected anteriorly and superiorly as needed. It is essential to complete the surgery by performing a dacryocystorhinostomy or at least vertical marsupialization of the nasolacrimal duct remnants to prevent nasolacrimal duct obstruction. A 70-degree nasal endoscope is often necessary for inspecting the anterior and lateral wall of the maxillary sinus, providing a detailed view of the surgical field.

At the end of the procedure, the comprehensive opening allows for thorough exploration and treatment of the sinus, including hard-to-reach areas such as the lateral recesses and the floor of the maxillary sinus. In cases of chronic recurrent sinusitis, the goal maybe to auto-obliterate the sinus by eradicating the maxillary sinus mucosa, thereby triggering the natural obliteration process and preventing future disease recurrence. This approach provides long-term relief for patients suffering from chronic and recalcitrant sinusitis [7].

4.4 Endoscopic Denker approach

Despite advances in endoscopic endonasal approaches, the anterolateral and anteroinferior aspects of the maxillary sinus and the infratemporal fossa remain challenging to reach through a purely transnasal approach. The Endoscopic Denker approach addresses these limitations by removing parts of the anterior wall of the maxillary sinus, providing more lateral access. This method combines the open procedures described by Denker and Sturmann/Canfield, performed entirely under endoscopic visualization. It offers the same port of entry into the maxillary sinus as the Caldwell-Luc procedure with similar indications [8].

The procedure begins with an incision in the mucosa at the junction of the nasal floor and lateral nasal wall, extending through the periosteum. A second mucosal incision is made superiorly along the lateral nasal wall, carried anteroinferiorly in front of the anterior head of the inferior turbinate, overlying the edge of the pyriform aperture. Following this, a subperiosteal dissection is performed to expose the anterior aspect of the maxilla, the infraorbital foramen, and its neurovascular bundle. A high-speed drill creates a bony window into the anterior wall of the maxillary sinus, staying inferior to the infraorbital nerve. Burrs connect the window to the inferior bony cut of the medial maxillectomy, allowing access to the anteroinferior corner of the maxillary sinus.

Once the tumor or diseased tissue is dissected free from the surrounding structures, the remaining bony and mucosal cuts are completed with burrs, osteotomes, and endoscopic scissors. Specifically, a superior cut is made at the level of the roof of the maxillary sinus, an inferior cut at the junction of the nasal floor and medial maxillary wall, and a posterior cut along the posterior wall of the maxillary sinus. At the end of the procedure, the nasolacrimal duct is identified, preserved, and sharply cut at an oblique angle to prevent stenosis (**Figure 3**), (Video 1, https://youtu.be/GXyKCt5UfGg).

5. Extended frontal sinus procedures

The primary surgical management of frontal sinus disease that has failed maximum medical treatment involves a frontal sinusotomy to reestablish the drainage

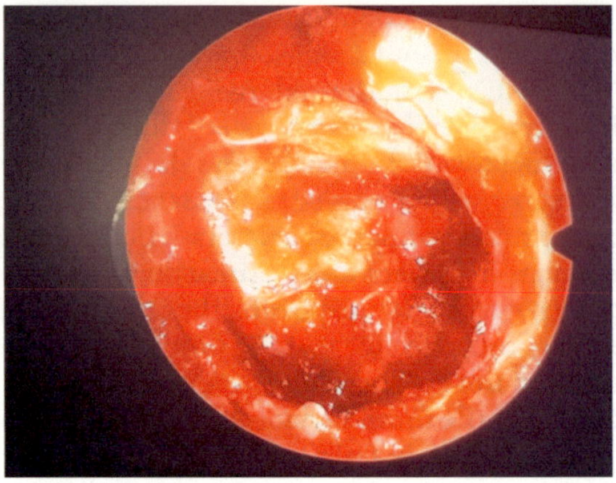

Figure 3.
70-degree nasal endoscope showing all the walls of the left maxillary sinus following and endoscopic Denker's procedure.

pathway. This approach can be extended to involve dilating the sinus by limited removal of the floor, as in the Draf IIa procedure, extending the opening from the middle turbinate to the lamina papyracea. When this procedure is inadequate, further management is required. Multiple extended frontal sinus procedures have been described, effectively replacing previously popular open procedures such as frontal sinus obliteration [9].

Extended frontal sinus surgery is indicated for various conditions, including recalcitrant sinus disease that failed primary sinus surgery, access to the lateral frontal sinus, severe stenosis due to osteoneogenesis, and remodeling in the frontal sinus recess that makes it unsafe to enter through the ipsilateral side. It is also indicated in cases of failed frontal sinus obliteration and cranialization, where failure often results from leaving mucosa behind, leading to the formation of a mucocele. Open approaches for these procedures are technically challenging, as the mucosa tends to adhere to the dura. In the case of benign and some malignant tumors, extended sinus surgery can offer the surgeon access to completely remove the tumor. For

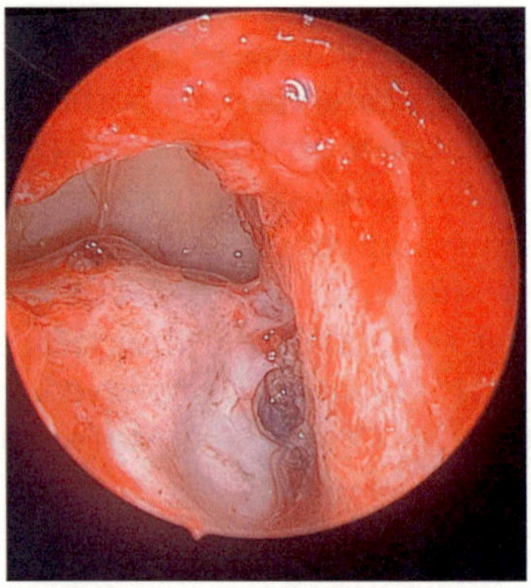

Figure 4.
A modified Lothrop procedure was used for access and repair of a left frontal sinus posterior table meningocele.

meningoencephaloceles, the extended approach allows for removal and repair while keeping the drainage of the frontal sinus patent (**Figure 4**) [10].

6. Key anatomical landmarks

Identifying the key anatomical landmarks is crucial for the success of extended frontal sinus procedures. These landmarks guide the surgeon during the procedure, ensuring precision and minimizing the risk of complications.

- *Anterior one-fourth of the middle turbinate vertical lamella:* Essential for orienting the surgeon during the initial stages of the procedure, its removal is necessary for completing the extended procedure and for the complete drilling of the floor of the frontal sinus.

- *Anterior olfactory cleft and olfactory fibrils:* Located medial to the middle turbinate vertical lamella, these structures are critical for avoiding skull base injury during Draf IIb or III procedures as they define the posterior limit of the resection.

- *Frontal sinus infundibulum posterior wall:* A critical reference point for performing extended frontal sinusotomies, helping maintain orientation and avoid inadvertent damage to the anterior skull base and cribriform plate.

- *Anterosuperior perpendicular plate and septectomy:* Facilitates exposure and allows the introduction of instrumentation from both sides of the nasal cavity, crucial for creating a common frontal sinus drainage pathway.

- *Frontal intersinus septum:* Necessary for combining the left and right frontal sinuses into a single, large cavity, a crucial step in Draf III (Modified Lothrop) procedures.

- *Nasal process of the frontal bone:* Drilling in this area helps enlarge the lateral dimensions of the frontal sinus opening, providing greater access to the supraorbital recesses.

- *Nasion:* The dense bone at the nasion is removed to enlarge the common frontal ostium, ensuring complete transillumination and visualization of the frontal sinus.

7. Extended approaches to the frontal sinus

7.1 Draf IIb procedure

This approach involves removing the frontal sinus floor from the nasal septum medially to the lamina papyracea laterally, creating a unilateral maximal opening of the frontal sinus outflow tract. This is particularly useful in cases of severe stenosis and for accessing the lateral frontal sinus (**Figure 5**) [11].

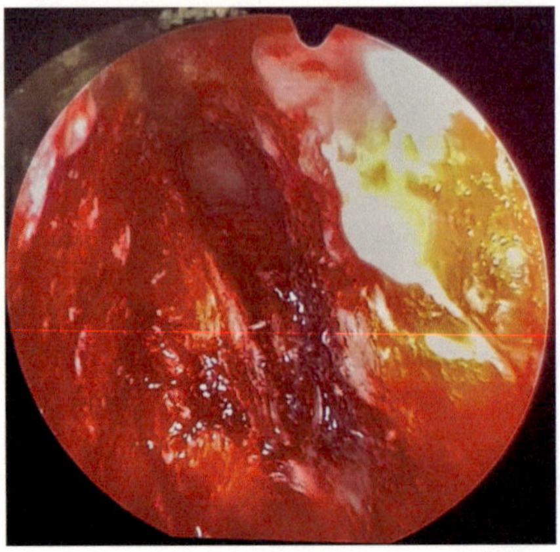

Figure 5.
The left frontal sinus following a Draf 2b procedure looked at using a 70- degree nasal endoscope.

7.2 Draf III (modified Lothrop) procedure

The Draf III procedure, or Modified Lothrop Procedure, is an endoscopic modification of the original procedure described by Lothrop in 1914. It involves the bilateral removal of the frontal sinus floor from one lamina papyracea to the other, including a superior septectomy and intersinus septectomy, creating a large common frontal sinus drainage pathway. This procedure is ideal for severe bilateral frontal sinus disease, failed prior frontal sinus surgeries, or extensive osteoneogenesis [12]. This procedure can also provide access for the repair of a posterior table cerebrospinal fluid (CSF) leak (**Figure 4**).

There are several modifications of the Draf III procedure tailored to specific anatomical and pathological challenges: [7].

- *Modified hemi-Lothrop procedure:* Adds a superior septectomy to a Draf IIb to increase access to the far lateral frontal sinus recess or a supraorbital ethmoid cell from the contralateral side (video 2, https://youtu.be/pHw8hR-vC0O).

- *Modified mini-Lothrop procedure:* Combines resection of the contralateral frontal sinus floor with a frontal intersinus septectomy, preserving the ipsilateral frontal sinus floor. This is suitable for cases where the ipsilateral frontal sinus cannot be safely accessed from the same side due to extensive scarring, previous medial orbital wall decompression, significant intraorbital contents obstructing the ipsilateral frontal sinus outflow tract, complex anatomy, severe stenosis, or pronounced osteoneogenesis.

- *Modified subtotal-Lothrop procedure:* Unilaterally removes the frontal sinus floor with a superior septectomy and intersinus septectomy, preserving as much normal sinonasal architecture as possible.

- *Modified central-Lothrop procedure:* Bilaterally resects the medial frontal sinus floor with a superior septectomy and intersinus septectomy, providing access and visualization to both frontal sinuses. This procedure is effective for addressing frontal sinus disease near the midline, particularly suitable for cases involving isolated disease or mucoceles in a frontal intersinus cell, central neoplastic processes, or skull base defects. The frontal sinus recesses and middle turbinates are left undisturbed.

8. Procedure details

8.1 Draf IIb procedure

The anterior one-fourth of the middle turbinate vertical lamella is removed, starting at its anterior insertion and heading superiorly in the direction of the posterior wall of the frontal sinus infundibulum or the superior aspect of the nasoseptal angle. This step is crucial for orienting the surgeon away from the skull base and anterior cribriform plate. The bony medial margin of the frontal infundibular opening is the superior point of insertion of this portion of the middle turbinate vertical lamella, which is removed with curettes or powered instruments to identify the junction of the perpendicular plate with the frontal intersinus septum. The mucosa on the posterior wall of the frontal infundibular area is preserved whenever feasible. However, significant fibrosis or osteoneogenesis in the area often makes this preservation impossible, and in that case, a Draf III procedure should be done.

8.2 Draf III (modified Lothrop) procedure

For a midline approach to access the frontal sinus, the anterior olfactory filaments should first be identified, located medial to the vertical lamella of the middle turbinate and posterior to the coronal plane of the posterior wall of the frontal infundibulum. Only the anterior quarter of the middle turbinate vertical lamella forms the medial margin of the frontal infundibulum. The rest is situated lateral to the cribriform plate, connecting with the lateral lamella of the cribriform plate and the medial fovea ethmoidalis.

Intraoperative navigation is beneficial for determining the correct trajectory during a transnasal midline frontal sinus approach and is typically used in all Draf III procedures. Elevating a posteriorly based flap starting in the wider septal angle area anterior to the cribriform plate, at the coronal plane just in front of the anterior attachment of the middle turbinate, helps identify the narrow V-shaped groove of the anterior cribriform plate and the first olfactory filaments, as well as the posterior wall of the frontal infundibulum. This step ensures that the posterior limit of the dissection is under direct vision, preventing inadvertent skull base injury and cerebrospinal fluid leaks. A 70-degree endoscope can be used for superior visualization.

The anterosuperior septectomy is performed anterior to the level of the anterior cribriform plate using cutting forceps or powered instrumentation, aligning with the coronal plane of the posterior wall of the frontal sinus infundibulum as viewed through a transethmoid or transseptal frontal sinusotomy. Posterior to this plane, there is a heightened risk of inadvertent intracranial penetration or damage to the cribriform plate or olfactory nerve fibers. The coronal plane of the posterior wall of the frontal sinus should always be kept in view to prevent accidental injury to the skull base and subsequent cerebrospinal fluid leaks. The size of the septal window should be adequate to enhance exposure and allow instrument insertion from both sides of the nose. Once the septal window is created, even straight instruments can access the common frontal cavity from either side of the nose. If this is not possible, the septal window should be lowered until bilateral access to the common frontal sinus is feasible.

The dense bone at the nasofrontal beak area, corresponding to the nasion, is removed using angled cutting burs, progressively enlarging the common frontal ostium and enhancing access to the superior and anterior aspects of the frontal sinus. Additionally, lateral enlargement toward the orbit is performed to increase the lateral dimensions of the opening, providing greater access to the supraorbital recesses of the frontal sinus from the contralateral side. The limits of the procedure are the dermis at the level of the nasion anteriorly and the supracanthal dermis on either side. Drilling the supracanthal bone of the orbit is usually done under endoscopic visualization, with the drill introduced through the contralateral nostril. Intermittent palpation of the nasion dermis and supracanthal dermis helps determine the anterior and lateral limits of drilling. Generally, limited exposure of the nasion or supracanthal dermis does not result in adverse cosmetic effects. The additional enlargement of the common frontal sinus ostium reduces the chances of failure due to restenosis, and maximizing the width of the opening should be attempted.

It is not uncommon to encounter one or more frontal intersinus cells that need to be removed to effectively enlarge the common frontal ostium and ensure adequate outflow. The final common extended frontal sinusotomy opening is a horseshoe-shaped opening measuring approximately 8–10 mm anteroposteriorly at the nasofrontal bone and 20–26 mm laterally from the orbital wall to the orbital wall. The posterior, lateral, and anterior walls of the common frontal ostium are composed of the posterior wall of the frontal sinus, the supracanthal lateral nasal dermis, and the anterior wall of the frontal sinus at the nasion, respectively. The final ostium should allow complete transillumination and visualization of the entire frontal sinus (**Figures 6** and 7) [4, 13].

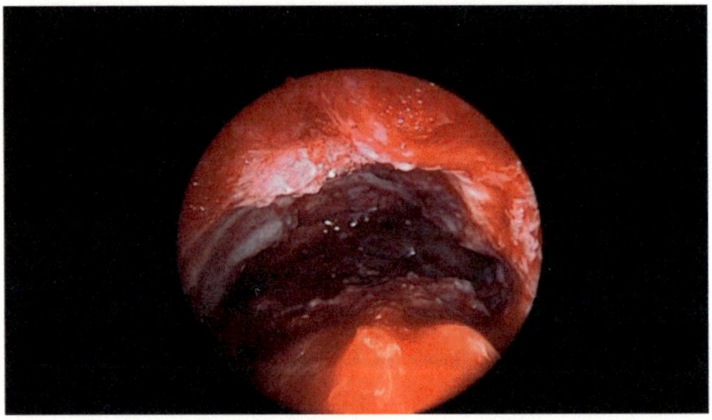

Figure 6.
Endoscopic view after a completed modified Lothrop procedure. The boundaries were reached to the limits to the medial canthi laterally and to the nasion anteriorly.

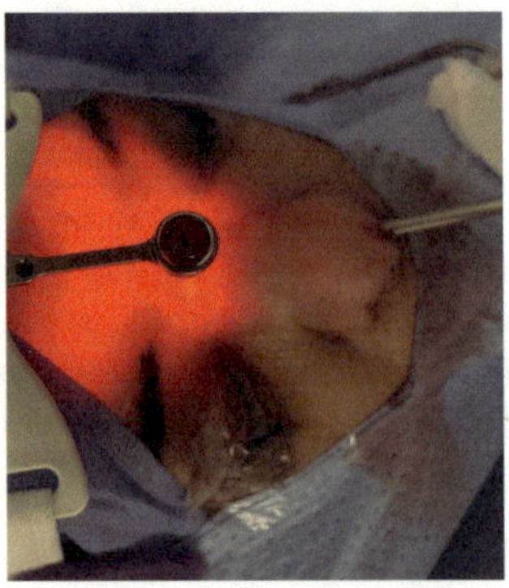

Figure 7.
Transillumination of both sides of the frontal sinus following a modified Lothrop procedure.

9. Extended sphenoid sinus procedures

Extended sphenoid sinusotomy is likely the most frequently performed extended sinus procedure as it is pivotal in transsphenoidal approaches to pituitary gland tumors, providing essential access to the skull base. This procedure creates a common sphenoid sinus cavity by removing the intersinus septum along with performing a posterior septectomy. Like other extended sinus procedures, extended sphenoid sinusotomy is crucial for managing recalcitrant sinus disease that has failed both appropriate medical therapy and primary sinus procedures. It is also indicated in cases of severe osteoneogenesis, which can obstruct entry to the ipsilateral sphenoid sinus through the natural ostium, as well as for the drainage of mucoceles and the resection of tumors [14].

The objective of performing this procedure, other than the common indication of gaining access to the skull base, is to minimize the chances of restenosis in chronic sinus disease, ensuring better long-term outcomes. Additionally, this procedure enhances access for postoperative examinations, debridement, and the administration of topical medications. Conceptually, it is similar to a frontal sinus Draf III (Lothrop) procedure, as it provides maximal access to the sphenoid sinus, facilitating improved sinus drainage and allowing for effective postoperative management and in-office debridement. By establishing a larger, more accessible cavity, it allows for thorough monitoring and treatment of any residual or recurring disease. This is particularly important for patients with complex sinus conditions, where meticulous postoperative care and the ability to apply topical medications are essential to prevent further recurrence and ensure sustained relief from symptoms.

10. Key anatomical landmarks

Identifying the key anatomical landmarks is crucial for the success of extended sphenoid sinus procedures. These landmarks provide reference points that guide the surgeon during the procedure, ensuring precision and minimizing the risk of complications.

- *Superior turbinate:* Acts as a key landmark for locating the sphenoid ostium. Identifying the superior turbinate and following it posteriorly often leads to the sphenoid ostium.

- *Sphenoid ostium:* The natural opening of the sphenoid sinus into the nasal cavity, essential for initial access.

- *Sphenoid rostrum:* The anterior bony wall of the sphenoid sinus, often removed or drilled to enlarge the sphenoidotomy and gain better access to the sinus cavity.

- *Optic nerve canal:* The bony canal through which the optic nerve runs, located in the lateral wall of the sphenoid sinus.

- *Internal Carotid Artery (ICA):* Runs laterally and inferiorly to the sphenoid sinus. Identifying and preserving this artery is critical to avoid catastrophic hemorrhage or stroke.

- *Vidian canal:* The bony canal containing the vidian nerve, located at the base of the sphenoid sinus. It serves as an important landmark for the ICA in complex skull base procedures.

- *Floor of the sella:* Important for transsphenoidal approaches to the pituitary gland.

- *Middle third of the clivus:* Inferior to the floor of the sella and providing a reference point for posterior sphenoid sinusotomy and access corridor in skull base surgeries.

11. Procedure details

The patient is placed in a supine position with the head slightly elevated, and the nose is decongested. A 30-degree or 0-degree endoscope is inserted into the nasal cavity. The natural sphenoid ostium is located, typically found superior to the choana and medial to the superior turbinate. The ostium is then enlarged using a microdebrider or Kerrison rongeurs to facilitate better access. The sinus that is opened first is typically the one easier to access and leads to a more pneumatized sphenoid sinus.

The sphenoid rostrum is removed completely using a high-speed drill or rongeurs to create a wider sphenoidotomy. This step involves careful drilling to avoid damage to surrounding structures, including the optic nerve and ICA. Subsequently, a posterior septectomy is performed by removing the posterior portion of the nasal septum, creating a common cavity between the left and right sphenoid sinuses, and enhancing surgical access and drainage.

Following the septectomy, the intersinus septum between the left and right sphenoid sinuses is removed, creating a single, large sphenoid cavity. The common sphenoid sinus opening is enlarged superiorly to the level of the planum sphenoidale and inferiorly to the floor of the sphenoid sinus, and laterally until the sphenoid cavity is contiguous with the medial orbital wall. This facilitates access and improves postoperative management, reducing the risk of restenosis. Identifying the optic nerve canal and ICA is crucial. The intersinus septum is removed using true-cutting instruments or a diamond burr. It is important to avoid pulling the intersinus septum using non-true-cutting instruments, as it may be attached to the parasellar or paraclival ICA, and pulling on it may result in catastrophic bleeding.

If pathologies such as mucoceles, tumors, or severe osteoneogenesis is present, the diseased tissue is carefully resected using various surgical instruments. The goal is to remove diseased tissue while preserving healthy mucosa. If the procedure is done as part of access to the skull base and a septal flap is needed, the flap should be raised before the procedure and tucked in the nasopharynx for later use in skull base repair.

12. The ethmoid sinus and the management of the middle turbinate

Anterior or total ethmoidectomy is the standard surgical procedure for managing ethmoid sinus disease that has not responded to medical treatment. This procedure emphasizes the concept of mucosal preservation. However, more aggressive approaches have been described for cases where standard ethmoidectomy fails. Jankowski et al. presented results of a more radical approach known as nasalization, or radical ethmoidectomy. This procedure involves removing all the bony lamellae and mucosa within the ethmoid sinus and includes performing a large maxillary antrostomy, sphenoidotomy, frontal sinusotomy, and middle turbinectomy.

In their study, Jankowski and colleagues compared nasalization to conventional ethmoidectomy. Patients who underwent nasalization showed better outcomes in terms of nasal symptoms compared to those who had conventional ethmoidectomy. Notably, olfactory improvement remained stable for three years in the nasalization group, whereas it worsened after 2 years in the ethmoidectomy group [15]. In a 2006 follow-up study, Jankowski and colleagues further found that nasalization was superior in terms of overall symptoms, as assessed by a questionnaire, endoscopic appearance of the mucosa, appearance on CT scan, and recurrence rate. This evidence supports the use of nasalization in more severe cases of ethmoid sinus disease that do not respond to standard surgical treatments [16].

13. Postoperative care

Postoperative care is a critical component of managing patients undergoing extended sinus surgeries. Proper care and monitoring following surgery can significantly impact the overall success and long-term outcomes of the procedure. Initially, patients may be prescribed antibiotics based on the cultures taken at the time of surgery. Saline irrigations mixed with topical corticosteroids, depending on the pathology, will help promote healing and reduce crusting [17].

Regular follow-up visits are crucial for assessing the surgical site and identifying any early signs of complications. During these visits, endoscopic examinations are performed to monitor the healing process, remove any crusts or debris, and ensure that the surgical cavities remain patent. Debridement may be necessary to remove any granulation tissue or crusts and to release any synechiae that could obstruct the sinus openings [18, 19].

Patients are advised to avoid activities that may increase nasal pressure, such as nose blowing or heavy lifting, to prevent disruption of the surgical site. They are also instructed to avoid exposure to irritants like smoke or dust, which can exacerbate inflammation and impede healing.

Long-term follow-up is necessary for patients with chronic conditions to monitor for any disease recurrence and to manage it as indicated.

Author details

Ghassan Alokby
King Faisal Specialist Hospital and Research Center, Alfaisal University, Riyadh, Saudi Arabia

*Address all correspondence to: galokby@icloud.com

References

[1] Snyderman CH et al. Technologic innovations in Neuroendoscopic surgery. Otolaryngologic Clinics of North America. 2009;**42**(5):883-890, x. DOI: 10.1016/j.otc.2009.08.019

[2] Alokby G et al. Endoscopic reconstruction of Skull Base defects using Tutoplast. Allergy & Rhinology (Providence, R.I.). 2021;**12**:21526567211009200. DOI: 10.1177/21526567211009200

[3] Eloy JA, Marchiano E, Vázquez A. Extended endoscopic and open sinus surgery for refractory chronic Rhinosinusitis. Otolaryngologic Clinics of North America. 2017;**50**(1):165-182. DOI: 10.1016/j.otc.2016.08.013. PMID: 27888912

[4] Casiano RR, Herzallah IR, Eloy JA, editors. Endoscopic Sinonasal Dissection Guide: Including Orbit and Skull Base. Second ed. New York: Thieme; 2018

[5] Costa ML et al. Long-term outcomes of endoscopic maxillary mega-Antrostomy for refractory chronic maxillary sinusitis. International Forum of Allergy & Rhinology. 2015;**5**(1):60-65. DOI: 10.1002/alr.21407

[6] Konstantinidis I, Constantinidis J. Medial Maxillectomy in recalcitrant sinusitis: When, why and how? Current Opinion in Otolaryngology & Head and Neck Surgery. 2014;**22**(1):68-74. DOI: 10.1097/MOO.0000000000000009

[7] Eloy JA, Marchiano E, Vázquez A. Extended Endoscopic and Open Sinus Surgery for Refractory Chronic Rhinosinusitis. Otolaryngol Medical Clinics of North America. Feb 2017;**50**(1):165-182. DOI: 10.1016/j.otc.2016.08.013. PMID: 27888912

[8] McCormick JP et al. Modified endoscopic Denker's approach for Management of Anterior Maxillary Sinus Tumors: Multicenter series of 58 cases. American Journal of Rhinology & Allergy. 2022;**36**(6):763-772. DOI: 10.1177/19458924221110652

[9] Turner JH, Vaezeafshar R, Hwang PH. Indications and outcomes for Draf IIB frontal sinus surgery. American Journal of Rhinology & Allergy. 2016;**30**(1):70-73. DOI: 10.2500/ajra.2016.30.4268

[10] Casiano RR, Livingston JA. Endoscopic Lothrop procedure: The University of Miami experience. American Journal of Rhinology. 1998;**12**(5):335-339. DOI: 10.2500/105065898780182444

[11] Draf W. Endonasal micro-endoscopic frontal sinus surgery: The Fulda concept. Operative Techniques in Otolaryngology-Head and Neck Surgery. 1991;**2**(4):234-240. DOI: 10.1016/S1043-1810(10)80087-9

[12] Gross CW, Gross WE, Becker DG. Modified Transnasal endoscopic Lothrop procedure: Frontal Drillout. Operative Techniques in Otolaryngology-Head and Neck Surgery. 1995;**6**(3):193-200. DOI: 10.1016/S1043-1810(06)80012-6

[13] Wormald P-J, Valentine R. Endoscopic Sinus Surgery: Anatomy, Three-Dimensional Reconstruction, and Surgical Technique. 3rd ed. New York Stuttgart: Thieme; 2013

[14] Simmen D, Jones N. Manual of Endoscopic Sinus Surgery and its Extended Applications. Stuttgart: Thieme; 2005

[15] Jankowski R, Pigret D, Decroocq F. Comparison of functional results after

Ethmoidectomy and nasalization for diffuse and severe nasal polyposis. Acta Oto-Laryngologica. 1997;**117**(4):601-608. DOI: 10.3109/00016489709113445

[16] Jankowski R et al. Comparison of radical (nasalisation) and functional Ethmoidectomy in patients with severe Sinonasal polyposis. A retrospective study. Revue De Laryngologie - Otologie - Rhinologie. 2006;**127**(3):131-140

[17] Rudmik L et al. Early postoperative care following endoscopic sinus surgery: An evidence-based review with recommendations. International Forum of Allergy & Rhinology. 2011;**1**(6):417-430. DOI: 10.1002/alr.20072

[18] Bugten V, Nordgård S, Steinsvåg S. The effects of debridement after endoscopic sinus surgery. The Laryngoscope. 2006;**116**(11):2037-2043. DOI: 10.1097/01.mlg.0000241362.06072.83

[19] Nilssen ELK et al. A randomized control trial of post-operative care following endoscopic sinus surgery: Debridement versus No debridement. The Journal of Laryngology and Otology. 2002;**116**(2):108-111. DOI: 10.1258/0022215021910041

Chapter 7

Dacryocystitis and Its Management

Sindhuja Bhanwala

Abstract

The lacrimal apparatus drains into the inferior meatus of the nose via the nasolacrimal duct. Blockage of this drainage pathway leads to a consistent and continuous epiphora which tends to become sticky and thick over time. Thus, it can lead to recurrent episodes of acute infections which present with a swelling in the medial canthal region (acute dacryocystitis). Dacryocystorhinostomy (DCR) is recognized as the most suitable treatment for patients with obstruction of the lacrimal system at the level of the sac or the nasolacrimal duct. This operation aims to create a bypass between the lacrimal sac and the nasal cavity. Traditionally, this was achieved by performing an external DCR using an external skin incision. But over the past two decades, we have seen advances in rigid endoscopic equipment, which have led to the development of a safer and less-invasive technique, namely, endoscopic DCR for treating symptomatic lacrimal apparatus disorders. This chapter provides a comprehensive overview of the lacrimal system, dacryocystitis etiology and management. It delves into the indications for surgical intervention and compares traditional external DCR with the less-invasive endoscopic DCR technique. By detailing postoperative care, this chapter aims to offer a streamlined approach to understanding and treating symptomatic lacrimal apparatus disorders.

Keywords: lacrimal apparatus diseases, dacryocystorhinostomy, endoscopic, epiphora, dacryocystitis

1. Introduction

Dacryocystitis, an inflammation of the nasolacrimal sac, often stems from an obstruction within the nasolacrimal duct (NLDO), resulting in tear stagnation and the hallmark symptom of epiphora. Additionally, it may manifest as a tender swelling near the eye's inferomedial canthus, periorbital cellulitis, erythema and purulent discharge. Timely diagnosis and effective management are pivotal in averting potential complications. Treatment typically encompasses antibiotics, wound care and addressing the underlying obstruction, often requiring surgical intervention such as dacryorhinocystostomy in chronic cases [1].

Collaborative efforts amongst ophthalmologists, otorhinolaryngologists, nurses, physicians and pharmacists are indispensable for comprehensive patient care and favorable outcomes.

2. Anatomy of the lacrimal system and its drainage pathway

The lacrimal sac, nestled within the lacrimal fossa, drains into the nasolacrimal duct. Lacrimal puncta, situated at the medial ends of the upper and lower eyelids, connect to the lacrimal sac via the upper and lower canaliculi. These canaliculi converge to form the common canaliculus, which opens high on the lateral wall of the lacrimal sac. It extends about 9 mm above the axilla of the middle turbinate. The thick anterior limb of the medial canthal tendon wraps along the anterior upper half of the lacrimal sac to insert onto the anterior lacrimal crest, and then the posterior limb passes behind the sac to insert onto the posterior lacrimal crest. The nasolacrimal duct travels within a bony canal formed by the maxillary and lacrimal bones, terminating at the inferior meatus of the nose.

Tear production by the lacrimal gland, along with accessory glands of Wolfring and Krause, nourishes and lubricates the eye. Tears flow through the puncta, ampullae, canaliculi and, finally, the lacrimal sac, before descending through the nasolacrimal duct and reaching the nasal cavity (**Figure 1**) [2, 3].

3. Classification and etiology

Dacryocystitis, resulting from obstruction in the lacrimal drainage pathway, manifests as either acute or chronic, congenital, or acquired.

Acute dacryocystitis, an infectious state, can be caused by various Gram-positive and Gram-negative bacteria including *Staphylococcus aureus*, *Streptococcus pneumoniae*, *Haemophilus influenzae*, *Serratia marcescens* and *Pseudomonas aeruginosa* [1].

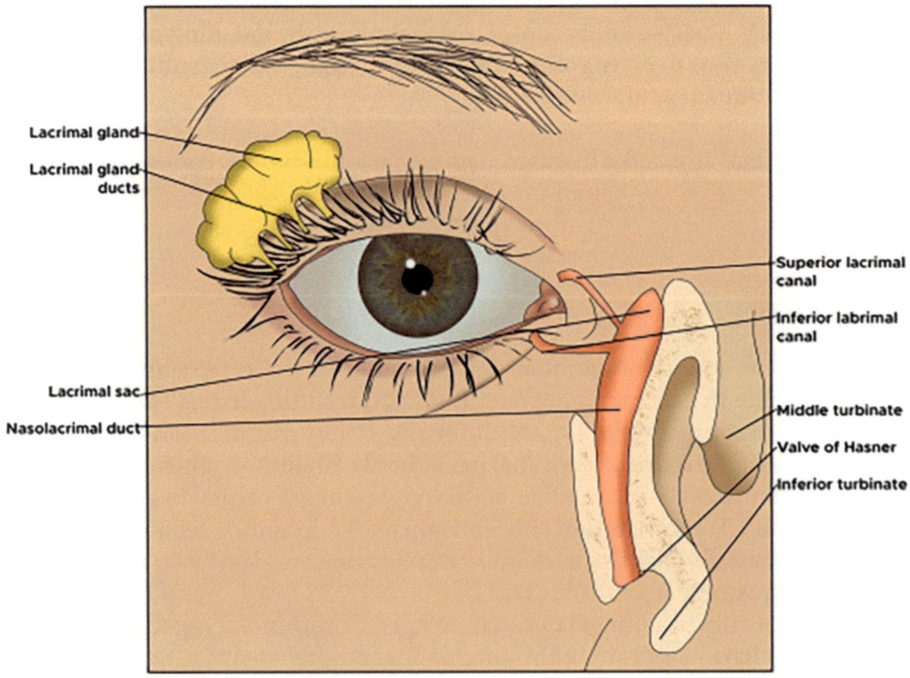

Figure 1.
Anatomy of the lacrimal apparatus and its drainage pathway.

Chronic dacryocystitis ensues from persistent obstruction, attributed to recurrent infections, dacryoliths, or systemic diseases like sarcoidosis and systemic lupus erythematosus.

Congenital dacryocystitis may stem from infection, failure of membrane rupture at the valve of Hasner, or persistent epithelial plugs [4].

Acquired dacryocystitis may be primary, characterized by chronic inflammation and fibrosis of the NLDO, or secondary, resulting from trauma, dacryolithiasis, tumors, or certain medications such as timolol, flourouracil and docetaxel.

4. Epidemiology

Dacryocystitis exhibits a bimodal distribution, with most cases occurring shortly after birth (congenital) or beyond 40 years of age. Predominantly affecting white individuals, over 75% of cases occur in females [1].

5. Pathophysiology

Obstruction within the nasolacrimal drainage system sets the stage for dacryocystitis, leading to the accumulation of tears in the lacrimal sac. Subsequent bacterial colonization incites infection, characterized by erythema, swelling and tenderness over the lacrimal sac. Untreated, it can progress to complications such as orbital abscess, cellulitis, necrotizing fasciitis and even meningitis [5].

6. Risk factors

Various factors contribute to the development of dacryocystitis, including gender (females have narrower ducts), older age (due to narrowed punctual openings and slower tear drainage), dacryoliths, nasal septal deviation, rhinitis, turbinate hypertrophy, idiopathic or iatrogenic trauma to the nasolacrimal system, systemic disorders and certain medications.

7. Clinical features

Acute dacryocystitis presents with sudden pain, erythema, fever and edema around the medial canthus and orbit. Symptoms progress over hours to days, often accompanied by purulent discharge. Chronic dacryocystitis typically manifests as epiphora, sometimes accompanied by conjunctival chemosis and changes in visual acuity due to altered tear film production (**Figure 2**).

8. Complications

Without intervention, dacryocystitis can extend to adjacent orbital tissues, leading to preseptal cellulitis, orbital cellulitis, orbital abscess, optic nerve compression and eventual vision loss.

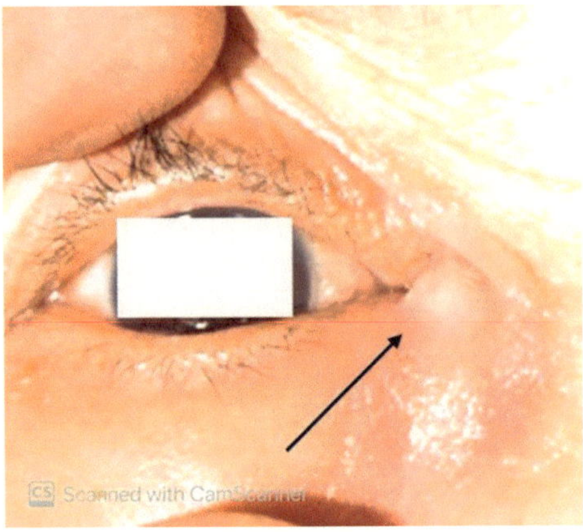

Figure 2.
Acute dacryocystitis with swelling and erythema at the inferomedial aspect of the eye.

9. Evaluation and investigations

Diagnosis of dacryocystitis relies primarily on clinical assessment and patient history. Imaging and blood work are reserved for acutely ill patients. Diagnostic procedures include tear duct massage for microbiological evaluation, serologic testing for systemic diseases suspicion and various tests like syringing, diagnostic probing and nasal endoscopy for surgical planning. In recurrent cases, additional tests such as the Jones test, fluorescein dye disappearance test, dacryocystography and imaging modalities like CT/MRI may be employed. Secretory tests like Schirmer's test may also be performed occasionally to rule out dry eye syndromes.

The following diagnostic tests are performed routinely if a surgical intervention is being planned:

- Syringing: used to check for the presence of any pre-sacral obstruction. Sterile saline solution is injected into the punctum, and if after 15–30 seconds, there was no regurgitation and if the patient appreciates a salty taste with awakening of the swallowing reflex, then the lacrimal passage is considered to be fully patent. If some saline regurgitated from either puncta and some saline passed into the nose, then the lacrimal passage is partially obstructed. A functional blockage is said to be present if there is no regurgitation of saline but a swelling developed in the lacrimal sac region (**Figure 3**) [6].

- Diagnostic probing: a lacrimal probe is passed through the punctum towards the canaliculus. If the common canaliculus is patent, the probe touches the medial wall of the sac and its underlying bone, giving a hard and bony feel (Hard stop). Hard stops are usually amenable to surgical correction. If there is a blocked common canaliculus, we encounter a soft, spongy feel.

- Diagnostic nasal endoscopy: is used to assess the nasal septum for deviation, middle meatus for any sinus disorders, osteomeatal complex and turbinates. It helps to exclude any rhinological pathology as the underlying etiology for dacryocystitis as well as helps in surgical planning.

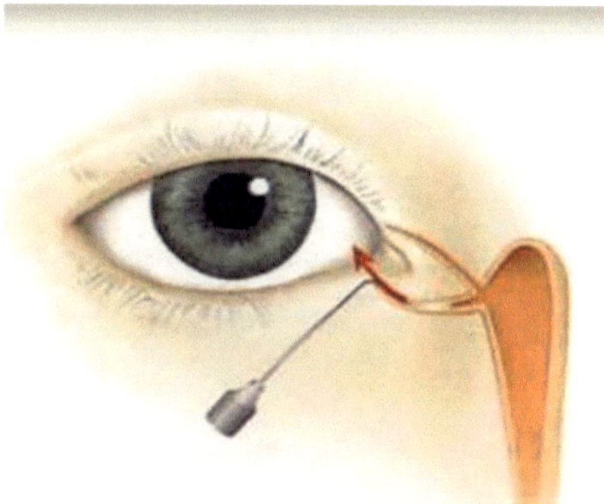

Figure 3.
Lacrimal syringing which shows regurgitation and an incomplete canalicular block.

Additionally, in recurrent cases, the following additional tests may be required to rule out any other causes for excessive tearing:

- Jones Test: *Primary Jones Test*—2% fluorescence dye is instilled in the conjunctival sac, and a cotton tip applicator moistened with 4% lignocaine is inserted into the inferior meatus. If the applicator is stained with the dye, then the nasolacrimal passage is patent. *Secondary Jones Test*—it is only performed if the primary test is suggestive of an obstructed nasolacrimal passage. After 5 minutes, the conjunctival sac is flushed with normal saline to remove any residual dye, and then clear normal saline is injected into the sac. If the dye appeared in the nose after injecting, then it is suggestive of a partial blockage of the passage. If no dye was recovered from the nose, then it is suggestive of a canaliculus block. If no fluid is recovered from the nose, then it is suggestive of a complete blockage of the nasolacrimal passage.

- Fluorescein dye disappearance test: a drop of 2% fluorescence dye is instilled into the conjunctival cul de sac. After 5 minutes, if all the dye had disappeared, then there is no obstruction. However, if there is presence of residual dye, it indicates an obstruction.

- Dacryocystography: it is a relatively simple, inexpensive and straightforward procedure in which a radiographic contrast material (either water based or oil based) is injected into the lower lid canalicular system using a lacrimal cannula. Serial postero-anterior and lateral X-ray films of the orbit are obtained immediately after the injection and also after 30 minutes. A delay in emptying time, or failure of the dye to appear in the nasal cavity or nasopharynx, is indicative of an obstruction (**Figure 4**).

- CT/MRI: is usually reserved of traumatic cases or if there is extensive infection with a suspicion of orbital cellulitis. It helps in surgical planning especially if an endonasal and endoscopic intervention is being planned. These are very useful in evaluating infants with congenital dacryocystitis as they delineate the potential aetiologies of a medial canthal mass such as congenital encephaloceles or vascular malformations [7].

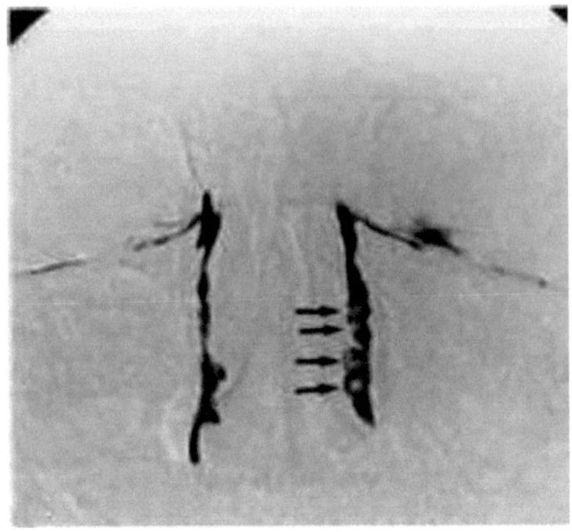

Figure 4.
Dacryocystogram showing left lacrimal dacryolithiasis.

- Secretory tests: are occasionally performed to rule out dry eye syndromes. Schrimer's test, Bengal rose staining test and lysozyme lysis tests are some examples of secretory tests which are occasionally performed.

10. Management

Dacryocystitis necessitates a combined approach of medical and surgical interventions. While acute cases typically respond well to conservative management, surgical intervention becomes more favorable for chronic cases.

Congenital dacryocystitis often resolves spontaneously within the first year of life in about 80–100% cases, with a significant percentage achieving resolution within the initial 6 months [8]. For symptomatic cases not showing spontaneous resolution, lacrimal sac massage with high-pressure irrigation can be attempted. Probing, once considered the primary intervention, is now reserved for children over 1 year of age, after failed spontaneous resolution, with a success rate of approximately 70%. However, it carries risks of creating false passages and epithelial injury leading to fibrosis and scarring, thus reserved for children older than 1 year [9]. If primary probing fails, alternative surgical interventions such as silicone tube intubation, dacryocystorhinostomy, balloon catheter dilation and inferior turbinate fracture may be explored. Topical antibiotics are usually reserved for any acute flare ups.

Acute dacryocystitis is managed with oral antibiotics covering both Gram-positive and Gram-negative organisms (preferably with antistaphylococcal activity), warm compresses and Crigler massage. Lacrimal probing is discouraged during acute episodes. Unresponsive cases are managed with incision and drainage with direct antibiotic application to the lacrimal sac, to provide immediate pain relief and rapid infection control [10, 11].

Chronic dacryocystitis typically necessitates surgical management, with probing often performed initially in outpatient settings. However, most cases eventually require surgical intervention to prevent disease progression. Techniques such as balloon dilation, silicone tube intubation and nasolacrimal stenting may be attempted. However, the primary management remains dacryocystorhinostomy (DCR), either through external or endoscopic endonasal approaches.

11. Dacryocystorhinostomy (DCR)

DCR creates an anastomosis between the lacrimal sac and nasal mucosa, bypassing the obstructed drainage pathway. External DCR, described by Adeo Toti in 1904, historically served as the standard surgical approach until the late twentieth century. But was associated with external scarring and a failure rate of 3–15% [12]. In pursuit of a less invasive, more physiological and more effective technique, endoscopic endonasal DCR (EE-DCR) emerged. Initially described by Rice in 1988 [13] and first performed on live patients by McDonough and Meiring in 1989 [14], EE-DCR offers a more physiological approach with potentially higher success rates compared to traditional external DCR.

- Indications: persistent congenital dacryocystitis not responding to conservative measures, congenital lacrimal duct obstruction associated with mucocele, dacryocystitis, acquired nasolacrimal duct obstruction

- Contraindications: acute dacryocystitis, malignant lacrimal sac mass, dry eye syndromes, bleeding dycrasias

- Surgical prerequisites: confirmation of diagnosis and clinical features, blood investigations like a complete hemogram, serological evaluation for HIV, HBV and HCV, coagulation profile, additional anesthetic investigations such as blood pressure measurement and random blood sugars based on the patient's age and comorbidities.

- Equipment and instruments: toothed forceps, needle holder, punctual dilator, Bowman's probe, Freer's periosteal elevator, blades (number 15, crescent knife), Kerrison punch, sutures (6-0 vicryl and 6-0 silk) and lacrimal (Crawford) stents.

- Anesthesia: can be performed under local or general anesthesia, but usually local anesthesia is preferred. Local infiltration to block the infratrochlear nerve in conjunction with topical anesthesia of the nasal mucosa is the preferred technique.

- Steps:

  A. A curvilinear incision, measuring about 1 cm in length, is given along the anterior lacrimal crest, 3–4 mm short of the medial canthus (**Figure** 5).

  B. Blunt subcutaneous dissection is done to expose the periosteum overlying the lacrimal fossa. Care is taken to avoid injury to the medial canthal tendon (**Figure 6**).

  C. Bone punches are made at the junction of the lamina payracea and the lacrimal bone to create a bony ostium which is then sequentially enlarged (**Figure** 7).

  D. Sac flaps are then created anteriorly and posteriorly using a Bowma's probe.

  E. H-shaped incision is then made across the sac funds till the nasolacrimal duct. The flaps are then raised, and the posterior one is cut.

F. Nasal mucosal flaps are then fashioned using an 11 blade along the bony ostium (**Figure 8**).

G. The nasal mucosal and the sac flaps are then anastomosed edge to edge.

H. Once the anastomosis is secure, the orbicularis is sutured back with 6-0 vicryl followed by skin with 6-0 silk (**Figure 9**).

I. Additionally, if there are adhesions within the lacrimal sac, 0.04% Mitomycin C can be applied intraoperatively. Intubation can be done simultaneously especially in case of inadequate flaps [15–17].

J. Nasal packing is optional, but may be required at times to achieve hemostasis.

- Postoperative care: postoperatively, the patient is prescribed analgesics, topical and oral antibiotics. Nasal decongestants and saline nasal drops are prescribed as well for wound care. The sutures are removed after 1 week, and the patient is regularly followed up in the clinic at 6 weeks, 12 weeks and 6 months.

- Complications:

A. Early (1–4 weeks): wound dehiscence, infection, tube displacement, nasal crusting and intranasal synechiae.

B. Intermediate (1–3 months): granulomas, tube displacement (**Figure 10**), synechiae, punctual cheese wiring, facial scar and nonfunctional DCR

C. Late (>3 months): rhinostomy fibrosis, webbed facial scar, medial canthal distortion and failed DCR

11.1 Endoscopic endonasal DCR

- Indications: epiphora caused by anatomic or functional obstruction of the lacrimal sac or the nasolacrimal duct, chronic relapsing dacryocystitis with purulent discharge, active infection of the lacrimal sac with infection of the overlying skin, acute dacryocystitis, nasolacrimal duct injury, dacryolithiasis, dacryocele and benign lacrimal sac mass

- Contraindications: dry eye syndromes, chronic epiphora due to presaccal obstruction, lagophthalmos with facial nerve palsy, ectropion of the lower lid, malignancy of the lacrimal system, extensive synechiae between the nasal septum and lateral nasal wall, sinonasal malignancy, Wegener's granulomatosis, significant collapse of the nasal dorsal due to trauma, bleeding dycrasias

- Surgical prerequistes: confirmation of diagnosis and clinical features, syringing and diagnostic probing, diagnostic nasal endoscopy, blood investigations like a complete hemogram, serological evaluation for HIV, HBV and HCV, coagulation profile, additional anesthetic investigations such as blood pressure measurement and random blood sugars based on the patient's age and comorbidities.

- Equipment and instruments: endoscopic camera system, video monitor, suction apparatus, antifog solution, light source and cable, 0° and 30° rigid nasal endoscopes, Blakesley forceps, endoscopic scissors, through cut forceps, Freer's elevator, ball probe, sickle knife, suction cannulas, Kerrison rongeur and burr tips (optional).

- Anesthesia: general anesthesia is the preferred modality for performing an endoscopic end-nasal DCR as it facilitates correction of any simultaneous nasal pathology which contributes to NLDO. However, in cases of severe and uncontrolled comorbidities, it can be performed under local anesthesia as well.

- Steps:

A. Nasal cavity is packed with cottonwoods soaked in 4% lignocaine with 1 in 10,000 adrenaline for about 5–10 minutes to decongest the nasal mucosa. If required, a septoplasty is then done to improve surgical exposure.

B. Lateral wall of the nose is infiltrated with 2% lignocaine with adrenaline, just anterior to the uncinate process.

C. A C-shaped mucosal flap measuring about 1.5 × 1 cm is then fashioned using a 15 number surgical knife. The first incision starts 1 cm above the axilla of the middle turbinate and runs forward by 1 cm, the blade is then turned vertically, a 1.5 cm incision is made downwards, the blade is then turned posteriorly and a 1 cm horizontal incision is made to create a posteriorly bases mucosal flap (**Figure 11**).

D. The mucosal flap is then elevated and is then reflected or excised, to expose the underlying lacrimal bone and the frontal process of the maxilla.

E. The thin lacrimal bone is elevated with a Freer elevator and removed with a forceps.

F. Kerrison rongeur is then used to take off the frontal process of the maxilla (some surgeons prefer to drill it out with microdrills and burr attachments, especially if the bone is very thick)

G. This exposes the medial wall of the lacrimal sac in its entirety.

H. The medial wall of the sac is incised with a sickle knife and excised with a Blakesly forceps or a through cut forceps.

I. Patency confirmed with saline irrigation via the inferior canaliculus.

J. The mucosal flap is then repositioned to cover any exposed bone. Care should be taken to prevent the flap from covering the sac incision.

K. Silicone stents can then be inserted, particularly in revision cases, from the upper and lower canaliculus. Once both the ends are visible inside the nasal cavity, they are held together using a Watzke sleeve or are gently tied together. The excess length of the tube is then excised intranasally.

L. 0.04% Mitomycin C can then be applied to prevent granuloma formation.

M. Nasal packs can be inserted to ensure hemostasis.

- Postoperative care: the patient is usually prescribed oral antibiotics, analgesics and saline nasal drops for 1 week in conjunction with topic antibiotic eye drops. Lacrimal syringing is performed twice daily on first and second postoperative days, and then the patients are discharged. They are followed up in the outpatient clinic at 1 week, 3 weeks and 3 months after the surgery. Nasal endoscopy is usually performed in the follow-up visits to ensure lacrimal passage latency and to remove any excessive crusting. In such cases, it is worthwhile to give saline nasal rinses and douches as well.

- Complications:

  A. Early (1–4 weeks): hemorrhage and bleeding into the orbit (in 5–10% cases), ecchymosis of the lower lid, emphysema of the lower lid or cheek, injury to lamina papyracea exposing the orbital fat, medial rectus injury leading to diplopia

  B. Intermediate (1–3 months): granulations, granulomas, synechiaea, stent migration, cheese wiring (excessively tight stent cuts through the canaliculus and the skin between them at the medial canthus) (**Figure 12**)

  C. Late (>3 months): synechiaea, granuloma formation, stent migration and cheese wiring, lacrimal sump syndrome (block of the common canaliculus intranasally due to mucus accumulation and accompanying sac inflammation, usually responds to external sac massage which helps to clear the accumulated mucus)

11.2 Ex-DCR vs. EE-DCR

EE-DCR has got several advantages over an Ex-DCR. It is a more physiological procedure as it avoids injury to the orbicularis oculi muscle, thereby preserving its pumping function. It avoids an external facial scar and any concomitant nasal pathologies

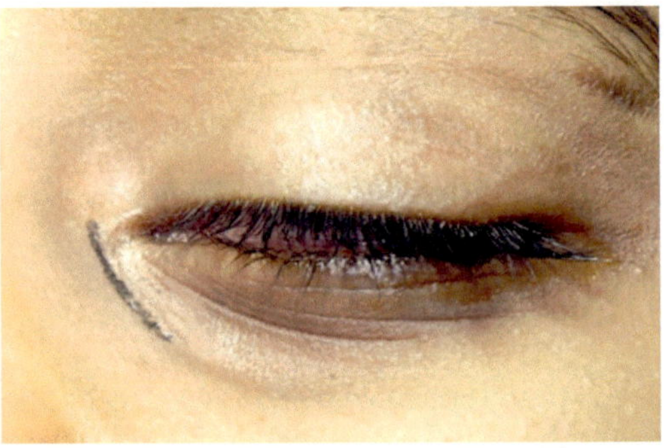

Figure 5.
Curvilinear incision used for ExDCR.

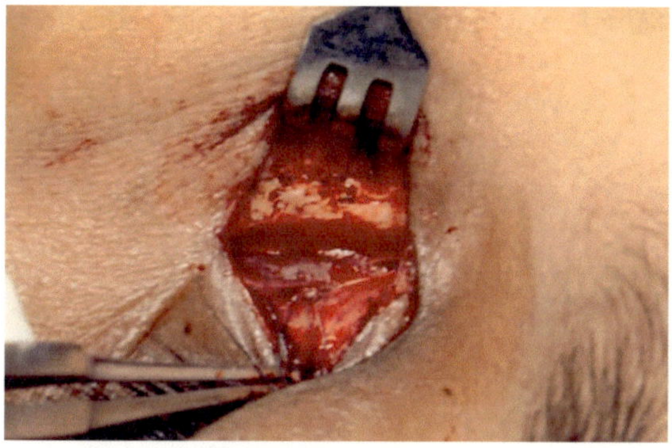

Figure 6.
Dissection of the lacrimal sac to expose the bony lacrimal fossa.

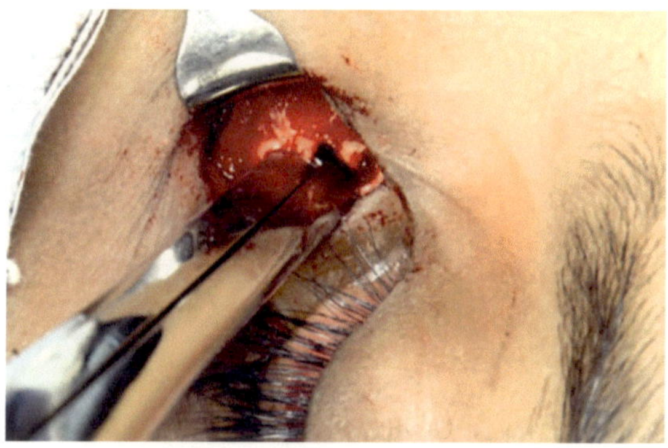

Figure 7.
Creation of bony ostium using a Kerrison punch.

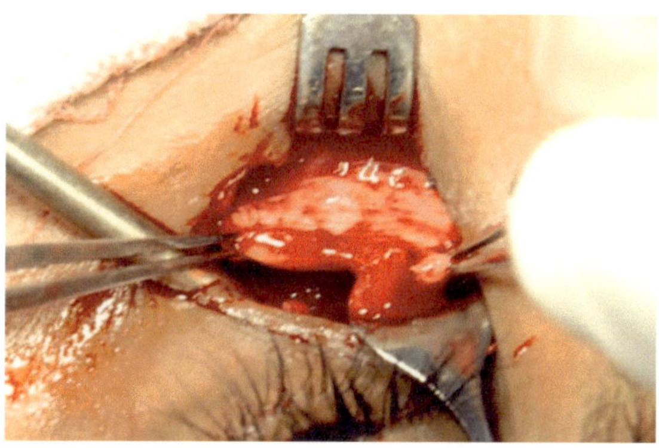

Figure 8.
Raising the nasal mucosal flap after making an incision in the sac.

can be corrected in a single procedure. However, it does require specialized training for performing procedures with endoscopic assistance, and moreover, the endoscopic equipment is expensive as well.

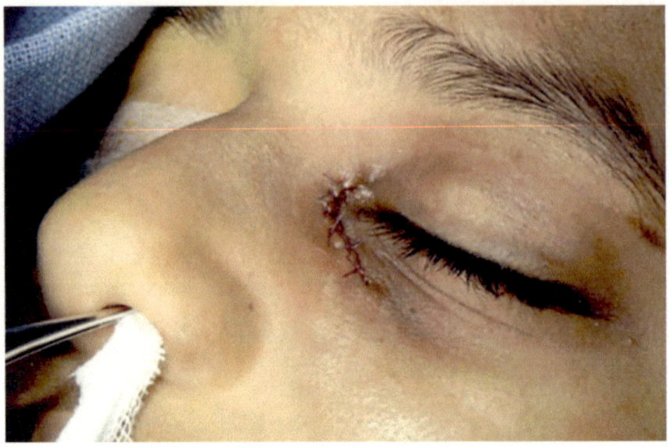

Figure 9.
Wound closure with sutures showing an external facial scar.

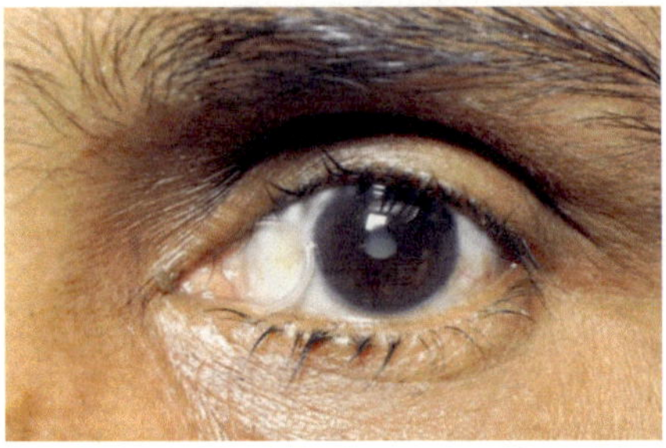

Figure 10.
Postoperative complication showing tube displacement.

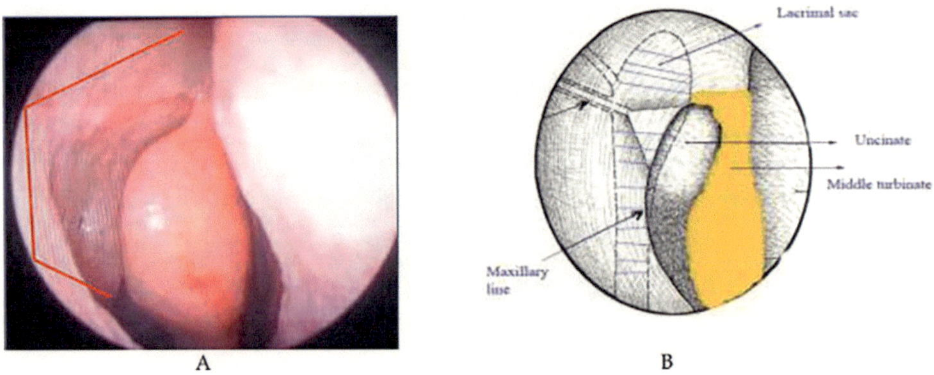

Figure 11.
Endoscopic view of the left lateral nasal wall showing the maxillary line and uncinate process (A) and a diagrammatic illustration demonstrating relationship of lacrimal sac to uncinate process and middle turbinate (B).

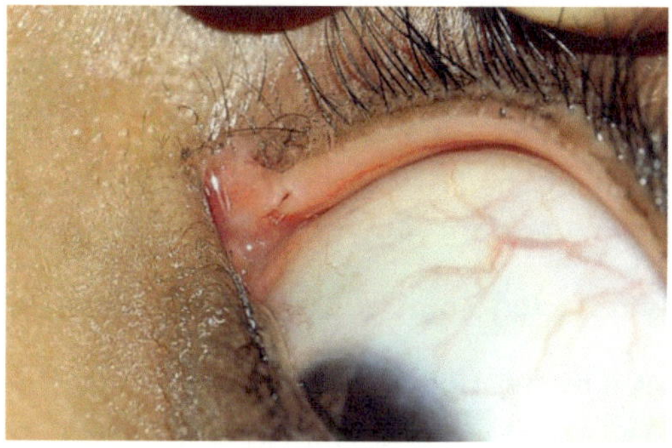

Figure 12.
Postoperative complication showing cheese wiring.

Aspect	EX-DCR	EE-DCR
Approach	External	Endoscopic, endonasal
Physiological consideration	Less physiological as orbicularis oculi gets severed	More physiological as orbicularis oculi and its pumping mechanism are preserved
Scar	External facial scar present	Avoids a facial scar
Correction of nasal pathologies	Cannot be corrected	Can be corrected in a single procedure
Anesthesia	Local anesthesia is preferred	General anesthesia is preferred
Training requirement	No additional training	Requires specialized training for using endoscopes
Equipment	Inexpensive	Relatively expensive
Active infection	Cannot be done in cases of active infection	Active infection is not a contraindication

Table 1.
Comparison between external and endoscopic DCR.

The major differences between the two procedures are summarized below in a tabular manner (**Table 1**).

12. Balloon catheter dilation of NLD/balloon dacryoplasty

It is a minimally invasive procedure to establish lacrimal system patency in patients with congenital NLDO [18]. It is typically performed in children above 12 months of age.

- Indications: failed probing, restenosis after failed silicone intubation
- Anesthesia: performed under general anesthesia
- Steps:

  A. Superior and inferior puncta are dilated using a punctual dilator

B. The balloon catheter is lubricated and inserted into the upper lacrimal system till the 15 mm mark. It is slowly advanced towards the nose and can be grasped via the nostril once it is in place.

C. The balloon is then inflated to 8 atmospheres for 90 seconds, then deflated, and then reinflated to 8 atmospheres for 60 seconds.

D. It is then gently retracted 5 mm where the same steps are repeated.

E. It is then removed by twisting in an anticlockwise manner.

- Postoperative care: topical antibiotic eye drops are prescribed for a week after the procedure. Usually analgesics are not required.

13. Recent advances

A new, novel lacrimal ostium stent (LOS) is currently being evaluated for use in patients with a small lacrimal sac, who are undergoing EE-DCR. Most cases of EE-DCR with a failed result, are attributed to the closure of the nasal ostium. In such cases, LOS comes in handy as it prevents the closure of the nasal ostium and thus, can enhance the success rate of the procedure.

The LOS is a tripartite structure composed of silicone. It has a smooth surface and has a central hollow tube which facilitates lacrimal drainage. It has also got an elliptical repositioning plate and four buckles which facilitate a strong fixation [19].

14. Conclusion

Dacryocystitis stands as a primary culprit behind epiphora and periorbital swelling, stemming from either nasolacrimal duct obstruction or lacrimal passage blockage, culminating in infection. It presents as either congenital or acquired, with the latter further categorized into primary or secondary forms. A comprehensive assessment involving physical examination and diagnostic tests is imperative for accurate diagnosis, emphasizing collaborative efforts between ophthalmologists and otorhinolaryngologists for optimal patient care. While acute cases typically receive conservative management, surgical intervention emerges as the cornerstone for chronic cases. Although external DCR (Ex-DCR) historically served as the gold standard surgical intervention, the tide is shifting towards Endoscopic Endonasal DCR (EE-DCR) owing to its myriad advantages.

Acknowledgements

I would like to express my sincere gratitude to my professors who have always been a beacon of wisdom and support in my professional journey. I would also like to thank my senior colleagues and coworkers, with whom I have had the pleasure to learn and grown both personally and professionally.

I want to thank authors Mohammed Javed Ali et al., Balwant Singh Gendeh and Roger S Taylor et al. whose wonderful illustrations have helped me gain a deeper understanding into the topic and hopefully, by using some of their illustrations, I will be able to reach out to a wider audience who will greatly benefit and learn, just like I had.

Conflict of interest

None.

Notes

I would like to express my gratitude to my family and friends who have always supported me and encouraged me to keep going, especially when things got tough. I want to humbly thank my teachers and professors, who have always motivated me and instilled in me a sense of curiosity which enables me to embark on new academic ventures. Lastly, I want to thank my pets—Taco and Jordan—for always giving me love and keeping stress at bay.

Author details

Sindhuja Bhanwala
University College of London NHS Foundation Trust, London, United Kingdom

*Address all correspondence to: sindhuja1495@gmail.com

References

[1] Chen L, Fu T, Gu H, Jie Y, Sun Z, Jiang D, et al. Trends in dacryocystitis in China. Medicine. 2018;**97**(26):e11318

[2] Singh S, Ali MJ. Congenital dacryocystocele: A major review. Ophthalmic Plastic & Reconstructive Surgery. 2019;**35**(4):309-317

[3] Sáenz González AF, Busquet I, Duran N, Arámbulo O, Badal Alter JM. Chronic dacryocystitis caused by sarcoidosis. Archivos De La Sociedad Espanola De Oftalmologia. 2019;**94**(4):188-191. Available from: https://www.pubmed.ncbi.nlm.nih.gov/30558969/

[4] Takahashi Y, Kakizaki H, Chan WO, Selva D. Management of congenital nasolacrimal duct obstruction. Acta Ophthalmologica. 2009;**88**(5):506-513

[5] Engelsberg K, Sadlon M. First-onset dacryocystitis: Characterization, treatment, and prognosis. Ophthalmology and therapy. 2022;**11**(5):1735-1741

[6] DrPA S. Chronic dacryocystitis—It's evaluation and management by various investigative and diagnostic test. IOSR Journal of Pharmacy (IOSRPHR). 2013;**03**(10):28-33

[7] Philadelphia LK, Beirut, Lebanon, and William R. Katowitz, MD. Managing Dacryoceles and Dacryocystitis. Newark, New Jersey: Jobson Medical Information LLC; [Internet]. 2018. Available from: https://www.reviewofophthalmology.com/article/managing-dacryoceles-and-dacryocystitis

[8] Vagge A, Ferro Desideri L, Nucci P, Serafino M, Giannaccare G, Lembo A, et al. Congenital nasolacrimal duct obstruction (CNLDO): A review. Diseases. 2018;**6**(4):2-6. Available from: https://www.ncbi.nlm.nih.gov/pmc/articles/PMC6313586/

[9] JDH Y, CJ ME, Ogston SA. Congenital nasolacrimal duct obstruction in the second year of life: A multicentre trial of management. Eye. 1996;**10**(4):485-491. Available from: https://www.nature.com/articles/eye1996107

[10] Prat D, Magoon K, Revere KE, Katowitz JA, Katowitz WR. Management of pediatric acute dacryocystitis. Ophthalmic Plastic and Reconstructive Surgery. 2021;**37**(5):482-487. Available from: https://www.pubmed.ncbi.nlm.nih.gov/33782322/

[11] Cahill KV, Burns JA. Management of acute dacryocystitis in adults. Ophthalmic Plastic & Reconstructive Surgery. 1993;**9**(1):38-42

[12] Metson R. The endoscopic approach for revision dacryocystorhinostomy. The Laryngoscope. 1990;**100**(12):1344-1347. Available from: https://www.pubmed.ncbi.nlm.nih.gov/2243532/

[13] Rice DH. Endoscopic intranasal dacryocystorhinostomy—A cadaver study. American Journal of Rhinology. 1988;**2**(3):127-128

[14] McDonogh M, Meiring JH. Endoscopic transnasal dacryocystorhinostomy. The Journal of Laryngology & Otology. 1989;**103**(6):585-587

[15] Dresner SC, Klussman KG, Meyer DR, Linberg JV. Outpatient dacryocystorhinostomy. Ophthalmic Surgery. 1991;**22**(4):222-224. Available from: https://www.pubmed.ncbi.nlm.nih.gov/1772483/

[16] Emmerich KH, Busse H, Meyer-Rüsenberg HW. Dacryocystorhinostomia externa. Technique, indications and results. Der Ophthalmologe: Zeitschrift Der Deutschen Ophthalmologischen Gesellschaft. 1994;**91**(3):395-398. Available from: https://www.pubmed.ncbi.nlm.nih.gov/8086760/

[17] Çokkeser Y, Evereklioglu C, Comparative EH, Dacryocystorhinostomy EVE. Results in 115 patients (130 eyes). Otolaryngology-Head and Neck Surgery. 2000;**123**(4):488-491

[18] Robbins Tien D, Young D. Balloon dilation of the nasolacrimal duct. Journal of American Association for Pediatric Ophthalmology and Strabismus. 2005;**9**(5):465-467

[19] Yu B, Zhou GM, Shi JL, Wu ED, Wu WC. Management of chronic dacryocystitis cases after failed external dacryocystorhinostomy using endoscopic technique with a novel lacrimal ostium stent. International Journal of Ophthalmology. 2022;**15**(3):413-419

Chapter 8

Advancements in the Application of Medical Laser in Rhinology

Lingling Zhou and Yu Huang

Abstract

With the development of laser technology and the advancement of otolaryngology, medical laser is playing increasingly important roles in the treatment of otolaryngology diseases, especially in early-stage laryngeal cancer by trans-oral endoscopic surgery, which significantly improved patients' survival and quality of life. Medical laser in the treatment of nasal diseases started, respectively, late, which is still under exploration. To advocate the application of medical laser in rhinology and provide inspiration for exploring new ideas of medical laser in this field, current application of medical laser in the treatment of nasal diseases is discussed, combined with our experience in clinical practice. Medical laser can be applied in various rhinologic practices. Laser surgery technology is playing increasing roles in rhinology, and it is of great potential that medical laser will bring new expansion to the field of minimally invasive nasal surgery.

Keywords: medical laser, rhinology, dacryocystorhinostomy, turbinate reduction, nostril atresia, rosacea

1. Introduction

Medical laser refers to utilization of advanced laser technology in a broad area to meet challenges in clinical diagnosis and therapy and to address healthcare issues that impact broad populations [1]. With the development of laser technology as well as the advancement of otolaryngology surgery conceptions and skills, new applications of medical laser have been explored to help diagnosis and therapy for otolaryngology diseases. Medical laser is playing increasingly important roles in the treatment of otolaryngology diseases, for instance, especially in the treatment of early and middle-stage laryngeal cancer by trans-oral endoscopic minimally invasive surgery, which has modified surgical approaches and significantly improved patients' survival and quality of life. Medical laser in the treatment of nasal diseases started, respectively, late, which is still in the stage of exploration. In this chapter, we will discuss the application of medical laser in the treatment of nasal diseases, combined with our experience of diagnosis and treatment in clinical practice, in order to advocate the applications of medical laser in rhinology and promote inspiration for exploring new applications of medical laser in this field.

2. Laser characteristics for medical application

Laser is abbreviated terminology of light amplification by stimulated emission of radiation. In 1960, physicist Theodore Maiman invented the world's first laser device, a ruby laser [2].

Laser has high monochromaticity, high directivity, high brightness, and good coherence. In medical procedures, these characteristics are utilized to convert laser energy into thermal effect on the focused area, so that the instantaneous temperature can reach up to 200 ~ 1000°C, leading to coagulation, incision, and vaporization of the target tissue.

Coagulation: The thermal effect of laser causes the temperature rising of target tissue, and when it exceeds 56°C, the tissue protein becomes denatured and coagulated, which can destroy the lesion, or achieve hemostasis by closing the blood vessels.

Incision: Laser energy and thermal effect are enhanced compared to coagulation mode, to achieve incising effect by rapidly burning up of target tissue, leaving necrotic areas on both sides of the incision.

Vaporization: The laser energy and thermal effect are further enhanced, and the local high temperature makes the tissue instantaneously transform into smoke and vapor, so as to eliminate the lesion.

3. Classification of commonly used medical lasers

Classification of medical lasers is typically according to the main materials they are composed of. Commonly used medical lasers include solid-state laser, gas laser, excimer laser, dye laser, semiconductor laser, and so on, and each of them have diverse characteristics, respectively.

3.1 Solid-state laser

The main material is glass or crystal material, adding ionic elements such as ruby, sapphire, neodymium-doped yttrium aluminum garnet (Nd:YAG), erbium-doped garnet (Er:YAG), neodymium-doped glass (Nd:Glass), KTP laser, holmium laser, thulium laser, and so on. The solid-state laser has the merits of high optical power, excellent beam quality, and wide tuning range [3].

3.2 Gas laser

The main material to release gas laser is the gas inside the medium such as helium-neon (He-Ne), argon ion, CO_2, copper vapor laser, and so on. Gas lasers were the first continuous wave (CW) lasers [4].

3.3 Excimer laser

The reactive gas is mixed with the inert gas as the medium to generate the laser such as chlorine/fluorine - argon/krypton/xenon laser. Excimer lasers have the characteristics of short wavelength and high power, belonging to the pulse laser.

3.4 Dye laser

Organic dyes are used as laser media in solution or suspensionsuch as PDL laser, high power, and good optics, but low efficiency, need to replace the dye. They generally allow for a wider range of wavelengths compared to solid-state or gas lasers [4].

3.5 Semiconductor laser

Generally refers to diode laser, which has low price, high efficiency, low environmental requirements, and long service life. The p–n junction of a semiconductor diode forms the active medium [4].

Among them, solid-state laser and gas laser are considered to have higher output power and energy efficiency. The commonly used lasers in otolaryngology include neodymium-doped yttrium aluminum garnet (Nd:YAG), KTP laser, holmium laser, helium-neon (He-Ne), CO_2, PDL laser, and so on.

4. History of medical laser application in otolaryngology

The technology of laser was first introduced by Theodore Maiman in Hughes Aircraft Laboratories (HRL) in 1960 [5]. The 33-year-old physicist created a kind of light that had not been observed previously from any natural origin, by stimulating specific cubic ruby material to emit red pulsating light with high brightness and high strength. Thereafter, a variety of materials have been gradually found presenting similar characteristics.

Ophthalmologist Charles J Campbell from Columbia Presbyterian Medical Center, who noticed the potential value of laser for clinical application, performed a retinal laser surgery in 1961, which was the first laser surgery in human body [6], resecting retinal neoplasm using a ruby laser.

In 1963, Kumar Patel invented CO_2 laser, which was considered a most appropriate solution designed for clinical use in otolaryngology [7].

In 1967, Polanyi and Bredemeier invented an optical transmission system suitable for otolaryngological endoscopic procedures [8]. The equipment was composed of quite similar components like lasers we use today.

In 1972, Strong and Jako performed the first laser surgery of laryngeal cancer resection [9]. Continuous CO_2 laser was coupled to surgical microscope, to resect lesions on vocal cord in a highly controllable mean.

Medical laser has played a significant role in changing surgical methods, improving patients' quality of life and prognosis in the treatment of early and middle-stage laryngeal cancer through oral endoscopic minimally invasive surgery. Applications in the ear include laser tympanostomy, stapedotomy, etc. Medical laser in remedy for nasal diseases started late, which is still in the stage of exploration.

Application of medical laser in rhinology started in late 1970s [10], as H Lenz from Germany utilized argon-ion laser in carbonization of the respiratory mucosa of the lower turbinates in vasomotor rhinitis [11]. Argon, KTP, and diode lasers were used in managing epistaxis resulting from hereditary hemorrhagic telangiectasia [12]. A variety of rhinology scenarios have been explored to attempt applicability of medical laser till now, which will be introduced in this chapter.

5. Laser commonly used in otolaryngology

Based on the classification of lasers referred above, there are various lasers for medical use in different medical fields. Among them, lasers commonly used currently in otolaryngology are listed below.

5.1 CO_2 laser

Of 10,600 nm wavelength, it is easily absorbed by water components in soft tissue, with rapid gasification and cutting effect on tissue. CO_2 laser causes mild damage to peripheral tissue. The classic optical guide arm mode is mostly used, and the optical fiber mode has appeared in recent years, as supplementary component to access particular surgical field that could not be reached by optical guide arm mode. CO_2 laser is the most commonly used laser in otorhinolaryngology.

5.2 Nd:YAG laser

Of 1064 nm wavelength, it is mainly absorbed by cellular proteins, so as to take thermal effect on deep tissue. On the other hand, this type of laser is of higher possibility in inducing postoperative scar and blisters.

5.3 Ho:YAG laser

Of 2014 nm wavelength, it is easily absorbed by water in the tissue, producing micro-bubbles to separate and cut hard tissue, like bones or calculi. The cutting effect on tissue is relatively superficial.

5.4 KTP laser (potassium titanium oxide phosphate laser, green laser)

Of 532 nm wavelength, it can be selectively absorbed by hemoglobin in the tissue, exhibiting good hemostatic effect. It produces less vapor in surgical field, providing a better view for surgeons.

5.5 PDL laser (pulsed dye laser)

Of 585/595 nm wavelength, it is mainly absorbed by hemoglobin and melanin in tissue, which can effectively treat a variety of vascular diseases.

5.6 He-Ne laser (red laser)

Of 632.8 nm wavelength, it is the most commonly used gas laser in medical field, which can be non-specifically absorbed by hemoglobin and proteins in tissue.

6. The application of medical laser in nasal diseases

Medical laser can be applied in various rhinologic clinical practices, ranging from inflammatory situations as allergic or nonallergic rhinitis, sinusitis, structural disease as turbinate hypertrophy, nostril atresia, nasal adhesion, nasal-orbital disease as dacryocystitis and abscess, epistaxis, and selected situation of neoplasms. Specific applications mainly include dacryocystorhinostomy, resection of external nasal masses and pigmented nevus, resection of benign and malignant tumors of nasal cavity and sinuses, hemostasis of nasal hemorrhage, and laser resection of nasal mucosal hemangioma, partial turbinectomy, sinusotomy and resection of nasal polyps, choanal plasty for posterior nostril atresia, nasal adhesion separation, and treatment of rosacea.

6.1 Dacryocystorhinostomy

Dacryocystorhinostomy is indicated in dacryocystitis and abscess caused by chronic lacrimal duct obstruction resulting from various reasons. Laser is utilized to incise the dacryocyst projection area on the lateral wall of the nasal cavity, including the bony structures of lacrimal bone and agger nasi, to achieve the effect of fusion of the dacryocyst cavity and the nasal cavity, as treatment of lacrimal cyst obstructive inflammation (See **Figure 1**). High power laser such as Ho-YAG laser is proposed [13, 14].

In clinical practice, the authors' experience is that after ablating the frontal process of maxilla and the lacrimal bone covering the dacryocyst area, a perpendicular incision could be made along the posterior margin of the cyst to incise into it, in order to avoid damaging the common lacrimal duct. The fusion of the dacryocyst cavity and the nasal cavity should be sufficient, so as to access full drainage.

6.2 Resection of external nasal masses and pigmented nevus

Regarding the size and depth of external nose, lesions like epithelial masses and pigmented nevus can be completely removed by cutting or vaporizing with the laser's photothermal effect. Commonly used lasers are CO_2 laser, and Nd:YAG laser [15, 16]. In case some lesion should have potentiality of malignancy, cutting with laser and keeping certain safe margins is abiding by the principle of en-bloc resection for tumors.

Furthermore, local treatment with CO_2 laser and pulsed dye laser is reported being a new and effective method for skin flap necrosis after basal cell carcinoma resection [17].

6.3 Resection of benign and malignant tumors of nasal cavity and sinuses

Laser has a unique role in the resection of benign and malignant tumors in the nasal cavity and sinuses, such as the resection and photocoagulation of the original

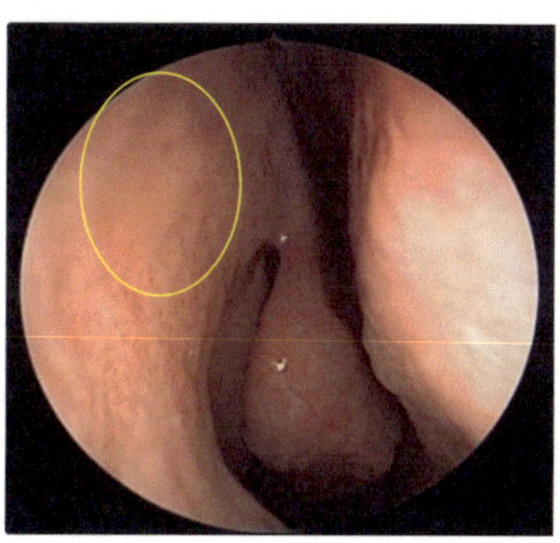

Figure 1.
Laser for dacryocystorhinostomy. Yellow circle: Dacryocyst projection area.

site of inverted papilloma (See **Figure 2**), the resection of malignant tumors regarding the safe margin, and the coagulation treatment of the tumor bed. Suggested types are KTP laser, PDL laser, etc. [14]. Some tumors are associated with possibility of bone invasion, while bone drilling should be performed along with laser resection and coagulation, to confirm acquisition of safe margins.

6.4 Hemostasis of nasal hemorrhage and laser resection of nasal mucosal hemangioma

For nasal bleeding of small vessel rupture, or mucosal hemangioma, the bleeding site and the surrounding area can be coagulated by laser in a non-contact way, and the small hemangioma was coagulated and blocked (See **Figure 3**). Suggested types are KTP laser, PDL laser, etc. [14]. The non-contact coagulation on the mucosa can minimize irritation to nasal normal condition, so as to help the patients recover with minimized nasal discomfort including nasal blockage, rhinorrhea, or subsequent risk of adhesion.

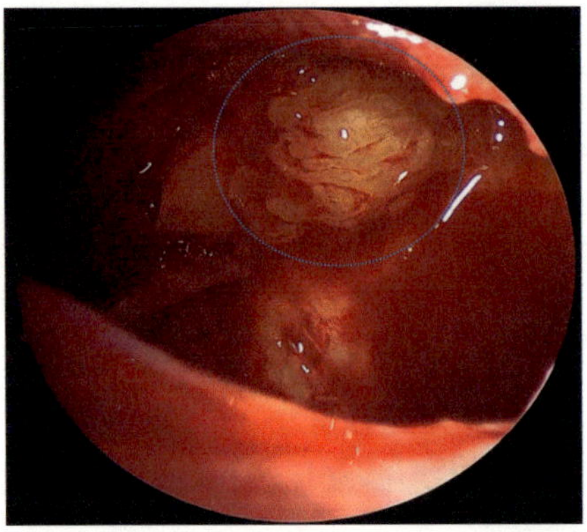

Figure 2.
Resection of stage 1 inverted papilloma. Blue dotted circle: Laser applied region.

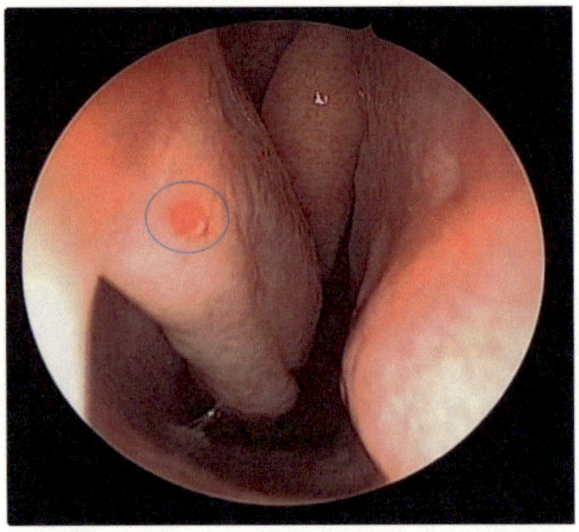

Figure 3.
Hemostasis of nasal hemorrhage by laser. Blue circle: Laser applied region.

DOI: http://dx.doi.org/10.5772/intechopen.1006193

6.5 Partial turbinectomy

For some patients suffering from nasal blockage by turbinate hypertrophy due to allergic rhinitis, vasomotor rhinitis, hypertrophic rhinitis, etc., nasal spray and oral medication effect are sometimes not satisfying. By laser surgery, partial removal of the turbinate tissue, coagulation, and postoperative scarring of the tissue can achieve the effect of reducing the volume of the turbinate, improving nasal ventilation and drainage, as well as reducing nasal secretions, which is rapid, effective, and with few complications. On the other hand, posterior nasal nerve ablation by laser can relieve nasal irritation symptoms of AR and VMR (See **Figure 4**) [14, 18]. CO_2 laser, KTP laser, and He-Ne laser can be used.

Raj Tajamul Hussain et al. conducted a prospective observational study of 53 symptomatic inferior turbinate hypertrophy patients who had poor effect by medication and found that CO_2-laser turbinoplasty improved nasal symptom scores, including nasal congestion symptom rating scale (NOSE) and visual analog scale (VAS) scores [19].

6.6 Sinusotomy and resection of nasal polyps

For patients with chronic rhinosinusitis and nasal polyps, laser can be used in functional nasal endoscopic sinus surgery, so as to reduce the sinus scarring, ensure

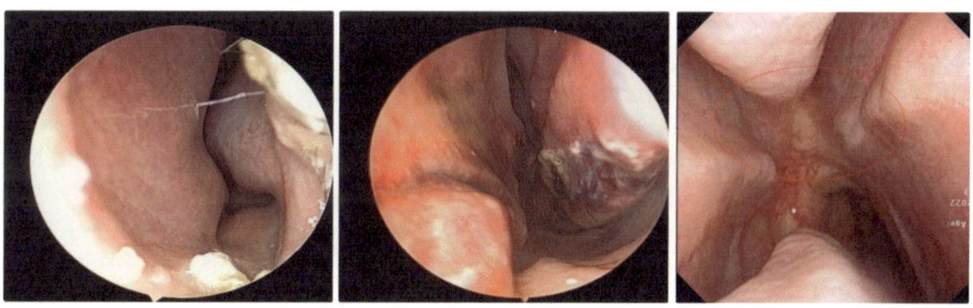

(A) (B) (C)

Figure 4.
Turbinate hypertrophy due to allergic rhinitis, vasomotor rhinitis, hypertrophic rhinitis, etc. A. Turbinate hypertrophy before laser. B. Nasal cavity view 1.5 weeks after laser surgery. C. Laser applied region for posterior nasal nerve ablation (Red oval).

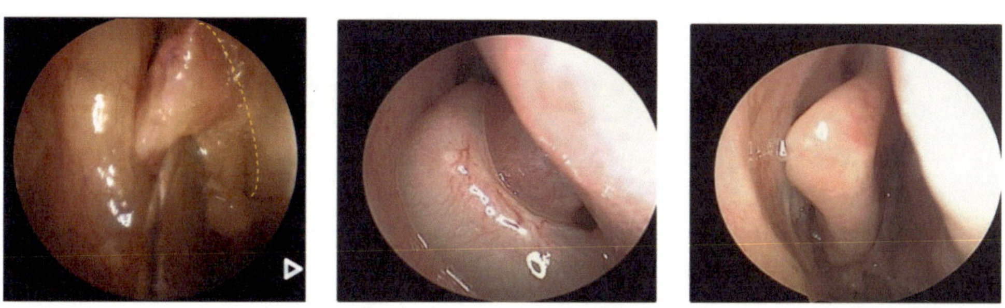

(A) (B) (C)

Figure 5.
A. Resection of polyps by laser. Yellow dotted line: Laser applied region. B. Polyp before laser surgery. C. Nasal cavity view 2 months after polyps laser surgery.

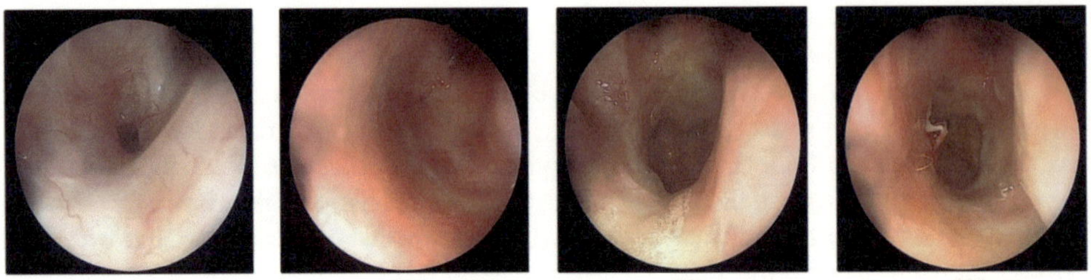

(A) (B) (C) (D)

Figure 6.
Endoscope view of a patient with bilateral posterior nostril atresia treated by laser choanal plasty. A. Right side, before surgery. B. Left side, before surgery. C. Right side, 1 month after surgery. D. Left side, 1 month after surgery.

long-term drainage, and carry out vaporization of nasal polyps. Laser ablation of the base of nasal polyposis can promote tissue scarring and reduce recurrence (See **Figure 5**) [14]. Fiber laser such as KTP laser and Ho:YAG laser can be used. As development of more effective instruments like powered microdebrider and plasma ablation, laser is relatively less suitable for sinusotomy and resection of nasal polyps, while it could still be an alternative in particular scenarios.

6.7 Choanal plasty for posterior nostril atresia

Posterior nostril atresia can be caused by congenital malformation/trauma/post-radiotherapy scar adhesion. The atretic tissue is different according to etiology, mainly of membrane/bone/mixed composition. According to the specific atretic tissue, appropriate type of laser is selected for excision, with less trauma and less scar contracture, and can be formed in one stage [20]. Suggested types are CO_2 laser, KTP laser, Ho:YAG laser, etc., depending on the atretic tissue. As shown in the figure, the author treated a patient with bilateral posterior nostril atresia after radiotherapy for nasopharyngeal carcinoma and achieved good results after CO_2 laser treatment, who has fluent nasal airflow and is under further follow-up (See **Figure 6**).

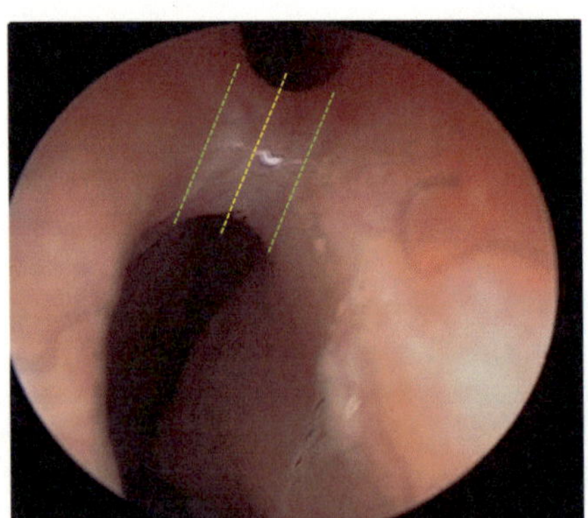

Figure 7.
Nasal adhesion to be separated by laser surgery. Yellow dotted line: Laser incision. Green dotted line: Laser ablation on proliferated scar tissue.

6.8 Nasal adhesion separation

Nasal adhesion is a usual complication after endoscopic surgery, nasal trauma, or inflammation. Fiber laser can be used to separate nasal adhesions, and local combination with mitomycin can enhance the efficacy and prevent re-adhesions [21]. CO_2 laser and KTP laser can be selected.

In clinical practice of the authors' institute, a 'incise-coagulate' method is utilized. The adhesion tissue is incised in the middle, then both ends are coagulated in endoscopic vision, to ensure sufficient coagulation of scar tissue, as well as avoid extensive injury to normal mucosa and vessels (See **Figure 7**).

6.9 Treatment of rosacea

Rosacea is a common chronic progressive skin inflammation characterized by persistent flushing, skin erythema, subcutaneous telangiectasia, papules, pustules, and nodoid changes in nasal appearance [22]. The laser of 500–600 nm wavelength can be selectively absorbed by the target tissue of rosacea lesion, to eliminate erythema and capillary dilatation, and the combination with hydroxymethazoline/hydroxychloroquine can enhance the efficacy [23]. Options include PDL laser, KTP laser, etc. Local skin pain, purpura, and pigmentation may occur after treatment. PDL laser significantly reduces nasal erythema and swelling.

There are some cautions after rosacea laser surgery. After laser surgery, patients may experience purple skin discoloration and discomfort as the area heals. Ice packs and steroids can reduce postoperative pain and swelling. Keeping the treatment area covered with vaseline ointment may support healing. Pain can usually be controlled with over-the-counter medications. It is recommended to use sunscreen outdoors during the healing period [24].

7. The development of new laser technology

With the development of new laser technology, we expect to apply some new technologies to the field of nasal surgery in the future. Femtosecond laser technology and lattice laser technology are introduced here. Furthermore, new materials are continuously being found capable for laser mediums, which have various new characteristics, presenting potentials for expanding the utilization of medical lasers.

7.1 Femtosecond laser technology

Femtosecond laser (FSL) works in ultrafast pulses with short duration of each pulse (10^{-15} second level), operates in the range of tens of to hundreds of femtoseconds, which can be focused in a very small space volume with a very high instantaneous power, generating little effect on the surrounding tissue. As high instantaneous power, focused precisely on the target lesion, a large number of microbubbles are generated through photolysis to separate tissues. FSL can easily achieve a spatial resolution of less than 100 nm to assure accuracy in operation [25]. The current application of FSL is mainly in the field of ophthalmic precise surgeries. It may be developed and applied in the field of rhinology surgery as necessity for specific scenarios requiring higher precision.

7.2 Lattice laser technology

Lattice laser was introduced by Dr. Rox Anderson of Harvard University in 2004. A lattice microtherapy area (MTZ) is arranged by the fusion of superpulse laser technology and computer graphics technology, as tissue vaporization caused by laser to produce many three-dimensional tiny thermal damage strips of uniform size and uniform arrangement surrounded by otherwise normal tissue, to stimulate the initiation of target tissue repair and reconstruction. The treatment depth, density, and treatment range can be preset, with high precision and small collateral damage [26–28]. It is mainly used in the treatment of skin hemangioma, pigmented nevus, vitiligo, scar, and so on. It may be developed and applied in the field of rhinology plastic surgery or therapy for other superficial lesions in epithelium or mucosa.

7.3 New laser types for candidate application in rhinology

There are continuously new laser types emerging along with the advancement of technology. Here, the author takes thulium fiber laser and autofocusing Bessel beams for example to introduce.

7.3.1 Thulium fiber laser

A novel thulium fiber laser (TFL) has been reported successfully moved forward from the preclinical trials into clinical practice and now is being widely used in clinics around the world. The wavelength of TFL is closer to the water absorption peak. It decreases carbonization and retropulsion ,which makes TFL an advantage in urologic surgery [29]. Thulium fiber laser has been subsequently compared with holmium laser in urologic clinical practice, to find TFL performing effective quick dusting and producing smaller fragments, as well as working at a lower power, leading to higher stone-free rate and lower risk of complications [30, 31]. TFL is currently mainly utilized in urology, while some situations similar to lithotripsy occurring in rhinology, such as dacryocystorhinostomy or other possible scenarios, should have the potential for application of TFL.

7.3.2 Autofocusing Bessel beams

A new family of abruptly autofocusing waves named autofocusing Bessel beams (ABBs) was introduced by Ding Z et al. in 2023, which only strongly influence the area near the focus. Such beams have a unique property: their maximum intensity changes slowly during propagation and suddenly increases by orders of magnitude near the focus [32]. This new kind of laser is of definite potential in medical field, playing profound roles in precise surgery meeting the need of medical frontiers, which may affect rhinology in unimaginable means.

8. Advantages of medical laser in rhinology

According to the authors' clinical experience, medical laser has some obvious advantages when applied in rhinology, mainly as listed below.

8.1 Non-contact resection or ablation

Nasal mucosa composed the lining surface of nasal cavity and sinuses, where extra touch applied by contacting devices such as microdebrider or coblation always causes extensive irritation to nasal mucosa, affecting comparatively more depth and/or area of the surgical region, inducing more scab formation followed by prolonged recovery and scar formation. Laser applied in rhinology works in a non-contact way to accomplish resection or ablation, reducing the extra risk attributed by contacting.

8.2 Instant closure of the wound

Laser energy absorbed by tissue cells or hemoglobin instantly degenerates the protein of the focused region, closing the wound and blood vessels in an efficient way, to reduce hemorrhage and tissue injury. Blood vessels less than 1 mm in diameter can be closed by laser.

8.3 Precise targeting

Laser beams applied in rhinology are usually less than 1 mm in diameter, restricting the target area of thermal effect. Comparing with microdebrider (about 4 mm in diameter) and coblation (about 4–5 mm in diameter for resecting, about 1 mm in diameter for submucosal coblation but invisible beneath the mucosa), laser has advantage in precise targeting and manipulation.

9. Potential risks and hazards of medical laser in rhinology

The most dangerous risk of medical laser in otolaryngology is airway fire [33, 34], which is a disastrous complication that should be carefully prevented, especially in trans-oral laryngeal surgeries. Laser in rhinology, as introduced above, also has potential risks of airway fire complication in some particular scenarios, resulting from the consistency of airway, which should be kept in mind.

Particulate matter exposures from medical lasers have been noticed, which was furthermore reported being able to cause dose-dependent respiratory inflammatory response in laboratory animals, and may contain viable bacteria and viruses, including human papilloma virus. A two-zone model was applied by Ramon Lopez, etc., in 2015 to estimate medical laser-generated particulate matter (LGPM) exposures in laser operations. Comparing to near-field from the simulated treatment room, far-field was detected to find significantly lower concentrations of LGPM. Larger room volume and higher air exchange rate were associated with lower LGPM concentrations in the near-field [35].

Another potential hazard is unexpected laser radiation to eyes of either operators or patients, which may cause eye injuries. Laser eyewear protection against laser radiation was suggested by Mike Regan on the 40th Annual Conference of the British Medical Laser Association in 2023 [36]. Appropriate eyewear protection for both operators and patients, which may be of different shapes or visibilities from each other, can avoid laser radiation injuries to their eyes.

10. Prospects

Laser surgery technology is playing increasing roles in rhinology, and it is of great potential that medical laser will bring new expansion to the field of minimally invasive nasal surgery. According to the different characteristics of different existing lasers, the application of these medical lasers in rhinology needs to be further explored and expanded in clinical practice (new application of existing technology), combining their characteristics and surgical need of rhinology to emerge new junctions. On the other hand, according to the needs of nasal surgery, new applicable laser hand-pieces will be developed, such as different integrated instrument functions (navigation, positioning, aspiration, etc.), to deliver laser beams in adequate forms and angles, or other specific desired ways. Laser surgery technology has been widely used in minimally invasive oral surgery, and it is of great potential that medical laser will bring new expansion, new surgical methods, and new surgical concepts, to the field of rhinology.

Acknowledgements

This chapter introducing advancements in the application of medical laser in rhinology was originally inspired by Dr. Jing Zhang of Department of Otolaryngology and Laser Surgery, Eye, Ear, Nose and Throat Hospital, Fudan University, Shanghai, China, who is chief of otolaryngology group of Laser medicine branch of Shanghai Medical Association, China. Thank her for inviting the author to talk about this very topic at annual meetings of otolaryngology group of Laser medicine.

Conflict of interest

There is no conflict of interest for this chapter.

Author details

Lingling Zhou[1] and Yu Huang[1,2*]

1 Department of Otorhinolaryngology, Head and Neck Surgery, Huadong Hospital Affiliated to Fudan University, Shanghai, China

2 Otolaryngology Group of Laser Medicine Branch of Shanghai Medical Association, China

*Address all correspondence to: huangyu1220@hotmail.com

References

[1] Sroka R, Stepp H, Hennig G, et al. Medical laser application: Translation into the clinics. Journal of Biomedical Optics. 2015;**20**(6):061110. DOI: 10.1117/1.JBO.20.6.061110

[2] Hecht J. A short history of laser development. Applied Optics. 2010;**49**(25):F99-F122. DOI: 10.1364/AO.49.000F99

[3] Qiao S, He Y, Sun H, et al. Ultra-highly sensitive dual gases detection based on photoacoustic spectroscopy by exploiting a long-wave, high-power, wide-tunable, single-longitudinal-mode solid-state laser. Light: Science & Applications. 2024;**13**(1):100. DOI: 10.1038/s41377-024-01459-5

[4] Mirza FN, Khatri KA. The use of lasers in the treatment of skin cancer: A review. Journal of Cosmetic and Laser Therapy. 2017;**19**(8):451-458. DOI: 10.1080/14764172.2017.1349321

[5] Maiman TH. Stimulated optical radiation in ruby. [J]. Nature. 1960;**187**:493-494

[6] Koester C, Campbell CJ. The first clinical application of the laser. In: Fankhauser F, Kwasniewska S, editors. Lasers in Ophthalmology: Basic, Diagnostic, and Surgical Aspects: A Review. The Hague: Kugler Publications; 2003. pp. 115-117

[7] Yan Y, Olszewski AE, Hoffman MR, et al. Use of lasers in laryngeal surgery. Journal of Voice. 2010;**24**(1):102-109. DOI: 10.1016/j.jvoice.2008.09.006

[8] Polanyi TG, Bredemeier HC, Davis TJ. A CO_2 laser for surgical research. Medical & Biological Engineering. 1970;**8**(6):541-548. DOI: 10.1007/BF02478228

[9] Strong MS, Jako GJ. Laser surgery in the larynx. Early clinical experience with continuous CO_2 laser. The Annals of Otology, Rhinology, and Laryngology. 1972;**81**(6):791-798. DOI: 10.1177/000348947208100606

[10] Karkos PD, Koskinas IS, Triaridis S, et al. Lasers in omicrontolaryngology: A laser odyssey from carbon dioxide to true blue. Ear, Nose, & Throat Journal. 2021;**100**(1_suppl):1S-3S. DOI: 10.1177/0145561320951681

[11] Lenz H, Preussler H. Histologic changes in the epithelium of the respiratory mucosa of the lower turbinates following argon laser strip carbonization in vasomotor rhinitis. Laryngologie, Rhinologie, Otologie. 1986;**65**(8):438-444

[12] Abiri A, Goshtasbi K, Maducdoc M, et al. Laser-assisted control of epistaxis in hereditary hemorrhagic telangiectasia: A systematic review. Lasers in Surgery and Medicine. 2020;**52**(4):293-300. DOI: 10.1002/lsm.23147

[13] Goel R, Nagpal S, Kamal S, et al. Study of microbial growth on silicone tubes after transcanalicular laser-assisted dacryocystorhinostomy and correlation with patency. Nepalese Journal of Ophthalmology. 2016;**8**(16):119-127. DOI: 10.3126/nepjoph.v8i2.16992

[14] Levine HL. Endonasal laser surgery: An update. Otolaryngologic Clinics of North America. 2006;**39**(3):493-501, viii. DOI: 10.1016/j.otc.2006.01.015

[15] Forbat E, Al-Niaimi F. Fractional radiofrequency treatment in acne scars: Systematic review of current evidence. Journal of Cosmetic and Laser Therapy. 2016;**18**(8):442-447. DOI: 10.1080/14764172.2016.1225964

[16] Delaney SW, Zhang P. Systematic review of low-level laser therapy for adult androgenic alopecia. Journal of Cosmetic and Laser Therapy. 2018;**20**(4):229-236. DOI: 10.1080/14764172.2017.1400170

[17] Nistico SP, Bennardo L, Sannino M, et al. Combined CO(2) and dye laser technique in the treatment of outcomes due to flap necrosis after surgery for basal cell carcinoma on the nose. Lasers in Surgery and Medicine. 2022;**54**(4):523-529. DOI: 10.1002/lsm.23502

[18] Krespi YP, Wilson KA, Kizhner V. Laser ablation of posterior nasal nerves for rhinitis. American Journal of Otolaryngology. 2020;**41**(3):102396. DOI: 10.1016/j.amjoto.2020.102396

[19] Hussain RT, Ahmad R. Role of CO(2) laser in the management of symptomatic hypertrophy of inferior nasal turbinates. Indian Journal of Otolaryngology and Head & Neck Surgery. 2022;**74**(Suppl. 2):826-834. DOI: 10.1007/s12070-020-01876-z

[20] Kubba H, Bennett A, Bailey CM. An update on choanal atresia surgery at Great Ormond Street Hospital for Children: Preliminary results with mitomycin C and the KTP laser. International Journal of Pediatric Otorhinolaryngology. 2004;**68**(7):939-945. DOI: 10.1016/j.ijporl.2004.02.010

[21] Hesham A, Fathi A, Attia M, et al. Laser and topical mitomycin C for management of nasal synechia after FESS: A preliminary report. European Archives of Oto-Rhino-Laryngology. 2011;**268**(9):1289-1292. DOI: 10.1007/s00405-011-1587-x

[22] Husein-ElAhmed H, Steinhoff M. Laser and light-based therapies in the management of rosacea: An updated systematic review. Lasers in Medical Science. 2021;**36**(6):1151-1160. DOI: 10.1007/s10103-020-03200-1

[23] Shaker G, Shaker A. A unique case of facial hypervascularity responding to pulsed-dye laser. Cureus. 2022;**14**(9):e28726. DOI: 10.7759/cureus.28726

[24] DeHart AN, Richter GT. Laser treatment of vascular anomalies. Dermatologic Clinics. 2022;**40**(4):481-487. DOI: 10.1016/j.det.2022.06.002

[25] Yan Q, Han B, Ma ZC. Femtosecond laser-assisted ophthalmic surgery: From laser fundamentals to clinical applications. Micromachines (Basel). 2022;**13**(10):1653. DOI: 10.3390/mi13101653

[26] Srisong H, Daduang S, Lopata AL. Current advances in ant venom proteins causing hypersensitivity reactions in the Asia-Pacific region. Molecular Immunology. 2016;**69**:24-32. DOI: 10.1016/j.molimm.2015.11.003

[27] Khan A, Al-Kathiri WH, Balkhi B, et al. The burden of bites and stings management: Experience of an academic hospital in the Kingdom of Saudi Arabia. Saudi Pharmaceutical Journal. 2020;**28**(8):1049-1054. DOI: 10.1016/j.jsps.2020.07.004

[28] Jin W, Li Z, Jin Z, et al. A novel technique for treating atrophic facial scars in Asians using ultra-pulse CO(2) laser. Journal of Cosmetic Dermatology. 2020;**19**(5):1099-1104. DOI: 10.1111/jocd.13335

[29] Taratkin M, Azilgareeva C, Cacciamani GE, et al. Thulium fiber laser in urology: Physics made simple. Current Opinion in Urology. 2022;**32**(2):166-172. DOI: 10.1097/MOU.0000000000000967

[30] Traxer O, Pearle M. Thulium fiber laser versus holmium:yttrium aluminum garnet for lithotripsy: Which

is the winner? European Urology. 2024;**85**(6):541-542. DOI: 10.1016/j.eururo.2024.02.023

[31] Uleri A, Farre A, Izquierdo P, et al. Thulium fiber laser versus holmium:yttrium aluminum garnet for lithotripsy: A systematic review and meta-analysis. European Urology. 2024;**85**(6):529-540. DOI: 10.1016/j.eururo.2024.01.011

[32] Ding Z, Yu Y, Li X, et al. Focusing property of autofocusing Bessel beams. Optics Express. 2023;**31**(20):33228-33240. DOI: 10.1364/OE.500383

[33] Leemann B, Heidegger T, Grossenbacher R, et al. A severe complication after laser-induced damage to a transtracheal catheter during endoscopic laryngeal microsurgery. Anesthesia and Analgesia. 2004;**98**(6):1807-1808. DOI: 10.1213/01.ANE.0000117144.16826.61

[34] Hermsen HW, Snijdelaar DG, Marres HA. Airway fire, a serious complication during laryngeal laser surgery. Nederlands Tijdschrift voor Geneeskunde. 2002;**146**(9):427-431

[35] Lopez R, Lacey SE, Jones RM. Application of a two-zone model to estimate medical laser-generated particulate matter exposures. Journal of Occupational and Environmental Hygiene. 2015;**12**(5):309-313. DOI: 10.1080/15459624.2014.989361

[36] Mike R. Update on laser eyewear standards. Selected abstracts from the 40th Annual Conference of the British Medical Laser Association: BMLA 2023 Annual Conference. Lasers in Medical Science. 2023;**38**(1):197. DOI: 10.1007/s10103-023-03842-x

Chapter 9

Implants and the Maxillary Sinus: Modern Oral Implantology Changed Paradigms

Stefan Ihde and Fadia Awadalkreem

Abstract

The maxillary sinus is the largest paranasal sinus located within the maxilla's body. The health of the maxillary sinus is of significant importance for the body's functions and health. Sinus pneumatization is a physiological process in which the volume of the sinus increases over time, exposing the roots of the teeth and leading to direct engagement of the roots with the sinus. Similarly, implants may protrude inside the maxillary sinus in cases with compromised ridge support, posing a challenge in implant treatment with respect to sinus health and implant survival rate. This chapter aims to elucidate the main aspect of the maxillary sinus's health and to analyse the various aspects of implant protrusion inside the maxillary sinuses.

Keywords: dental implant, implant protrusion, maxillary sinus, osseofixated implant, sinus health

1. Introduction

The paranasal sinuses are paired air-filled spaces, lined with mucosa, situated in the maxillofacial region and communicating with the nasal cavity [1]. The paranasal sinuses form an integral part of the respiratory tract in addition to the nose, tracheobronchial tree and lung.

Four pairs of paranasal sinuses have been elucidated, including the maxillary, ethmoid, frontal and sphenoid sinuses. Among them, the maxillary sinus is the largest, with a pyramidal shape with a variable size [2–11], not only between the different individuals [higher volume in male] [3, 11], but also between the two sides of the same individual, and an average volume of around 15–30 cubic centimetres (cc) per side 4–10 [3, 11]. It was first described by Leonardo da Vinci in 1489, and later in 1651, an English anatomist named Nathaniel Highmore documented it as *the antrum of highmore*.

2. Development of the maxillary sinus

The maxillary sinus is the first paranasal sinus to develop. At the 10th week of foetal life, it appeared as a mucosal invagination associated with the deeper cranial tip ethmoidal infundibulum; later, at the 11th week, it formed an oval-shaped space, indicating the primitive (early) maxillary sinus [3, 12]. A rapid maxillary sinus growth occurred during the 4th to 5th and 6th to 7th months to be measured at birth (10 mm × 3 mm × 4 mm). Additional episodes of slow and rapid growth continue after birth until it reaches its full size after the eruption of all the permanent teeth [13].

3. Anatomy of the maxillary sinus

The maxillary sinus is situated within the body of the maxillary bone with six wall boundaries: the superior, inferior, anterior, posterior, medial and lateral walls. The superior, anterior, lateral and medial walls are broad, while the posterior and inferior walls are narrow [1, 14–16].

By forming the orbital floor, the superior wall separates the maxillary sinus from the orbital content. The front wall has a noticeable infero-lateral focal convexity, a canine eminence and a slightly anteriorly concave surface. The narrow and thin posterior bony border correlates with the maxillary artery, maxillary vein and the maxillary division of the trigeminal nerve. On the buccal side, the lateral wall descends and meets the lower part of the alveolar ridge. It faces posterior-laterally towards the infratemporal fossa. The lateral wall of the nasal cavity forms the medial wall of the sinus, while the inferior wall of the sinus continues with the maxillary alveolar process and contains the roots of the maxillary dentition (**Figures 1, 2**) [1, 14–16].

A thin layer of compact bone typically separates the molar roots from the floor of the antrum. However, the roots of the first and second molars may form conical projections (recess) while projecting into the antral floor. In some cases, the first, second, pre-molar and third molar roots, rarely canine roots, may protrude into the sinus (**Figure 3**) [3, 17, 18].

The maxillary sinus had four recesses: the infraorbital recess superiorly, the zygomatic recess laterally, the alveolar recess inferiorly and finally, the palatine recess. The sinus opening is located superiorly, where the ostium communicates with the nasal cavity [19].

The inside of the maxillary sinus is lined with a pseudostratified columnar ciliated epithelium with many goblet cells and the *Schneiderian membrane* that provide

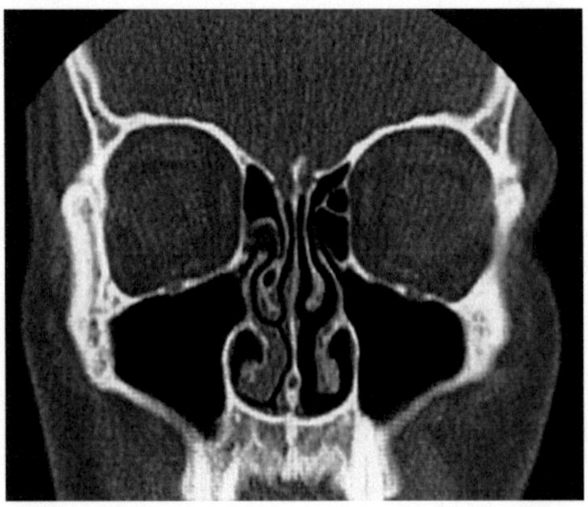

Figure 1.
A. cone beam CT showing a normal maxillary sinus.

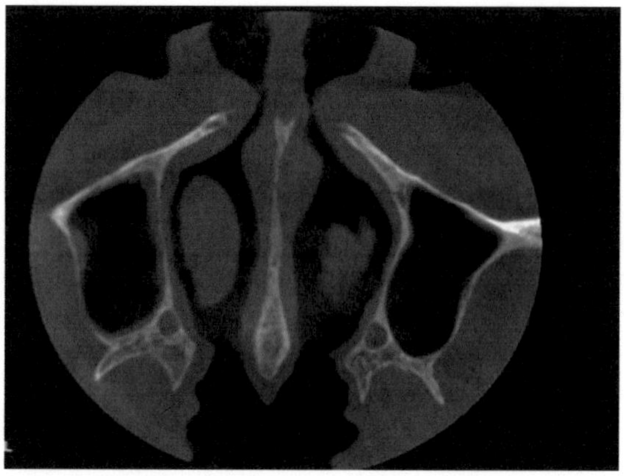

Figure 2.
A cone beam CT, axial view, illustrating a normal maxillary sinus.

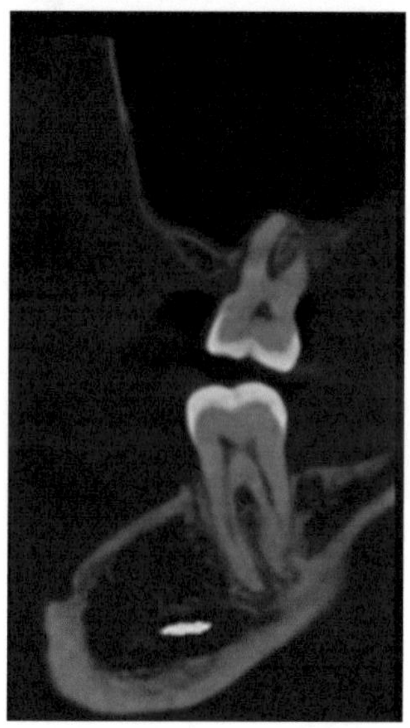

Figure 3.
A cone beam CT, sagittal view, showing the protrusion of the maxillary molar's root inside the sinus.

the necessary blood supply and serve thereby as a periosteum [1, 20–22]. Since the forces acting on the inside of the maxillary sinus are nil and almost no blood supply is required for the underlying bone, its secretorial function into the lumen of the maxillary sinus consumes most of the available blood.

4. Blood supply for the maxillary sinus

The maxillary artery, through the posterior superior alveolar artery, the infraorbital artery and the posterior lateral nasal artery, provides the core blood supply for the maxillary sinus [22–24].

5. Innervation of the maxillary sinus

The infraorbital, anterior, middle and posterior superior alveolar branches of the maxillary nerve provide the sinus with nerve sensation [22–25].

6. Detox-function of the maxillary sinus

The nasal mucosa provides the respiratory tract's primary defence mechanism against inhaled pollutants, allergens and pathogens. The maxillary sinus is protected against the intrusion of infective agents by the constant outflow of liquid. Each maxillary sinus produces more than 1.5 l liquid per day, and this liquid also rinses the nose. The mucociliary clearance mechanism provides the necessary layer of mucous and the direction of the clearance. The mucous will then glue all particles and pathogens inhaled by the nose to the mucus for natural excretion [26, 27].

Consequently, there is a direct correlation between the health of the paranasal sinuses and the nose and the efficacy of mucociliary clearance. Any condition that disturbs the normal mucociliary clearance function can result in rhinosinusitis [26, 27].

Just as the mucosa of the pharynx, the Schneiderian Membrane allows the ejection of exosomes that are transported through the lymphatic system, as a result, this mucosa in the pharynx and the maxillary sinus performs a very important function for body detoxification. The content of the exosomes will be forwarded by ciliae out of the maxillary sinus and into the nose [26, 27]. From there, this outflow of the maxillary sinus, including the toxins, is swallowed (or cleaned) and then excreted [26, 27]. If a massive toxic attack on the human body occurs, a large number of exosomes will require their own elimination, which may lead to symptoms of the flu.[1] The British medical researcher Herbert Sheldon clarified already in 1944 as follows: "There was never any medicine against the flu, the flu is the medicine". This sentence clarifies that therapy against detoxification must be done and that all medical efforts to reduce this natural process of detoxification are resulting in even more problems.

7. Sinus "pneumatization" and "contraction"

The bones surrounding the maxillary sinus have a large potential to undergo morphological changes. This is due to the fact that they undergo frequent waves of remodelling in response to tooth extractions and changes in the masticatory pattern. Both the inner and outer cortical of the maxillary sinus are not stable and are prone to atrophy due to ageing mechanisms as well as disuse atrophy.

Sinus "pneumatization" is a physiological process in which the volume of the paranasal sinuses increases over time through the progressive expansion of the sinus volume, a process that begins after birth and continues until the 2nd–3rd decades of life [28–31]. Pneumatization proceeds in an inferomedial direction into the hard

[1] This specific function of the soft tissues of the pharynx and maxillary sinus was (mis-)used during the recent "pandemic" to quantify toxins in the individual through a "PCR"test and to (wrongly) label them as effects of viruses.

palate, laterally into the zygomatic bone and posteriorly into the ethmoids [32]. The most pneumatized area, according to the literature, is the anteromedial wall of the maxillary sinus [33, 34].

Disuse atrophy as well as "Sinus pneumatization" can be extensive, exposing the roots of the respected teeth and leading to direct engagement of the roots of the posterior teeth within the sinus. Moreover, this disuse atrophy can affect both bone quality and quantity [35]. Furthermore, the loss of posterior maxillary teeth aggravates the situation [31, 36–42]. According to Elsayed et al. [28], sinus pneumatization significantly reduces the average bone height and density in the edentulous locations. A consequence that may lead to complications during extraction and present a challenge during implant placement [42, 43]. Additionally, cortically anchored implants may lose contact with the 2nd cortical as a result of sinus expansion or contraction.

Although tooth loss in the posterior maxilla leads to a collapse of the corticals that surround the roots, in our experience, immediate functional loading protocols using method of osseofication can overcome this collapse by generating new functions in these cortical themselves. Hence, emphasising the use of immediate functional loading over the sinus lifting and bone augmenting procedure (**Figures 4a** and **b** and **5a** and **b**).

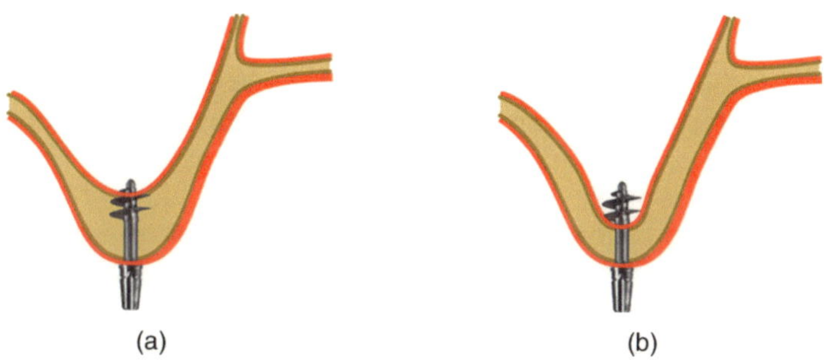

Figure 4.
a: baseline situation. b: due to bone loss and sinus "pneumatization" of the maxillary sinus, the load transmitting threads of the implants have lost contact with the 2nd cortical. The implant does not contribute to load transmission and more and a corrective intervention may be indicated.

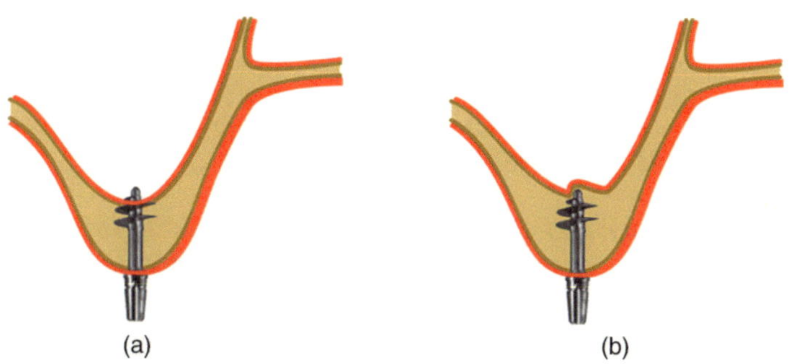

Figure 5.
a: baseline situation. b: Due to bone growth inside the maxillary sinus, the load-transmitting threads of the implants have lost contact with the 2nd cortical. The implant does not contribute to load transmission and more and a corrective intervention may be indicated.

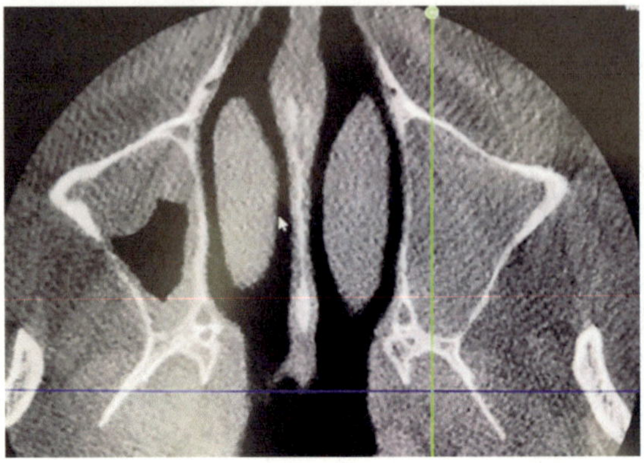

Figure 6.
A cone beam cross-section of the maxillary sinus with the anterior cortical wall of the maxillary sinus on the left side of the picture shows thickening due to the persistence of chronic granulation in that area (inside the maxillary sinus) that created changes in the cortical border of the maxillary sinus, including increased bone mass and mineralisation. While the opposite maxillary sinus (right side of the picture) shows signs of an acute infection.

8. Thickening of the bone structures around the sinus cavity

The literature demonstrates that mesenchymal stem cells derived from the sinus membrane possess the capacity to create bone and thicken the Schneiderian membrane using various techniques such as cadaver inspections, CT scans and CBCT imaging [43–46].

The normal thickness of the Schneiderian membrane is 1 mm [46, 47]–2 mm [44, 46], with many reported factors associated with a symptomatic increase in membrane thickness. Monje et al. [48] and Munakata et al. [49] categorised these factors into patient-related factors (age and smoking habits), teeth-related factors (like periodontitis, bone loss and lesions near the gums) and maxillary sinus variations (including sinus septa and nasal septum deviation).

Recently, Alghofaily et al. [50] highlighted the association between the increased thickness of the sinus membrane with male's gender, the periapical lesions, inadequate endodontic treatment and sinusitis.

Cone beam CT can be successfully used to distinguish between chronic and acute infections of the maxillary sinus, as chronic infections lead to a thicker and more mineralised cortical (**Figure 6**).

9. Diseases in the maxillary sinus

A wide spectrum of diseases can involve the maxillary sinus, originating either from or within the lining of the sinus; the other paranasal sinuses; the nasal space; dental and oral tissues; the adjacent bone; and hence extending to the sinus [51]. These diseases can be either inflammatory (the result of a bacterial infection, an allergy, or a result of detox function), neoplastic (benign or malignant, or metastases), odontogenic (infections, cysts, foreign body), mucocele and granulomatous vasculitis (**Figure 7**).

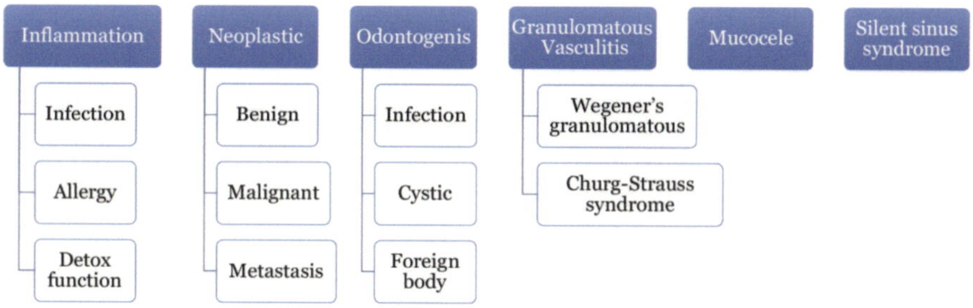

Figure 7.
Diseases of the maxillary sinus.

10. The odontogenic maxillary sinusitis (OMS)

Acute or chronic maxillary sinusitis is defined as symptomatic inflammation of the maxillary sinus due to bacterial, allergic or fungal rhinitis [52, 53]. Nowadays, we have to add symptoms of overly strong detox function of the maxillary sinus.

Odontogenic maxillary sinusitis (OMS) is a type of sinusitis caused by infections that originate from dental or dentoalveolar structures, affecting the floor of the maxillary sinus and the sinus membrane. It is slightly more common in females, with the commonest age group being 40–60 years [44, 54].

11. Historical background of the odontogenic maxillary sinusitis (OMS) and incidence

The concept of the odontogenic maxillary sinus (OMS) was first proposed by William H. Bauer in 1943 [54].

Brook [55] and Mehra and Jeong [56] documented that odontogenic infections account for approximately 10–12% of maxillary sinusitis cases, with the high prevalence of dental caries serving as a causative factor [53]. Later, Maillet et al. [43] reported an increase in incidence to 51.8% while Wuokko-Landén et al. [57] mentioned 15% and Patel and Ferguson [58] reported 40%.

The main causes of sinusitis can be either odontogenic infection and/or dentoalveolar surgery, which is associated with perforation of the Schneiderian membrane and oroantral fistula formation [52, 59–62]. Some authors reported some iatrogenic causes, such as impacted teeth after dental care, implants, dental amalgams and the oro-antral fistula [54, 62–66].

Lechien et al. [61] investigated the different causes of odontogenic sinusitis among 674 patients from January 1980 to January 2013. They reported 65.7% for iatrogenic aetiology and 25.1% for apical periodontal pathologies, which were distributed as follows: apical periodontitis—16%, apical granulomas—5%, odontogenic cysts—2.5 and 8.3% for marginal periodontitis. Molteni et al. [63] identified oroantral communication as the causative factor in 45.8% of patients. Troeltzsch et al. [60] reported that extractions with subsequent OAF or healing disturbances were the main cause of sinusitis (64%). Almost a similar percentage (57%) was reported by Zirk et al. [64], while Puglisi et al. [65] consider oroantral communication as a consequence of tooth extraction, sinus lift, third molar removal, improper implant placement, cyst excision and dislodged tooth fragments. Moreover, Troeltzsch et al. [60] reported 5.2% and 2.3% for dental implants and periimplantitis, while Lechien et al. [61] documented a lower rate (0.3%).

Furthermore, Craig et al. [67] reported a higher percentage of endodontic sources than an iatrogenic one. More recently, Yassin-Kassab et al. [68] documented the same observation (51.7% compared to 48.3%).

The percentage of older people has increased significantly in recent years as the world population grows. In 2019, 703 million people aged 65 years or older were documented, a number that may reach up to 1.5 billion [69, 70] by 2050. The growing percentage of elderly people and patients suffering from tooth loss is increasing the demand for dental rehabilitation [69, 70]. Today, dental implants [69–71] have become the gold standard treatment for dental rehabilitation in edentulous patients with an increasing percentage of use and a high reported success rate. Frequent and recurring problems with "sinus lifts" [69] have instilled justified fears in the field of conventional implantology, and this method, in our opinion, is becoming more and more left.

Several studies have proved that well-installed and healed implants are never the cause of maxillary sinusitis per se [72–84], independent of their penetration depth into the maxillary [72–75].

12. Implants and the maxillary sinuses

Cases with compromised ridge support present a challenge for implant rehabilitation due to the compromised bone quality and quantity, which adversely affect implant primary stability and diminish implant survival rate. Moreover, sinus pneumatizion may complicate the situation [75–77].

The literature describes numerous treatment alternatives, including sinus left and bone grafting procedures [75–77, 84–86], the use of short implants [69–71, 82, 84], the all-on-4, and even all-on-3 techniques [85, 87, 88], implants in remote bony areas such as zygomatic and tubero-pterygoid regions [72–74, 85, 89, 90] and basal implants [72–74, 91–93].

Several authors recommended using a bi-cortical or tri-cortical design to improve implant stability [73, 74, 88–96]. In this scenario, sinus membrane perforation and subsequent implant protrusion within the maxillary sinus could occur. Al-Salman and Almas [96] documented that perforation of the Schneiderian membrane (range 7–35%), while Stacchi et al. [97] reported that sinus membrane perforation and haemorrhage are the most common intraoperative complications associated with sinus floor elevation, with overall occurrences of 15.7 and 0.4%, respectively.

12.1 Implant protrusion inside the maxillary sinus and the maxillary sinus's health

In 2007, Jung et al. [75] studied the consequences of implant penetration in the maxillary sinus in mongrel dogs for 6 months and reported that none of the investigated sinus cavities displayed any features of inflammation.

Moreover, Tabrizi et al. [77] in 2012 retrospectively investigated the effects of the implant's tip penetration inside the maxillary sinuses. They documented no problem associated with implants protruding inside the sinuses; all the investigated implants remained successfully integrated, with no radiographic evidence of bone loss or other complications. They highlighted that in cases where no membrane penetrated, bone formation was noticed in the maxillary sinus floor.

Zhong et al. [78] investigated the effect of different depth implant penetrations (group A: 0 mm; group B: 1 mm; group C: 2 mm; group D: 3 mm) on maxillary sinus health in a dog mode. None of the maxillary sinuses showed any signs of sinus inflammation. At penetrating depths of 1–2 mm, newly formed membrane and partially formed bone completely covered the implants, while at depths of 3 mm or more, neither membrane nor bone were present. All the implants showed an obvious increase in bone-to-implant contact (BIC). The same observation was reported by Abi Najm et al. [79], Ghnaem and Gad [80], Elhamruni et al. [81], Shihab [82] and Ragucci et al. [72]. While Awadalkreem et al. [73] and Ahmad et al. [74] stated that implant penetration per se has no adverse effect either on maxillary sinus health or on the survival rate, moreover, Lazarov [76] reported only one maxillary sinus adverse reaction that does not necessitate the use of antibiotic among 131 maxillary sinuses penetrated with 217 implants, with none of the implants failing.

Sala et al. [35] conducted a recent systematic review and meta-analysis, which concluded that the prevalence of postoperative infection, including sinusitis, is low and has a reported correlation with the dimensions of the perforation and the anatomical predisposition.

On the other hand, some investigators reported the occurrence of sinusitis, including Nooh [84] (in one patient out of 63 patients), and Lazarov [76] (in one sinus out of 131, after Corticobasal® implants were used) Lazarov identified however retrospectively surgical error (too short implant chosen) and subsequent prosthetic mistake (to deep cementation zone inside the palatal mucosa in combination with effects of the open bone wound) as the origins of the problem.

12.1.1 The "bony seal"

Ihde postulated, based on observation of polished osseofixated implants, that it is not the implant's body inside the maxillary sinus that causes the infection, but the fact that the implant's endosseous surface is partially not in direct contact with the bone, which leads to a loss of the seal between the bacterial-loaded oral cavity and the potentially sterile maxillary sinus (**Figure 8a–f**).

12.2 Bone augmentation and sinusitis

Sinus grafting is a common procedure used to increase the vertical and horizontal amount of bone in the maxillary sinus, with a high reported implant success rate (after successful grafting). Several complications have been documented: membrane perforation, infections, bleeding, implant migration and dissemination or loss of the graft material [98–111].

Barone et al. [102] found that membrane perforation (60%), infection (21%), bleeding (9%) and implant migration were the most common problems that could happen in connection with a maxillary sinus graft.

Lu et al. [45] recently reported that the most common problems with sinus grafts were perforation of the Schneiderian membrane, bleeding, infection, persistent sinusitis and loss of the graft material.

Iușan et al. [69] documented membrane perforation and the formation of an oro-antral fistula; the occurrence of acute and chronic rhinosinusitis; facial congestion;

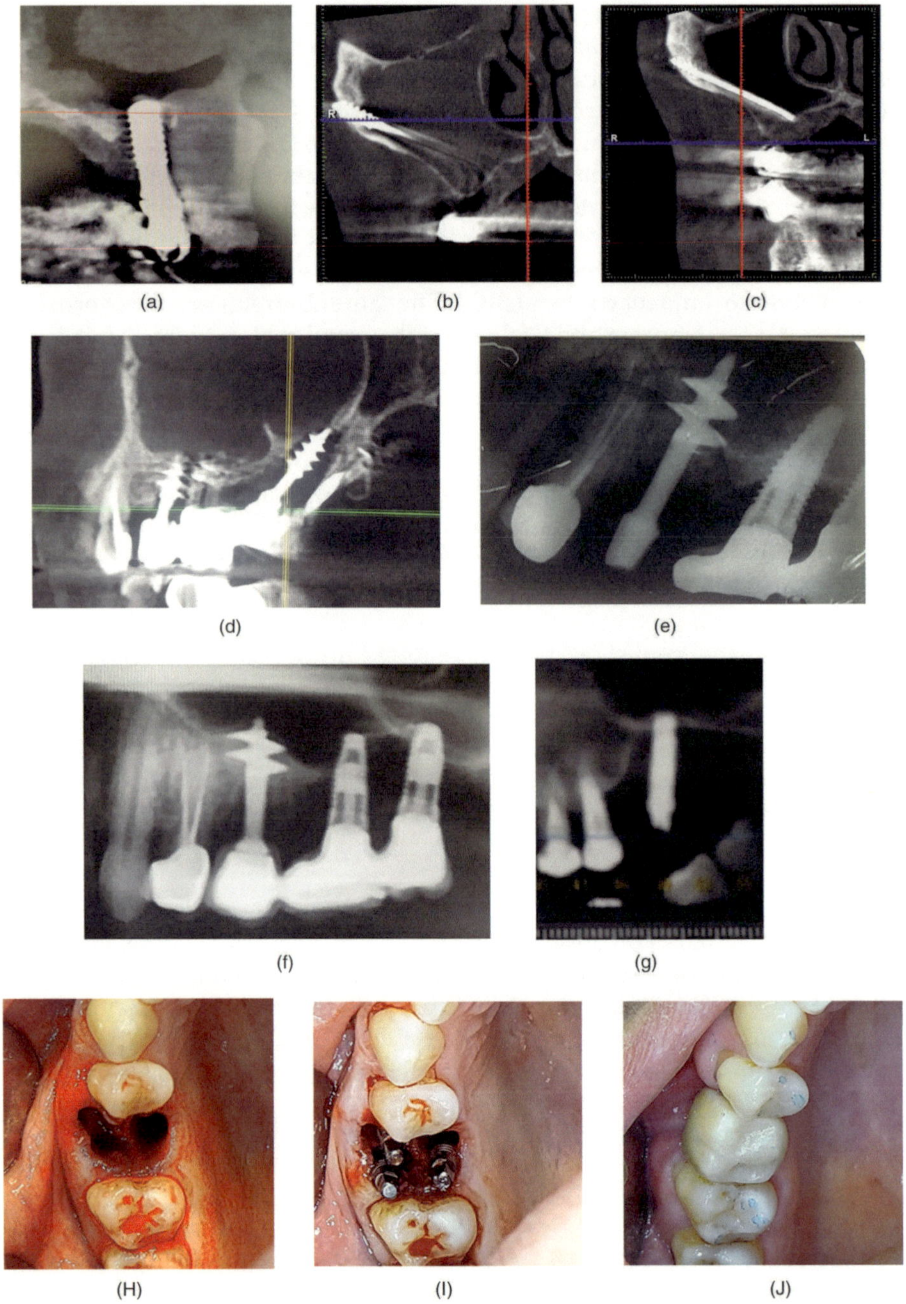

Figure 8.
a: Shows that the infection and detachment of a 2-stage endosseous surface from the bone happens from the nose down, and from the oral cavity up. b and c: A zygomatic implant penetrates the maxillary sinus, leading to subsequent maxillary sinusitis. The origin of the problem, in our opinion, was the missing seal between the oral cavity and the maxillary sinus. The management approach to the situation was as follows: the implant was left within the sinus; however, it was detached from the cementing abutment and the bridge by removing the abutment. After 6 weeks, no signs of infection were visible on the control scan, while the bridge was still in full function. d: shows the detachment of the peri-implant bone from the oblique ossefixated implant in area 26; the osteolytic zone has reached the maxillary sinus, causing maxillary sinusitis. The bony seal was replaced by connective tissue. This tissue has no sealing function. e and f: when the bone seal is ensured, the protrusion of osseofixated implants may even lead to a considerable increase in the size of the basal membrane of the maxillary

sinus (image after 19 months of follow-up). 8F. Note the increase in mineralised bone in area 25, while peri-implantitis progresses in areas 26 and 27. g. Peri-implantitis had resorbed all of the alveolar bone. Nevertheless, the basal and cortical bones of the floor of the maxillary sinus keep the implant stable. This portion of the cortical (around the implant's apex) even shows an increase in bone volume due to increased (cantilever) forces. h: After the extraction of an upper 1st molar, three root extraction sockets were available, allowing the placement of three osseofixated implants. i: Osseofixated implants (BCS® implant designs) of adequate diameters and length were placed into each of the three extraction sockets. j: After the temporary crown had been in place for 9 months, tooth 16 was replaced by a full zirconium crown. The high polish surface of both the implant's shafts and the zirconium crown ensures excellent periodontal health. Using a thin interdental brush, the patients can clean and maintain their oral hygiene.

the incidence of nasal obstruction, the possibility of implant displacement into the maxillary sinus; and sinus graft failure as the main complications associated with sinus grafting procedures. Moreover, they reported a positive correlation between peri-implant infection and sinus pathology.

In the same line, literature [83, 104–112] highlighted that membrane perforation, dislodgement of the graft material into the sinus, disrupting the normal sinus physiology, overfilling of the maxillary sinus with graft material, as well as graft infection, can be the main predisposing factors for sinusitis. This reported sinusitis may require antibiotic treatment or an evenly distributed surgical draining procedure [82].

In augmented sinuses, the cause of the infection cannot be diagnosed, because even if the implant is the cause, the augmented area will be affected by the infection **Figure 8g**.

In the sinus site where bone augmentation "sinus-lift" has been performed, the basal cortical of the maxillary sinus is, however, missing due to resorption. Therefore, the infection from peri-implantitis could potentially spread to the maxillary sinus, leading to the development of pan-sinusitis. An explanation that is in line with Iușan et al. [69].

A reported relation between implant surface tomography and an increased risk of peri-implantitis, including orthograde and retrograde peri-implantitis has been emphasised by many investigators [112, 113]. Hence, this infection affects both the first and the second cortical bones, resulting in implant mobility and consequence implant loss. A complication that presents a challenge for using this implant in severely resorbed posterior maxillary bone, especially with the increased demands of the patient for immediate loading.

Hence, there is an increase in demand for a treatment modality with a high success rate and fewer complications. Today the use of use of osseofixated implants in compromised ridge support areas, including the posterior maxilla, has been emphasised with high success and survival rates, numerous biological and mechanical advantages, limited complications and significant improvements in patient satisfaction and quality of life owing to their smooth surface, thin penetrating tip, small diameter, devoid of bone grafting, and most importantly, no risk of peri-implantitis [73, 76, 91–93]. From our experience, and as shown in **Figure 8(e and f)**, the cortical floor of the maxillary sinus tends to increase in thickness after being mechanically loaded by cortically anchored implants (osseofixated implants). This increase in both volume and mineralization is not astonishing: as the bone is loaded through the roots of the premolars and molars. However, following tooth extraction, the caudal cortical of the maxillary sinus receives close to zero functional stimuli and hence atrophies. As we do not know if the additional bone is going to be generated inside the maxillary sinus or inside the maxillary bone, the rationale of osseofixated implants is to place approximately half of the implant thread into the maxillary sinus and half of it into the maxillary bone.

Similarly, when osseofixated implants are placed into fresh extraction sockets of upper molars and premolars, the lamina cribrosa of these teeth remains under constant functional stimulus and it does not atrophy away (**Figure 8h–j**).

12.3 Implant protrusion inside the maxillary sinus and the peri-implant bone contact

Khairnar and Gaur [114] reported an increase in bone apposition and excellent implant stability in bi-cortical implants anchored in rabbits. Moreover, Kim et al. [115] reported a significant increase in peri-implant bone level when the initial bone height was less than 5 mm, but no increase in peri-implant level when it was 5 mm or more. Furthermore, after 12 months of follow-up, Tabrizi et al. [77] found no radiographic signs of bone loss in any of the penetrated implants.

In the same line, Awadalkreem et al. [73] reported the absence of any osteolytic reaction around the penetrated implants with increased bone-implant contact after 18 months of follow-up. The growth of bone was assumed to be a result of increased local function.

12.4 Implant protrusion inside the maxillary sinus and implant survival rate

Implant penetration inside the maxillary sinus has no significant effect on the implant survival rates. This observation has been emphasised by Awadalkreem et al. [73] who reported a 100% survival rate after 18 months of follow-up, and Kim et al. [116] who documented a high survival rate. Ragucci et al. [72] documented a 95.6% survival rate after a 52.7-month follow-up of implants. Furthermore, Ghnaem et al. [80] reported a survival rate of 100% after 6 years of follow-up, while Nooh [83] mentioned a survival rate of 98.4%.

12.5 Implant protrusion inside the maxillary sinus and the incidence of epistaxis

Despite the fact that implant penetration inside the nasal and maxillary sinuses does not adversely affect the health of the nasal and maxillary sinuses, some investigators [72, 83], reported the incidence as an immediate postoperative complication. Nooh [83] reported mild epistaxis as an immediate postoperative occurrence in 7 of the 63 patients in their study. Kim et al. [115] noted similar observations, reported by 3 of the 39 patients immediately postoperatively. Furthermore, according to Ragucci et al. [72], epistaxis is the most common complication associated with implant pentation inside the maxillary sinus. Tabrizi et al. [77] reported the incidence of transit epistasis in one patient.

12.6 Implant protrusion inside the maxillary sinus and the thickening in the Schneiderian membrane

As described previously, several factors can be interlinked with the thickening of the Schneiderian membrane [48, 49, 52, 72, 75]. Researchers found a significant relationship between the thickness of the membrane and the patient's gender, hypertension and smoking habits [48, 49, 52, 72, 75]. In the same line, Tabrizi et al. [77] reported a nonsignificant thickening of the sinus membrane radiologically in two patients with no clinical signs of sinusitis. Both patients who showed membrane thickness in the present study were men and smokers.

Ragucci et al. [72] found that the radiological complications associated with implant penetration inside the maxillary sinus were 14.8%, with no significant difference between the different implant penetration levels.

12.7 Implant protrusion inside the maxillary sinuses and its effect on patient satisfaction and quality of life

Now a day, implant-supported prostheses represent one of the most common treatment alternatives to treat edentulism, with significant improvements in patient's aesthetic and oral functions, including mastication and phonation, patient satisfaction and quality of life [116–118].

Implant protrusion inside the maxillary sinuses does not affect patient satisfaction and quality of life, as reported by Awadalkreem et al. [73] and Ahmad et al. [74]. All the investigated patients showed a substantial improvement following implant treatment, despite the fact that implants protruded inside the sinuses.

A recent consensus from 2022 [119], which included 31 experts (23 in implantology, 6 otolaryngologists and 2 radiologists), looked into what happens when implants stick out into the maxillary sinuses and nasal fossae. They featured significant statements emphasising that: implants protruding into the sinus do not always imply implant protrusion through the sinus mucosa is not necessarily related to sinus pathology; implants should be monitored clinically during the maintenance programme; but the implants should not be removed; and ENT consultation is not mandatory. Even when implants are associated with symptoms (foul smell, loss of smell, posterior nasal drainage, anterior nasal drainage, nasal obstruction and facial pressure) and radiographic signs of sinus or sinonasal pathology, the patient should be referred to an ENT specialist or surgical provider for evaluation; implants should not be removed, and pharmacological treatment should be attempted first.

13. Conclusion

Implant therapy is a successful treatment option for patients with complete and partial edentulous conditions. Implants can protrude inside the maxillary sinus without clinical or radiographic evidence of sinusitis, having a negative effect on implant survival as well as a positive effect on patient satisfaction. During the patient's follow-up schedule, we can regularly monitor protruded implants, both clinically and radiographically.

Acknowledgements

There are no acknowledgements to declare.

Conflict of interest

There are no conflicts of interest to declare.

Ethical approval

Institutional approval was not required.

Author details

Stefan Ihde[1,2]* and Fadia Awadalkreem[3]

1 Department for Prosthetics, Jaipur Dental College, Maharaj Vinayak Global University, Jaipur, Rajasthan, India

2 International Implant Foundation, Evidence and Research Department, Munich, Germany

3 RAK College of Dental Sciences, RAK Medical and Health Sciences University, Ras Al Khaimah, United Arab Emirates

*Address all correspondence to: ihde1962@gmail.com

References

[1] Whyte A, Boeddinghaus R. The maxillary sinus: Physiology, development and imaging anatomy. Dento Maxillo Facial Radiology. 2019;**48**(8):20190205. DOI: 10.1259/dmfr.20190205

[2] Aktuna Belgin C, Colak M, Adiguzel O, Akkus Z, Orhan K. Three-dimensional evaluation of maxillary sinus volume in different age and sex groups using CBCT. European Archives of Oto-Rhino-Laryngology. 2019;**276**(5):1493-1499. DOI: 10.1007/s00405-019-05383-y

[3] Somayaji K, Muliya VS, Kg M, et al. A literature review of the maxillary sinus with special emphasis on its anatomy and odontogenic diseases associated with it. Egypt The Journal of Otolaryngology. 2023;**39**:173. DOI: 10.1186/s43163-023-00536-7

[4] Testori T. Maxillary sinus surgery: Anatomy and advanced diagnostic imaging. Journal of Prosthodontics-Implant Esthetic and Reconstructive Dentistry. 2011;**2**:6-14

[5] Sargi ZB, Casiano RR. Surgical anatomy of the paranasal sinuses. In: Kountakis SE, Onerci TM, editors. Rhinologic and Sleep Apnea Surgical Techniques. New York: Springer; 2007. pp. 17-26

[6] Van den Bergh JP, ten Bruggenkate CM, Disch FJ, Tuinzing DB. Anatomical aspects of sinus floor elevations. Clinical Oral Implants Research. 2000;**11**(3):256-265. DOI: 10.1034/j.1600-0501.2000.011003256.x

[7] Gosau M, Rink D, Driemel O, Draenert FG. Maxillary sinus anatomy: A cadaveric study with clinical implications. The Anatomical Record (Hoboken). 2009;**292**(3):352-354. DOI: 10.1002/ar.20859

[8] Guo ZZ, Liu Y, Qin L, Song YL, Xie C, Li DH. Longitudinal response of membrane thickness and ostium patency following sinus floor elevation: A prospective cohort study. Clinical Oral Implants Research. 2016;**27**(6):724-729. DOI: 10.1111/clr.12655

[9] Rancitelli D, Borgonovo AE, Cicciù M, Re D, Rizza F, Frigo AC, et al. Maxillary sinus septa and anatomic correlation with the Schneiderian membrane. The Journal of Craniofacial Surgery. 2015;**26**(4):1394-1398. DOI: 10.1097/SCS.0000000000001725

[10] Kalyvas D, Kapsalas A, Paikou S, Tsiklakis K. Thickness of the Schneiderian membrane and its correlation with anatomical structures and demographic parameters using CBCT tomography: A retrospective study. International Journal of Implant Dentistry. 2018;**4**(1):32. DOI: 10.1186/s40729-018-0143-5

[11] Wallace JA. Transantral endodontic surgery. Oral Surgery, Oral Medicine, Oral Pathology, Oral Radiology, and Endodontics. 1996;**82**(1):80-83. DOI: 10.1016/s1079-2104(96)80381-7

[12] Nuñez-Castruita A, López-Serna N, Guzmán-López S. Prenatal development of the maxillary sinus: A perspective for paranasal sinus surgery. Otolaryngology and Head and Neck Surgery. 2012;**146**(6):997-1003. DOI: 10.1177/0194599811435883

[13] Bailey BJ. Head and Neck Surgery Otolaryngology. 2nd ed. Philadelphia: Lip-Pincott Raven Publishers; 1998. pp. 418-446

[14] Lovasova K, Kachlik D, Rozpravkova M, Matusevska M, Ferkova J, Kluchova D. Three-dimensional CAD/CAM imaging of the maxillary sinus in ageing process. Annals of Anatomy. 2018;**218**:69-82. DOI: 10.1016/j.aanat.2018.01.008

[15] Boeddinghaus R, Whyte A. Trends in maxillofacial imaging. Clinical Radiology. 2018;**73**:4-18

[16] Tian XM, Qian L, Xin XZ, Wei B, Gong Y. An analysis of the proximity of maxillary posterior teeth to the maxillary sinus using cone-beam computed tomography. Journal of Endodontia. 2016;**42**(3):371-377. DOI: 10.1016/j.joen.2015.10.017

[17] Schuh E, Schmiedl R, Vogel G. Anatomische Grenzen der enossalen zahnärztlichen implantologie [Anatomic limits of endosseous dental implantation]. Zeitschrift für Stomatologie (1984). 1984;**81**(2):81-90. German

[18] Wyler B, Mallon WK. Sinusitis update. Emergency Medicine Clinics of North America. 2019;**37**(1):41-54. DOI: 10.1016/j.emc.2018.09.007

[19] Ford RL, Barsam A, Velusami P, Ellis H. Drainage of the maxillary sinus: A comparative anatomy study in humans and goats. Journal of Otolaryngology—Head and Neck Surgery. 2011;**40**(1):70-74

[20] Lin Y-H, Yang Y-C, Wen S-C, Wang H-L. The influence of sinus membrane thickness upon membrane perforation during lateral window sinus augmentation. Clinical Oral implants Research. 2015;**00**:1-6. DOI: 10.1111/clr.12646

[21] Iwanaga J, Wilson C, Lachkar S, Tomaszewski KA, Walocha JA, Tubbs RS. Clinical anatomy of the maxillary sinus: Application to sinus floor augmentation. Anatomy and Cell Biology. 2019;**52**(1):17-24. DOI: 10.5115/acb.2019.52.1.17

[22] Standring S. Gray's Anatomy: The Anatomical Basis of Clinical Practice. 41st ed. London: Elsevier Health Sciences; 2015

[23] Kqiku L, Biblekaj R, Weiglein AH, Kqiku X, Städtler P. Arterial blood architecture of the maxillary sinus in dentate specimens. Croatian Medical Journal. 2013;**54**(2):180-184. DOI: 10.3325/cmj.2013.54.180

[24] Flanagan D. Arterial supply of maxillary sinus and potential for bleeding complication during lateral approach sinus elevation. Implant Dentistry. 2005;**14**(4):336-338. DOI: 10.1097/01.id.0000188437.66363.7c

[25] Duncavage J. The Maxillary Sinus: Medical and Surgical Management. New York: Thieme Medical Publishers; 2011

[26] Cohen NA. Sinonasal mucociliary clearance in health and disease. The Annals of Otology, Rhinology and Laryngology. Supplement. 2006;**196**:20-26. DOI: 10.1177/00034894061150s904

[27] Bustamante-Marin XM, Ostrowski LE. Cilia and Mucociliary Clearance. Cold Spring Harbor Perspectives in Biology. 3 Apr 2017;**9**(4):a028241. DOI: 10.1101/cshperspect.a028241

[28] Elsayed SA, Alassaf MS, Elboraey MO, Mohamado LL, Huwaykim DA, Albouq AK, et al. The impact of maxillary sinus pneumatization on the quality of the alveolar bone in dentated and edentulous patients: A cone-beam computed tomography radiographic analysis.

Cureus. 2023;**15**(9):e46005. DOI: 10.7759/cureus.46005

[29] Jun BC, Song SW, Park CS, Lee DH, Cho KJ, Cho JH. The analysis of maxillary sinus aeration according to aging process; volume assessment by 3-dimensional reconstruction by high-resolutional CT scanning. Otolaryngology and Head and Neck Surgery. 2005;**132**(3):429-434. DOI: 10.1016/j.otohns.2004.11.012

[30] Nimigean V, Nimigean VR, Măru N, Sălăvăstru DI, Bădiță D, Tuculină MJ. The maxillary sinus floor in the oral implantology. Romanian Journal of Morphology and Embryology. 2008;**49**(4):485-489

[31] Alqahtani S, Alsheraimi A, Alshareef A, Alsaban R, Alqahtani A, Almgran M, et al. Maxillary sinus pneumatization following extractions in Riyadh, Saudi Arabia: A cross-sectional study. Cureus. 2020;**12**(1):e6611. DOI: 10.7759/cureus.6611

[32] Lang J. Clinical Anatomy of the Nose, Nasal Cavity and Paranasal Sinuses. New York, US: Georg Thieme Verlag; 1989

[33] Mauri M, de Oliveira CO, Franche G. Pneumosinus dilatans of the maxillary sinus. Case report. The Annals of Otology, Rhinology, and Laryngology. 2000;**109**(3):278-280. DOI: 10.1177/000348940010900308

[34] Tovi F, Gatot A, Fliss DM. Air cyst of the maxillary sinus (pneumosinus dilatans, pneumocoele). The Journal of Laryngology and Otology. 1991;**105**(8):673-675. DOI: 10.1017/s0022215100116998

[35] Sala YM, Lu H, Chrcanovic BR. Clinical outcomes of maxillary sinus floor perforation by dental implants and sinus membrane perforation during sinus augmentation: A systematic review and meta-analysis. Journal of Clinical Medicine. 2024;**13**(5):1253. DOI: 10.3390/jcm13051253

[36] Lim HC, Kim S, Kim DH, Herr Y, Chung JH, Shin SI. Factors affecting maxillary sinus pneumatization following posterior maxillary tooth extraction. Journal of Periodontal and Implant Science. 2021;**51**(4):285-295. DOI: 10.5051/jpis.2007220361

[37] Yücesoy T, Göktaş TA. Evaluation of sinus pneumatization and dental implant placement in atrophic maxillary premolar and molar regions. The International Journal of Oral & Maxillofacial Implants. 2022;**37**(2):407-415. DOI: 10.11607/jomi.9215

[38] Levi I, Halperin-Sternfeld M, Horwitz J, Zigdon-Giladi H, Machtei EE. Dimensional changes of the maxillary sinus following tooth extraction in the posterior maxilla with and without socket preservation. Clinical Implant Dentistry and Related Research. 2017;**19**(5):952-958. DOI: 10.1111/cid.12521

[39] Sarilita E, Lita YA, Nugraha HG, Murniati N, Yusuf HY. Volumetric growth analysis of maxillary sinus using computed tomography scan segmentation: A pilot study of Indonesian population. The Anatology and Cell Biology. 2021;**54**(4):431-435. DOI: 10.5115/acb.21.051

[40] Li Y, Deng S, Mei L, Li J, Qi M, Su S, et al. Accuracy of alveolar bone height and thickness measurements in cone beam computed tomography: A systematic review and meta-analysis. Oral Surgery, Oral Medicine, Oral Pathology, Oral Radiology. 2019;**128**(6):667-679. DOI: 10.1016/j.oooo.2019.05.010

[41] Cavalcanti MC, Guirado TE, Sapata VM, Costa C, Pannuti CM, Jung RE, et al. Maxillary sinus floor pneumatization and alveolar ridge resorption after tooth loss: A cross-sectional study. Brazilian Oral Research. 2018;**32**:e64. DOI: 10.1590/1807-3107BOR-2018.vol32.0064

[42] Lawson W, Patel ZM, Lin FY. The development and pathologic processes that influence maxillary sinus pneumatization. The Anatomical Record (Hoboken). 2008;**291**(11):1554-1563. DOI: 10.1002/ar.20774

[43] Maillet M, Bowles WR, McClanahan SL, John MT, Ahmad M. Cone-beam computed tomography evaluation of maxillary sinusitis. Journal of Endodontia. 2011;**37**(6):753-757. DOI: 10.1016/j.joen.2011.02.032

[44] Shanbhag S, Karnik P, Shirke P, Shanbhag V. Association between periapical lesions and maxillary sinus mucosal thickening: A retrospective cone-beam computed tomographic study. Journal of Endodontia. 2013;**39**(7):853-857. DOI: 10.1016/j.joen.2013.04.010

[45] Lu Y, Liu Z, Zhang L, Zhou X, Zheng Q, Duan X, et al. Associations between maxillary sinus mucosal thickening and apical periodontitis using cone-beam computed tomography scanning: A retrospective study. Journal of Endodontia. 2012;**38**(8):1069-1074. DOI: 10.1016/j.joen.2012.04.027

[46] Nascimento EH, Pontual ML, Pontual AA, Freitas DQ, Perez DE, Ramos-Perez FM. Association between odontogenic conditions and maxillary sinus disease: A study using cone-beam computed tomography. Journal of Endodontia. 2016;**42**(10):1509-1515. DOI: 10.1016/j.joen.2016.07.003

[47] Abrahams JJ, Glassberg RM. Dental disease: A frequently unrecognized cause of maxillary sinus abnormalities? AJR. American Journal of Roentgenology. 1996;**166**(5):1219-1223. DOI: 10.2214/ajr.166.5.8615273

[48] Monje A, Diaz KT, Aranda L, Insua A, Garcia-Nogales A, Wang HL. Schneiderian membrane thickness and clinical implications for sinus augmentation: A systematic review and meta-regression analyses. Journal of Periodontology. 2016;**87**(8):888-899. DOI: 10.1902/jop.2016.160041

[49] Munakata M, Yamaguchi K, Sato D, Yajima N, Tachikawa N. Factors influencing the sinus membrane thickness in edentulous regions: A cone-beam computed tomography study. The International Journal of Implant Dentistry. 2021;**7**(1):16. DOI: 10.1186/s40729-021-00298-y

[50] Alghofaily M, Alsufyani N, Althumairy RI, AlSuhaibani A, Alfawzan F, AlSadhan L. Odontogenic factors associated with maxillary sinus Schneiderian membrane thickness and their relationship to chronic Sinonasal symptoms: An ambispective cohort study. Diagnostics (Basel). 2023;**13**(16):2710. DOI: 10.3390/diagnostics13162710

[51] Bell GW, Joshi BB, Macleod RI. Maxillary sinus disease: Diagnosis and treatment. British Dental Journal. 2011;**210**(3):113-118. DOI: 10.1038/sj.bdj.2011.47

[52] Ferguson M. Rhinosinusitis in oral medicine and dentistry. Australian Dental Journal. 2014;**59**(3):289-295. DOI: 10.1111/adj.12193. Epub 2014 Jul 30

[53] Psillas G, Papaioannou D, Petsali S, Dimas GG, Constantinidis J. Odontogenic maxillary sinusitis: A comprehensive review. Journal of Dental

Science. 2021;**16**(1):474-481. DOI: 10.1016/j.jds.2020.08.001

[54] Bauer WH. Maxillary sinusitis of dental origin. American Journal of Orthodontics and Oral Surgery. 1943;**29**(3):B133-B151

[55] Brook I. Sinusitis of odontogenic origin. Otolaryngology and Head and Neck Surgery. 2006;**135**(3):349-355. DOI: 10.1016/j.otohns.2005.10.059

[56] Mehra P, Jeong D. Maxillary sinusitis of odontogenic origin. Current Allergy and Asthma Reports. 2009;**9**(3):238-243. DOI: 10.1007/s11882-009-0035-0

[57] Wuokko-Landén A, Blomgren K, Välimaa H. Acute rhinosinusitis—Are we forgetting the possibility of a dental origin? A retrospective study of 385 patients. Acta Oto-Laryngologica. 2019;**139**(9):783-787. DOI: 10.1080/00016489.2019.1634837

[58] Patel NA, Ferguson BJ. Odontogenic sinusitis: An ancient but under-appreciated cause of maxillary sinusitis. Current Opinion in Otolaryngology and Head and Neck Surgery. 2012;**20**(1):24-28. DOI: 10.1097/MOO.0b013e32834e62ed

[59] Gaudin RA, Hoehle LP, Smeets R, Heiland M, Caradonna DS, Gray ST, et al. Impact of odontogenic chronic rhinosinusitis on general health-related quality of life. European Archives of Oto-Rhino-Laryngology. 2018;**275**(6):1477-1482. DOI: 10.1007/s00405-018-4977-5

[60] Troeltzsch M, Pache C, Troeltzsch M, Kaeppler G, Ehrenfeld M, Otto S, et al. Etiology and clinical characteristics of symptomatic unilateral maxillary sinusitis: A review of 174 cases. Journal of Cranio-Maxillo-Facial Surgery. 2015;**43**(8):1522-1529. DOI: 10.1016/j.jcms.2015.07.021

[61] Lechien JR, Filleul O, Costa de Araujo P, Hsieh JW, Chantrain G, Saussez S. Chronic maxillary rhinosinusitis of dental origin: A systematic review of 674 patient cases. International Journal of Otolaryngology. 2014;**2014**:465173. DOI: 10.1155/2014/465173

[62] Akhlaghi F, Esmaeelinejad M, Safai P. Etiologies and treatments of odontogenic maxillary sinusitis: A systematic review. Iranian Red Crescent Medical Journal. 2015;**17**(12):e25536. DOI: 10.5812/ircmj.25536

[63] Molteni M, Bulfamante AM, Pipolo C, Lozza P, Allevi F, Pisani A, et al. Odontogenic sinusitis and sinonasal complications of dental treatments: A retrospective case series of 480 patients with critical assessment of the current classification. Acta Otorhinolaryngologica Italica. 2020;**40**(4):282-289. DOI: 10.14639/0392-100X-N0457

[64] Zirk M, Dreiseidler T, Pohl M, Rothamel D, Buller J, Peters F, et al. Odontogenic sinusitis maxillaris: A retrospective study of 121 cases with surgical intervention. Journal of Cranio-Maxillo-Facial Surgery. 2017;**45**(4):520-525. DOI: 10.1016/j.jcms.2017.01.023

[65] Puglisi S, Privitera S, Maiolino L, Serra A, Garotta M, Blandino G, et al. Bacteriological findings and antimicrobial resistance in odontogenic and non-odontogenic chronic maxillary sinusitis. Journal of Medical Microbiology. 2011;**60**(Pt 9):1353-1359. DOI: 10.1099/jmm.0.031476-0

[66] George M, Noor A, Thorpe AR, Sritharan N, Riffat F. Odontogenic sinusitis: A literature review. Oral Surgery. 2024;**17**(2):170-178. DOI: 10.1111/ors.12861

[67] Craig JR, Poetker DM, Aksoy U, Allevi F, Biglioli F, Cha BY,

et al. Diagnosing odontogenic sinusitis: An international multidisciplinary consensus statement. International Forum of Allergy and Rhinology. 2021;**11**(8):1235-1248. DOI: 10.1002/alr.22777

[68] Yassin-Kassab A, Peterson EL, Craig JR. Total times to treatment completion and clinical outcomes in odontogenic sinusitis. American Journal of Otolaryngology. 2023;**44**(4):103921. DOI: 10.1016/j.amjoto.2023.103921

[69] Iușan SAL, Costache C, Lucaciu OP, Petrescu BN, Mirică IC, Toc DA, et al. Correlations between dental implant infectious pathologies and maxillary sinusitis: A review article. Journal of Clinical Medicine. 2023;**12**(15):5059. DOI: 10.3390/jcm12155059

[70] Kim WJ, Cho YD, Ku Y, Ryoo HM. The worldwide patent landscape of dental implant technology. Biomaterials Research. 2022;**26**(1):59. DOI: 10.1186/s40824-022-00307-0

[71] Carinci F, Lauritano D, Bignozzi CA, Pazzi D, Candotto V, Santos de Oliveira P, et al. A new strategy against peri-implantitis: Antibacterial internal coating. International Journal of Molecular Sciences. 2019;**20**(16):3897. DOI: 10.3390/ijms20163897

[72] Ragucci GM, Elnayef B, Suárez-López Del Amo F, et al. Influence of exposing dental implants into the sinus cavity on survival and complications rate: A systematic review. The International Journal of Implant Dentistry. 2019;**5**(1):6. DOI: 10.1186/s40729-019-0157-7

[73] Awadalkreem F, Ahmad AG, Ihde S, et al. Effects of corticobasal implant protrusion inside the nasal and maxillary sinus. Annals of Maxillofacial Surgery. 2020;**10**(1):114-121. DOI: 10.4103/ams.ams_42_20

[74] Ahmad AG, Awadalkreem F, Osman M, Palka L. Does the protrusion of corticobasal implants in the maxillary sinuses affect sinus health? A retrospective study. The Journal of Contemporary Dental Practice. 2023;**24**(6):357-363. DOI: 10.5005/jp-journals-10024-3521

[75] Jung JH, Choi BH, Jeong SM, Li J, Lee SH, Lee HJ. A retrospective study of the effects on sinus complications of exposing dental implants to the maxillary sinus cavity. Oral Surgery, Oral Medicine, Oral Pathology, Oral Radiology, and Endodontics. 2007;**103**(5):623-635. DOI: 10.1016/j.tripleo.2006.09.024

[76] Lazarov A. A prospective cohort study of maxillary sinus complications in relation to treatments with strategic implants® penetrating into the sinus. Annals of Maxillofacial Surgery. 2020;**10**(2):365-369. DOI: 10.4103/ams.ams_85_20

[77] Tabrizi R, Amid R, Taha Özkan B, Khorshidi H, Langner NJ. Effects of exposing dental implant to the maxillary sinus cavity. The Journal of Craniofacial Surgery. 2012;**23**(3):767-769. DOI: 10.1097/SCS.0b013e31824dbd19

[78] Zhong W, Chen B, Liang X, Ma G. Experimental study on penetration of dental implants into the maxillary sinus in different depths. Journal of Applied Oral Science. 2013;**21**(6):560-566. DOI: 10.1590/1679-775720130078

[79] Abi Najm S, Malis D, El Hage M, Rahban S, Carrel JP, Bernard JP. Potential adverse events of endosseous dental implants penetrating the maxillary sinus: Long-term clinical evaluation. The Laryngoscope. 2013;**123**(12):2958-2961. DOI: 10.1002

[80] Ghnaem A, Gad HM. The influence of exposing dental implants

to the maxillary sinus cavity on sinus complications and implant survival rate: 6 years prospective follow-up. Egyptian Dental Journal. 2014;**60**(3):2-9

[81] Elhamruni LM, Marzook HA, Ahmed WM, Abdul-Rahman M. Experimental study on penetration of dental implants into the maxillary sinus at different depths. Oral and Maxillofacial Surgery. 2016;**20**(3):281-287. DOI: 10.1007/s10006-016-0568-z

[82] Shihab OI. Intentional penetration of dental implants into the maxillary sinus: A retrospective study. Zanco Journal of Medical Sciences. 2017;**21**(1):1536-1539. DOI: 10.15218/zjms.2017.001

[83] Nooh N. Effect of Schneiderian membrane perforation on posterior maxillary implant survival. The International Journal of Oral Health. 2013;**5**(3):28-34

[84] Ali SA, Karthigeyan S, Deivanai M, et al. Implant rehabilitation for atrophic maxilla: A review. The Journal of Indian Prosthodontic Society. 2014;**14**(3):196-207. DOI: 10.1007/s13191-014-0360-4

[85] Chiapasco M, Casentini P, Zaniboni M. Bone augmentation procedures in implant dentistry. The International Journal of Oral and Maxillofacial Implants. 2009;**24**(Suppl):237-259

[86] Starch-Jensen T, Nielsen HB. Prosthetic rehabilitation of the partially edentulous atrophic posterior mandible with short implants (≤8 mm) compared with the sandwich osteotomy and delayed placement of standard length implants (>8 mm): A systematic review. Journal of Oral and Maxillofacial Research. 2018;**9**(2):e2. DOI: 10.5037/jomr.2018.9202

[87] Soto-Penaloza D, Zaragozí-Alonso R, Penarrocha-Diago M, et al. The all-on-four treatment concept: Systematic review. Journal of Clinical and Experimental Dentistry. 2017;**9**(3):e474-e488. DOI: 10.4317/jced.53613

[88] Bocklage R. Advanced alveolar crest atrophy: An alternative treatment technique for maxilla and mandible. Implant Dentistry. 2001;**10**(1):30-35. DOI: 10.1097/00008505-200101000-00009

[89] Goiato MC, Pellizzer EP, Moreno A, et al. Implants in the zygomatic bone for maxillary prosthetic rehabilitation: A systematic review. International Journal of Oral and Maxillofacial Surgery. 2014;**43**(6):748-757. DOI: 10.1016/j.ijom.2014.01.004

[90] Sharma A, Rahul GR. Zygomatic implants/fixture: A systematic review. The Journal of Oral Implantology. 2013;**39**(2):215-224. DOI: 10.1563/AAID-JOI-D-11-00055

[91] Ihde S, Kopp S, Maier T. Comparison of implant survival with implants placed in acceptable and compromised bone: A literature review. Journal of Oral and Maxillofacial Research. 2009;**8**(1):1-7. DOI: 10.1007/s12663-009-0001-3

[92] Ihde S, Ihde A. Immediate Loading Guideline to Successful Implantology. 2nd ed. Munich, Germany: International Implant Foundation Publishing; 2010. pp. 2-12

[93] Awadalkreem F, Khalifa N, Satti A, Suliman AM. Rehabilitation of patients with compromised ridge support using immediately loaded corticobasal implant-supported prostheses: A prospective observational study. The Journal of Contemporary Dental Practice. 2022;**23**(10):971-978. DOI: 10.5005/jp-journals-10024-3416

[94] Morgano AT. Functional load in oblique bicortical implants: Parasinusal implants and palatine implants. The Journal of Oral Implantology. 2013;**39**(4):467-474. DOI: 10.1563/AAID-JOI-D-09-00121.1

[95] Scortecci G. Basal Implantology. New York: Springer International Publishing; 2019. DOI: 10.1007/978-3-319-44873-2

[96] Al-Salman WT, Almas K. Maxillary sinus and success of dental implants: An update. General Dentistry. 2015;**63**(4):47-54

[97] Stacchi C, Andolsek F, Berton F, et al. Intraoperative complications during sinus floor elevation with lateral approach: A systematic review. The International Journal of Oral and Maxillofacial Implants. 2017;**32**(3):e107-e118. DOI: 10.11607/jomi.4884

[98] Chan HL, Wang HL. Sinus pathology and anatomy in relation to complications in lateral window sinus augmentation. Implant Dentistry. 2011;**20**(6):406-412. DOI: 10.1097/ID.0b013e3182341f79

[99] Shiffler K, Lee D, Aghaloo T, Moy PK, Pi-Anfruns J. Sinus membrane perforations and the incidence of complications: A retrospective study from a residency program. Oral Surgery, Oral Medicine, Oral Pathology, Oral Radiology. 2015;**120**(1):10-14. DOI: 10.1016/j.oooo.2015.02.477

[100] Kim J, Jang H. A review of complications of maxillary sinus augmentation and available treatment methods. Journal of the Korean Association of Oral and Maxillofacial Surgeons. 2019;**45**(4):220-224. DOI: 10.5125/jkaoms.2019.45.4.220

[101] Barone A, Santini S, Sbordone L, Crespi R, Covani U. A clinical study of the outcomes and complications associated with maxillary sinus augmentation. The International Journal of Oral & Maxillofacial Implants. 2006;**21**(1):81-85

[102] Nolan PJ, Freeman K, Kraut RA. Correlation between Schneiderian membrane perforation and sinus lift graft outcome: A retrospective evaluation of 359 augmented sinus. Journal of Oral and Maxillofacial Surgery. 2014;**72**(1):47-52. DOI: 10.1016/j.joms.2013.07.020

[103] Manor Y, Mardinger O, Bietlitum I, Nashef A, Nissan J, Chaushu G. Late signs and symptoms of maxillary sinusitis after sinus augmentation. Oral Surgery, Oral Medicine, Oral Pathology, Oral Radiology, and Endodontics. 2010;**110**(1):e1-e4. DOI: 10.1016/j.tripleo.2010.02.038

[104] Timmenga NM, Raghoebar GM, van Weissenbruch R, Vissink A. Maxillary sinus floor elevation surgery. A clinical, radiographic and endoscopic evaluation. Clinical Oral Implants Research. 2003;**14**:322-328

[105] Hainarosie R, Pietrosanu C, Cherecheanu A, Stoian AP, Stefanescu C, Pituru S. Risk factors for dental implant failure and medicolegal implications. Romanian Journal of Legal Medicine. 2019;**27**:52-56

[106] Park WB, Kim YJ, Herr Y, Lim HC. Functional endoscopic sinus surgery for paranasal sinusitis originating from a peri-implantitis-triggered infection in the augmented maxilla: A case report. The Journal of Oral Implantology. 2020;**46**(4):423-429. DOI: 10.1563/aaid-joi-D-19-00381

[107] Scarano A, Cholakis AK, Piattelli A. Histologic evaluation of sinus grafting materials after peri-implantitis-induced failure: A case series. The International Journal of Oral and Maxillofacial Implants. 2017;**32**(2):e69-e75. DOI: 10.11607/jomi.5303

[108] Hong SO, Shim GJ, Kwon YD. Novel approach to the maxillary sinusitis after sinus graft. Maxillofacial Plastic and Reconstructive Surgery. 2017;**39**(1):18. DOI: 10.1186/s40902-017-0115-3

[109] Park WB, Han JY, Oh SL. Maxillary sinusitis associated with peri-implantitis at sinus floor augmented sites: Case series. Implant Dentistry. 2019;**28**(5):484-489. DOI: 10.1097/ID.0000000000000922

[110] Rodriguez AE, Nowzari H. The long-term risks and complications of bovine-derived xenografts: A case series. Journal of Indian Society of Periodontology. 2019;**23**(5):487-492. DOI: 10.4103/jisp.jisp_656_18

[111] Park MJ, Park HI, Ahn KM, Kim JH, Chung YS, Jang YJ, et al. Features of odontogenic sinusitis associated with dental implants. The Laryngoscope. 2023;**133**(2):237-243. DOI: 10.1002/lary.30069

[112] Caccianiga G, Rey G, Caccianiga P, Leonida A, Baldoni M, Baldoni A, et al. Rough dental implant surfaces and peri-implantitis: Role of phase-contrast microscopy, laser protocols, and modified home oral hygiene in maintenance. A 10-year retrospective study. Applied Sciences. 2021;**11**:4985. DOI: 10.3390/app11114985

[113] Sinjab K, Sawant S, Ou A, Fenno JC, Wang HL, Kumar P. Impact of surface characteristics on the peri-implant microbiome in health and disease. Journal of Periodontology. 2024;**95**(3):244-255. DOI: 10.1002/JPER.23-0205

[114] Khairnar M, Gaur V. Evidence of bone formation in the nasal floor around polished surface bi-cortical screw implants after indirect nasal lift in an atrophied maxilla: Cone beam computed tomography-based case report. Journal of Indian Society of Periodontology. 2015;**19**(2):236-238. DOI: 10.4103/0972-124X.149941

[115] Kim HY, Yang JY, Chung BY, Kim JC, Yeo IS. Peri-implant bone length changes and survival rates of implants penetrating the sinus membrane at the posterior maxilla in patients with limited vertical bone height. Journal of Periodontal and Implant Science. 2013;**43**(2):58-63. DOI: 10.5051/jpis.2013.43.2.58

[116] Altayyar S, Al-Zordk W, Algabri R, Rajah E, Al-Baadani A, Alqutaibi AY, et al. Prospective evaluation of implants-supported, tooth-implant supported, and teeth-supported 3-unit posterior monolithic zirconia fixed restorations: Bite force and patient satisfaction. Cllinical and Experimental Dental Research. 2023;**9**(5):810-819. DOI: 10.1002/cre2.780

[117] Beuer F, Sachs C, Groesser J, Gueth JF, Stimmelmayr M. Tooth-implant-supported posterior fixed dental prostheses with zirconia frameworks: 3-year clinical result. Clinical Oral Investigations. 2016;**20**(5):1079-1086. DOI: 10.1007/s00784-015-1594-1

[118] Lemos CAA, Verri FR, de Luna Gomes JM, Santiago Junior JF, Miyashita E, Mendonca G, et al. Survival and prosthetic complications of monolithic ceramic implant-supported

single crowns and fixed partial dentures: A systematic review with meta-analysis. The Journal of Prosthetic Dentistry. 2022, 2022:S0022-3913(22)00736-3. DOI: 10.1016/j.prosdent.2022.11.013. Available from: https://www.sciencedirect.com/science/article/abs/pii/S0022391322007363

[119] Testori T, Clauser T, Saibene AM, Artzi Z, Avila-Ortiz G, et al. Radiographic protrusion of dental implants in the maxillary sinus and nasal fossae: A multidisciplinary consensus utilising the modified Delphi method. International Journal of Oral Implantology (Berlin). 2022;**15**(3):265-275. Erratum in: International Journal of Oral Implantology (Berlin). 2022;15(4):371-373